ELECTRICAL ENGINEERING DEVELOPMENTS

CMOS TECHNOLOGY

Electrical Engineering Developments

Additional books in this series can be found on Nova's website
under the Series tab.

Additional E-books in this series can be found on Nova's website
under the E-book tab.

ELECTRICAL ENGINEERING DEVELOPMENTS

CMOS TECHNOLOGY

MIN-JUN KWON
EDITOR

Nova Science Publishers, Inc.
New York

LIBRARY OF CONGRESS CATALOGING-IN-PUBLICATION DATA

CMOS technology / editor, Min-jun Kwon.
p. cm.
Includes bibliographical references and index.
ISBN 978-1-61761-325-8 (hardcover)
1. Metal oxide semiconductors, Complementary. I. Kwon, Min-jun.
TK7871.99.M44C58 2010
621.39'5--dc22
2010029837

Published by Nova Science Publishers, Inc. † New York

CONTENTS

PREFACE

Chapter 1 - The science and engineering of lithography has been the major driving force for continuous CMOS scaling from micro- down to nano-scale, well known in the semiconductor industry as Moore's Law which predicts that the density of transistor on a chip roughly doubles every 18-24 months. This scaling trend has continued for more than five decades and reshaped our information technology and global society in the 20th and 21st centuries. It is not expected to stop until it hits the economic or physical limit when further device shrinking will not create enough financial incentives or simply violates the fundamental physical principles. As the size of transistors decreases, more components can be crammed onto integrated circuits to enable larger memory capacity and faster computing speed at lower costs. The scaling pace of different semiconductor sectors varies significantly, depending on the design and integration complexity of devices. For example, flash memory NAND devices enjoy their 1-D array characteristics and can be fabricated with more aggressive shrinking processes (e.g., self-aligned spacer double/multiple patterning); while scaling of DRAM and logic devices lags behind due to the difficulty of fabricating more complicated device structures. The evolution of CMOS design and process will be shown as a reflection of lithographic difficulty in the semiconductor manufacturing. Several process techniques to overcome the optical resolution limit such as double/multiple patterning will also be introduced. The technical and cost issues of these sub-resolution spatial frequency multiplication techniques will be addressed. This chapter will be organized into a self-contained manner to not only give a historical review of CMOS lithography, but also provide a technical guidance for industrial/academic readers to understand the fundamental principles, critical challenges and future directions of lithography technology. The topics to be covered include conventional technologies such as optical and extreme ultraviolet (EUV) lithography, and emerging technologies such as maskless EUV lithography.

Chapter 2 - In this work, the approaches adopted at (Universitat Autònoma de Barcelona) UAB to evaluate the nanoscale sources of device variability related to the MOSFETs gate oxide and to analyse the impact of device variability and aging on circuit performance and reliability are described. First, a Conductive Atomic Force Microscope is used to evaluate the effects of processing on the morphological and electrical characteristics of gate dielectrics at the nanoscale. As example, the dependence of the electrical properties of Al2O3/SiO2 stacks on a thermal annealing at different temperatures is analyzed. Second, a reliability circuit simulation methodology to transfer the variability and aging effects in devices up to circuit level, which combines Montecarlo and SPICE simulations, is presented. The methodology is applied to evaluate the impact of threshold voltage time-dependent variability in differential amplifiers performance and reliability.

Chapter 3 - The differential voltage current conveyor (DVCC) has recently emerged as a versatile building block for the realization of a variety of analog signal processing functions including, but not limited to, analog filters, amplifiers, integrators, etc. Most of these applications can be classified as linear although non-linear applications like the voltage-mode comparator, function generators and oscillators are also reported in the technical literature.

This chapter presents several new and traditional applications of DVCC including both linear and non-linear ones. Amongst the linear applications, voltage-mode and current-mode analog filters, amplifiers and integrators are reviewed. Similarly, amongst the available non-linear ones, oscillators and negative resistance converter are reviewed whereas DVCC-based digital logic gates, linear equation solver and circuits for solving linear & quadratic programming problems are proposed. Real device simulations on some of the newly presented circuits are also given. The material presented is intended to explore full potential of DVCC covering a broad range of practical circuit applications.

Chapter 4 - A global view of today's world permits to notify that many hardware and software based systems, are characterized by complex behavior and often have special features and structures requiring high complexity approaches for modeling [1, 2]. In such situation, this may result in many unwanted phenomena disturbing the correct function particularly when taking into account the VLSI constraints. In order to avoid the heavy costs associated with maintenance and diagnostic operations [3], a deep understanding of the degradation process should be established so that the reliability of such systems can be enhanced significantly. Despite that the recent progress in different fields has made resulted products more powerful than ever, the increasing level of complexity implied that reliability problems not only will continue to exist, but also are likely to require ever more innovative solutions [4].

The demand for high-performance devices has been experiencing a steady rise in the sophistication degree of semiconductor manufacturing techniques, which in turn have driven the associated Metal Oxide Semiconductor Field Effect Transistor (MOSFET) size close to their physical limits. Theoretical analyses make it clear that we should be able to rearrange atoms and molecules one by one to get more efficient tools at a tiny scale [5]. Therefore, the initial stage consists in the seek of new design approaches for devices miniaturization scheme, the adopted approach based on keeping the internal electric fields constant had to be abandoned due to several practical reasons including [6]:

Loss of compatibility with previous generation circuits,

Decrease in noise margins because of the non-scaling of the threshold voltage and the subthreshold slope,

Decrease in operating speeds in sub-micron devices due to the non-scaling of parasitic capacitances.

As a result, the design of alternate scaling schemes becomes an inevitable choice to remedy the previous cited disadvantages. In these schemes, despite the obvious correlation between the consequent increase in internal electric fields in aggressively scaled MOSFETs and the benefit of increased carrier velocities leading to a considerable amelioration in switching speed, it also presents a major reliability concern for the long term operation. As device feature size moves into deep submicron region, the benefits of higher electric fields saturate while the corresponding reliability problems get worse [7]. Such tradeoff between both criteria has initiated a tremendous race towards developing strategies providing guidelines for devices downscaling including degradation effect, which is reflected by a shift in the main measured parameters describing the device operating under different regimes circumstances.

The authors will start this chapter by reviewing the various types of defects existing in CMOS-based devices. The authors will then present a surface-potential-based approach for the analytical modeling of main device parameters in subthreshold regime and later they will develop accurate models to handle the device operation in saturation regime. The obtained results provide comprehensive description of multigate MOSFETs behavior including hot-carriers effects, and offer new insights for future CMOS-based circuits design.

Chapter 5 - In the majority of high performance analog signal processors, power efficiency, speed, noise and dynamic range should be traded against one another to find an optimum device bias condition for a given application. Due to these constraints IC designers are forced to operate MOSFETs in moderate inversion, where the model accuracy is questionable. Considering that the model will always limit the simulation accuracy, it is of great importance the designer to be aware of the capabilities – limitations of the applied CMOS model. Concerning the modeling of a MOS device, the model should be simple enough that the results produced in a simulation to be reliable. The model should also be simple in order the simulation time to be minimal and the process for parameter extraction can be easily performed. A balance between the model simplicity and accuracy needs to be attained. Regarding Analog/RF IC design, inconsistencies have been observed mainly in relation to high frequency behavior, linearity-distortion estimation and noise performance. In this work, MOSFET model reliability issues are addressed in relation to the used MOSFET model, the operating mode, the frequency band of interest (baseband or wideband, RF) and the related applications.

Chapter 6 - From their introduction in 1968 by Smith and Sedra and subsequent reformulation in 1970 by them, current conveyors have proved to be functionally flexible and versatile, rapidly gaining acceptance as both a theoretical and practical analog building block. Differential Voltage Current Conveyors are undoubtedly the most widely accepted operational devices in continuous time, current mode and voltage mode analog signal processing. This chapter is based on the CMOS implementation and application of Differential Voltage Current Conveyor (DVCC) and its' modified version, a relatively new active element suited for differential signal processing. In this chapter, a new circuit topology for realizing several second-order versatile (current mode as well as voltage mode) quadrature oscillator (VQO) is proposed. Each circuit employs three differential voltage current conveyors and all grounded passive components, ideal for IC implementation. The circuits exhibit non-interactive frequency control and low THD. Effects of non-idealities are also analyzed. PSPICE simulations using 0.5μ CMOS parameters confirm the validity and practical utility of the proposed circuits.

Chapter 7 - In this chapter the authors present a comprehensive and up-to-date studyt on the feedthrough logic (FTL) concpt for designing high performance arithmetic circuits in CMOS technology.

The FTL logic family, for high speed and lower power CMOS applications, was instroduced by the authors in the recent past. FTL works successfully on the domino concept with the added feature that gates commence evaluation even before their input signals are vallid. This is accomplished by means of initial quasi evaluation time in the computational blocks for the final evaluation when inputs arrive. Furthermore, the well known problems of domino logic, such as the need for output inverters and charge redistribution are completely eliminated, thus reducing the chip are and delay, and improving the performance.

The FTL is well suited to arithmetic circuits where the critical path is made of a large cascade of inverting gates. Furthermore, FTL based circuits perfomr better in high fanout and high swithcing frequencies due to both lower delay and dynamic power consumption.

Experimental results, from the chip measurements, demonstrated superior performance of the FTL ripple carry adders (RCA) when compared with the dynamic domino and traditional CMOS logic styles. Our 14-bit low power implementation performs faster, (2.6 times smaller propagation time delay, an 1.85 times higher max-imun frequency), and provides a better energy efficiency (3.11 times or 67.9%), when compared with the dynamic domino style. On the other hand, an 18-bit high speed FTL design, working at maximum frequency, outperforms the dynamic domino logic in terms of the propagation delay (19.5 times less), maximum frequency (12.1 times more, nd energy efficiency per bit (29.97 times or 96.7% better). Moreover, the same 18-bit high speed FTL adder outperfoms other high performance adders, such as multilevel CSAs, in terms of both, energy efficiency (1.72 times) and propagation time delay (1.78 times).

However, FTL is very sensitive to the device mismatch, and the variations in the capacitive loads in the manufacturing proces. The authors show how the sensitivuty of the FTL based design can be improved through very clever design techniques.

This chapter also discusses the capabilities of the FTL logic in practical applications, and how to extend the use of this logic to larger word-length arithmetic circuits.

Chapter 8 - In this chapter the most important flicker noise models that have been proposed in the literature are analytically presented and discussed. The available flicker noise simulation models in software packages like SPICE and HSPICE and SPECTRE are also examined and selection criteria in relation to the type of MOSFET and the operating regime are proposed. The impact of scaling down technologies is discussed and extra analysis is presented in relation to the type of the bias in CMOS transistors. In particular the switched biasing effect in the MOSFET low frequency noise performance is presented. Additionally, low frequency noise measurement set-ups are presented and measurement techniques are suggested.

In: CMOS Technology
Editor: Min-jun Kwon, pp.1-79
ISBN: 978-1-61761-325-8

Chapter 1

PRINCIPLES, INTEGRATION AND CHALLENGES OF LITHOGRAPHY TECHNOLOGY FOR DEEP NANO-SCALE CMOS PATTERNING

***Yijian Chen*[1] *and Yashesh Shroff*[2]**
[1]Applied Materials, 3225 Oakmead Village Drive, M/S 1220,
Santa Clara, CA 95054, USA
[2]Intel Corp., 2200 Mission College Blvd., M/S SC1-01,
Santa Clara, CA 95052, USA

ABSTRACT

The science and engineering of lithography has been the major driving force for continuous CMOS scaling from micro- down to nano-scale, well known in the semiconductor industry as Moore's Law which predicts that the density of transistor on a chip roughly doubles every 18-24 months. This scaling trend has continued for more than five decades and reshaped our information technology and global society in the 20th and 21st centuries. It is not expected to stop until it hits the economic or physical limit when further device shrinking will not create enough financial incentives or simply violates the fundamental physical principles. As the size of transistors decreases, more components can be crammed onto integrated circuits to enable larger memory capacity and faster computing speed at lower costs. The scaling pace of different semiconductor sectors varies significantly, depending on the design and integration complexity of devices. For example, flash memory NAND devices enjoy their 1-D array characteristics and can be fabricated with more aggressive shrinking processes (e.g., self-aligned spacer double/multiple patterning); while scaling of DRAM and logic devices lags behind due to the difficulty of fabricating more complicated device structures. The evolution of CMOS design and process will be shown as a reflection of lithographic difficulty in the semiconductor manufacturing. Several process techniques to overcome the optical resolution limit such as double/multiple patterning will also be introduced. The technical and cost issues of these sub-resolution spatial frequency multiplication techniques will be addressed.

This chapter will be organized into a self-contained manner to not only give a historical review of CMOS lithography, but also provide a technical guidance for industrial/academic readers to understand the fundamental principles, critical challenges and future directions of lithography technology. The topics to be covered include conventional technologies such as optical and extreme ultraviolet (EUV) lithography, and emerging technologies such as maskless EUV lithography.

1. Introduction

CMOS (complementary metal-oxide-semiconductor field-effect transistor) inverter is the basic cell of most of modern digital circuits as arbitrarily complex logic circuits with low power dissipation can be built by suitably interconnecting inverters. In a conventional CMOS process (e.g., before introduction of 45-nm logic node), we normally pattern poly-Si gates first, followed by ion implantation and high-temperature annealing to activate dopants in the source and drain areas. The advantages of this gate-first technique are its less process complexity and easier scaling to smaller device nodes. On the other hand, to suppress the depletion effect of the gates as CMOS feature size continues to shrink, the poly-Si gate material is gradually replaced by metal whose functionality is nevertheless affected by the thermal budget in the following steps. It has been found in the gate-first CMOS process, patterning metal gate followed by a high-temperature annealing poses severe challenges to tight control of the work function of metal, threshold voltage and device performance. To overcome this problem, a more complicated gate-last approach was invented and introduced into manufacturing by Intel, and it has been adopted by other semiconductor manufacturers. A short description of the gate-last approach will be given below.

As demonstrated in step (1) of Figure 1, the sacrificial poly-Si and spacers are formed first, followed by source/drain doping and annealing, which is similar to the sequence of the gate-first process. However, the sacrificial poly-Si is only a dummy gate and it will be etched away in step (3). A thin film is deposited in step (2) to fill in the space between poly-Si lines and then polished to expose poly-Si. A highly selective poly-Si etching will remove the dummy gates without attacking other materials. In steps (4) and (5), metal A is formed, but remaining only in the P-MOS area by separate lithography and etching processes. A different metal B for N-MOS is required, as shown in step (6), to tune the metal work function and threshold voltage of N-MOS device. After metals A and B are formed in the P-MOS and N-MOS regions respectively, a conductive material (e.g., metal C) is finally used to fill the trenches and then planarized using a CMP process. Apparently, forming metal gates after the high-temperature annealing allows a significantly lower thermal budget to improve the metal functionality.

Lithographic patterning of gates is one of the most critical steps in a CMOS process as the gate CD (critical dimension) will directly impact the speed of transistors. In the CMOS literature, a word "node" is frequently used to indicate the minimum feature size of gates. For example, a 22-nm node normally refers to the minimum gate CD of about 22 nm. However, this number is not an accurate indication of the lithographic capability which is physically related to the pitch (or spatial period) of a periodic pattern. For patterns with a certain pitch, a trimming process (e.g., an isotropic plasma etching) can often be applied to reduce the resist (or BARC: bottom anti-reflective coating) line CD to be much smaller than the half pitch. Therefore, "node" sometimes is considered to be ambiguous from a lithography point of view, and "half pitch (HP)" is preferred to avoid the confusion about the resolution capability of a lithographic system. In Table 1, the predicted lithographic technology/wavelength, gate length, and half pitch of different generations of CMOS device are shown. "SE" and "DP" stand for single exposure and double patterning, respectively. As we can see from the table, the "half pitch" of CMOS devices is about twice of the gate CD (or node CD).

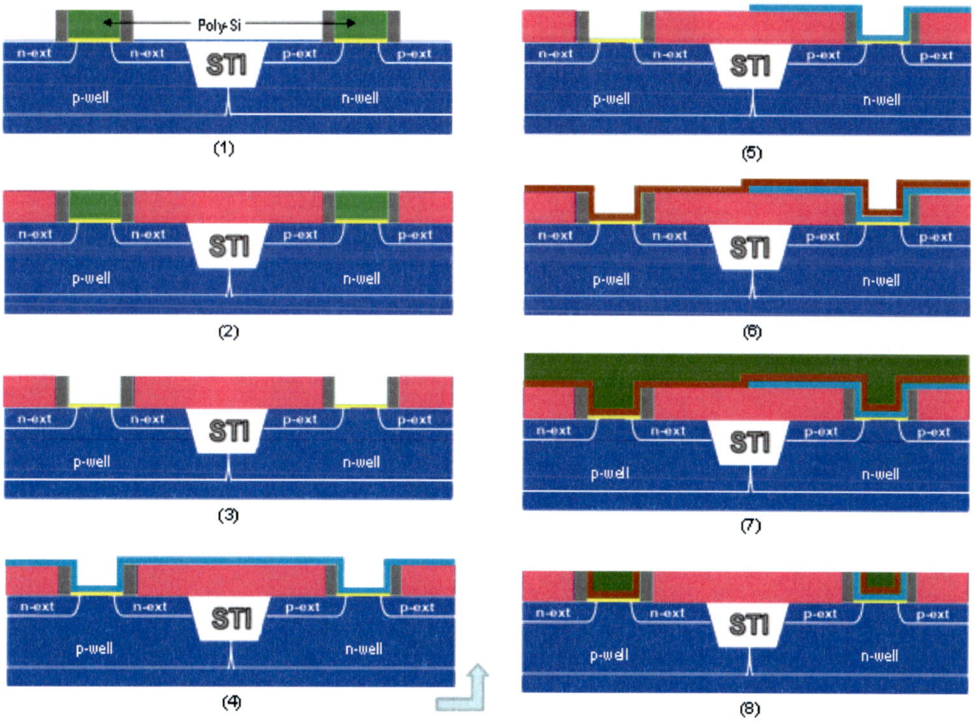

Figure 1. An example of the gate-last CMOS process flow

A brief description of a typical lithographic process flow is shown in Figure 2. A lithographic process usually starts from the bottom anti-reflective coating (BARC) to reduce the standing-wave effect, followed by the resist coating and soft (post-apply) bake. The main purpose of a soft bake is to remove the solvent in the chemically amplified resist. The alignment and exposure are done in a lithographic tool, often called scanner or stepper. After exposure, the wafer will be sent to a thermochemical post-exposure bake (PEB). PEB is one of the most critical process control steps that will activate the catalyst reaction to convert the exposed resist areas and also help to eliminate the standing-wave effect by the diffusion of photoactive compound. Once the PEB step is completed, the wafer will be developed to remove the exposed resist (in a positive tone resist) and form the final pattern. An optional post-development bake (hard bake) sometimes is applied to harden the resist. Before developing a lithographic process, an optical simulation to determine the optimal thickness of BARC to minimize the standing-wave effect is often needed. Normally, the resist aspect ratio of the minimum features should not exceed 3-4 to avoid the collapse problem.

Table 1. The predicted lithographic technology/wavelength, gate length, and half pitch of different generations of CMOS device

CMOS Node (nm)	Half pitch (nm)	Gate length (nm)	Lithography, wavelength (nm)
45	85-90	40-45	Dry/Immersion DUV (SE), 193
32	60-65	30-35	Immersion DUV (SE), 193
22	40-50	20-25	Immersion DUV (SE), 193
15	30-35	15-22	Immersion DUV (DP), 193
11	22-25	11-15	Immersion DUV (DP), 193
8	16-20	8-10	EUV, 13.5

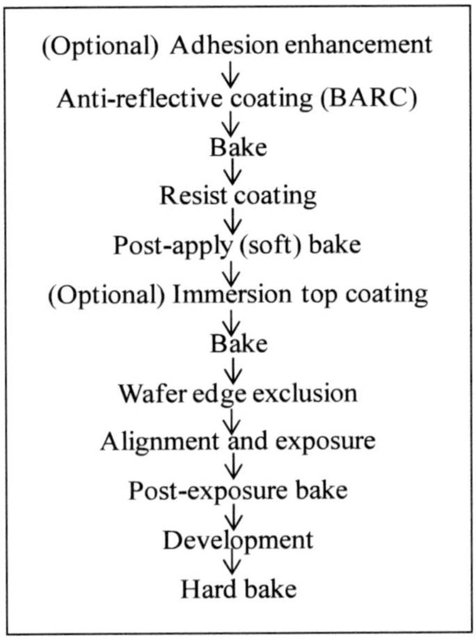

Figure 2. A typical lithographic process flow

193 nm deep ultraviolet (DUV) immersion lithography currently is the cutting-edge workhorse of the semiconductor industry, and 13.5 nm extreme ultraviolet (EUV) lithography is widely considered as the most promising next-generation technology for future advanced patterning needs [1]. However, compared to current and prior optical lithography technologies, the wavelength of EUV light is much shorter and the light energy will be absorbed by the materials it travels through. As a photon-based lithography technology, EUV systems have some modules similar to DUV scanners, which is an advantage over other next-generation lithography contenders such as e-beam and nanoimprint technologies. DUV immersion lithography is a mature technology and its infrastructural description is readily available in the lithographic literature. Therefore, this chapter will be focused on the introduction of EUV research and development, and the fundamental optical principles that are foundations of both DUV and EUV lithography.

Table 2. Comparison of an EUV lithography system with a DUV (193nm and above wavelength) optical lithography scanner

	EUVL Scanner	UV/DUV Scanner
Wavelength	13.5nm	193nm or higher
NA	≥ 0.25	0.93(dry)-1.35(wet)
Environment	Vacuum	N_2 purged (Atm)
Optical system	Reflective	Mostly refractive[1]
Reticles	Reflective	Transmissive

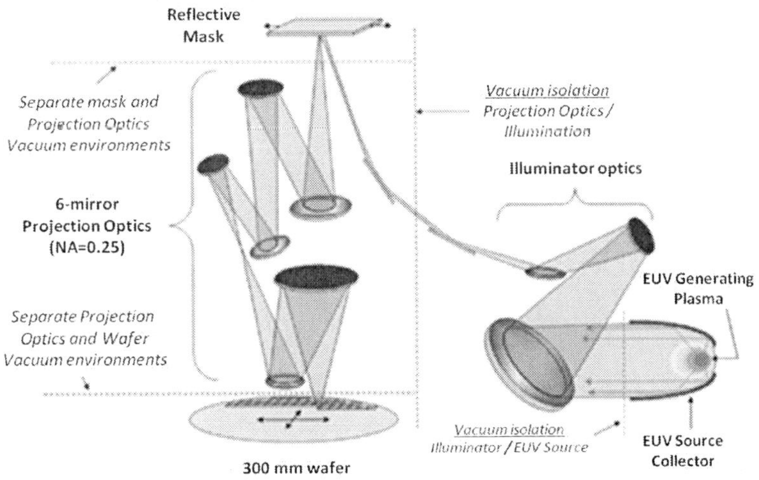

Figure 3. EUV reflective illumination and imaging systems [37]

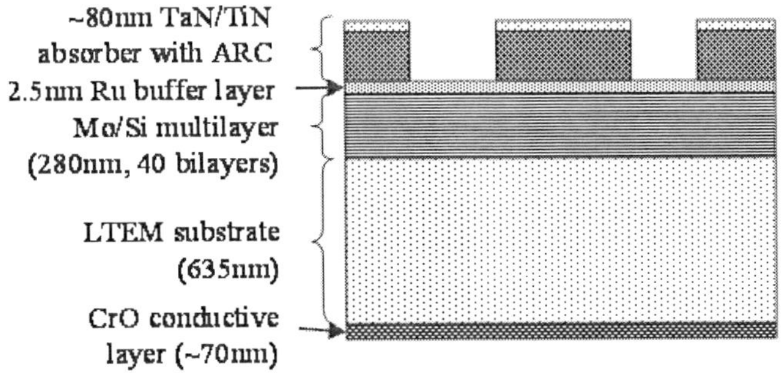

Figure 4. EUV multilayer mask with patterned absorber stack [30]

The EUV imaging system (including the light collector, reticle/mask, and projection unit) is reflective and is composed of a stack of Mo/Si multilayers with absorber stack for patterning. Schematic diagram of a reflective EUV optical system is shown in Figure 3 and a cross-section schematic of an EUV multilayer mask is shown in Figure 4. Critical differences between EUV and DUV technologies are highlighted in Table 2.

[1] Modern 193nm scanners use what's called a catadioptic system which utilizes reflective mirrors to reduce the number of optical elements in the projection optic, thereby reducing tool weight and dimension.

The resolution capability of any lithographic system is usually described as: $k_1\lambda/NA$. Process metric, k_1, is a factor that encompasses all the effects outside of exposure wavelength (λ) and numerical aperture (NA); thus it characterizes the complexity of a lithographic process, i.e., a lower k_1 indicates a more complex process. Lower k_1 is achieved by pushing all other elements (resist, reticle, and tool/illumination) to as close to perfect operation conditions as possible. A k_1 of 0.25 is the theoretical lower limit for single exposure imaging and a detailed analysis of the imaging theory will explain why later. Lower "k_1 equivalent" values are obtainable through the multiple patterning technique that will be discussed soon. Later in this section, we shall cover the major modules of a EUV system and understand their challenges and progress made. While the current 193 nm DUV technology dictates having complicated reticle sets and low-k_1 process engineering techniques, EUV lithography has the advantages of a 14x shorter wavelength and much higher k_1, thus providing larger process latitude using fewer expensive optical proximity correction (OPC) and resolution enhancement techniques (RETs).

2. EUV Tool Development: Status and Requirements

An EUV lithography (EUVL) system comprises of a plasma light source, an illumination system, a reflective mask and a projection system, and a wafer stage, all of which are enclosed in a vacuum environment. The vacuum environment is necessary because soft x-ray radiation at EUV wavelength is significantly attenuated by the atmosphere, particularly oxygen and water vapor. Typically the source and tool are separated at the point of intermediate focus (IF). The IF plane may contain a spectral-purity-filter (SPF) to block out any out-of-band radiation deemed harmful to EUVL imaging performance. All optics and mask are reflective, composed of Si/Mo multilayer with a protective buffer layer on top to prevent oxidation of the surface. Carbon contamination is a serious issue and is mitigated by constant flow of cleansing agents such as oxygen in the environment chamber.

Currently there are several EUV tools in operation at various research consortia and IC manufacturing facilities around the world. The creation of 10x-microsteppers was one of the earlier tools developed for the study of EUV imaging. These tools utilized a Schwarzschild projection optical system which contains a convex primary mirror and a concave secondary mirror. Schwarzschild optics design is notable because it can be designed to produce images free of spherical aberration, coma and astigmatism. An initial 0.088 NA 10x-microstepper [2] and a subsequent 0.10 NA Engineering Test Stand (ETS) [3] were supplemented with 0.3 NA micro-exposure tools (MET). The METs have a 5x demagnification and 4-5deg tilt of incident wavefront on the reticle. Located at Lawrence Berkeley National Lab (LBNL) and Intel Corporation, these tools are used extensively for studying defects, resist, flare, and ultimate resolution of the imaging system [4-7]. A small field exposure tool (SFET) has also been deployed at SELETE for aiding in EUVL research in Japan [8]. Design is underway to upgrade the LBNL MET to an NA of 0.5, with ultimate resolution capability of 8 nm half-pitch. The exposure field of an MET is about 600 um x 200 um [9].

Full-field alpha tools have also been developed. The alpha demo tool (ADT) by ASML at IMEC in Belgium has a tin based discharge produced plasma (DPP) source [10, 11]. There are two EUV1 tools manufactured by Nikon – both use a xenon based DPP light source [12,

68]. Full-field tools have a 6-mirror projection optics (PO) design and support a 26 mm ring width and can scan at least 33 mm long field. The alpha tools are precursor to the 0.32 NA, 6-mirror projection optics based pre-production tools currently in development at ASML. Besides that, 6 NXE (3100) tools with 0.25NA & 0.8σ targeted at 27 nm half pitch are also being built by ASML, although the throughput goal still lags behind at 60 wafers per hour assuming a resist sensitivity of 10mJ/cm2.

Beyond the need of the next two or three nodes, it will become necessary to increase the NA of the EUVL full-field tools to 0.40 or higher [13]. This is where the optical design becomes increasingly complicated. It requires to either introduce a central obscuration for designs with NA≥0.40 with six-mirrors or move to an unobscurated 8-mirror design. At NA>0.60, even 8-mirror systems would require obscuration [14]. The number of mirrors in the projection system needs to be even such that a practical design can have the object and the image plane in opposite planes, thereby not restricting the wafer stage motion.

All mirrors in modern EUVL PO systems are aspheric. The goal is to keep residual aberrations to a minimum in any PO design. By keeping the incident angle low (below 10-deg), aberrations are controlled to allow the tool to reach ultimate resolution capability for the given NA. Current systems such as Nikon's EUV1 keep the total aberrations to 0.40 nm RMS, an impressive feat by any account [15, 43]. Today's systems are well beyond the necessary aberration constraints, so it is conceivable that the potential exists to increase NA with existing mirror polishing technology or change the field size.

The total light transmission of any optical system is governed by its étendue. The area solid angle product or étendue is defined as:

$$E_{tool} = \pi \times Area_{wafer} \times NA_{wafer} \tag{1.1}$$

the total photon flux allowed by the system (a metric we want to continually optimize upward to increase wafer throughput) is the product of the transmission of light by the optics and étendue. The tool étendue is limited by the ability of the optics to create a distortion-free image at the wafer plane. To ensure that the maximum efficiency of the light source is utilized, it is important to match source étendue with the tool étendue.

3. EUV Lithography Challenges

According to ITRS roadmap [18], the semiconductor industry was poised to print 22 nm half-pitch patterns by 2013. However, certain memory industry such as NAND flash has scaled faster than what ITRS predicted, e.g., 24 nm half-pitch NAND flash memory has been reportedly available in 2010. It is expected the minimum NAND half-pitch will reach the limit of spacer double patterning at about 20 nm by 2012. After that, there are several lithographic candidates for the NAND application: self-aligned triple patterning (SATP), self-aligned quadruple patterning (SAQP), and EUVL. We shall discuss the SATP and SAQP processes in detail later in this chapter.

Table 3. EUV risks and infrastructure readiness overview

Module	Specification	Status	Concerns and options
Source power	>180W at intermediate focus (2% BW)	50W calculated[2]	Improve conversion efficiency, laser power for LPP. Medium risk.
Resist	<1.5nm 3σ LWR $<10mJ/cm^2$ sensitivity at below 20nm HP node	LWR ~ 4-5nm at 20nm HP node and dose around 15-20mJ/cm2	Mitigate resist outgas so as to not impact optic lifetime. High risk.
Defects	Zero printing defects	Varies	Establish mask blank and patterned inspection infrastructure. High risk.
Optics	<0.1nm rms figure <7% intrinsic flare	ca. 0.5nm rms and 15% flare	Steady progress. Low risk.
Optics contamination	>5 year lifetime	No consensus	Need validation from established toolsets. Medium risk.

Table 4. EUV power requirements for HVM

Throughput	100WPH
Exposure Time[1]	18 sec
Resist sensitivity	10mJ/cm2
Transmission	0.5%
In-band EUV power at IF	180W

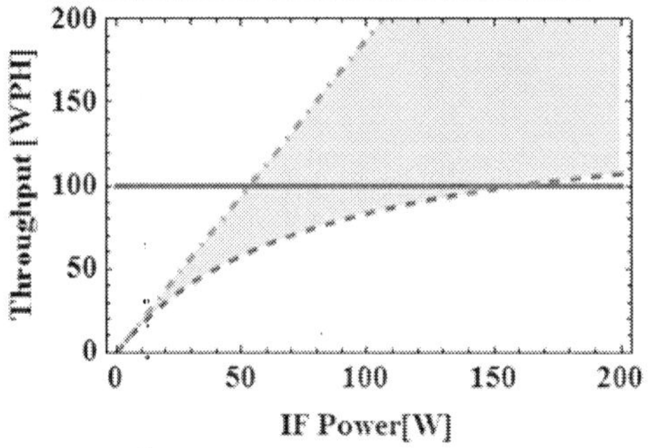

Figure 5. Throughput model assuming resist sensitivity of 10mJ/cm2. IF power of 160W is required to meet 100WPH high-volume requirement based on a realistic wafer overhead of 15sec and exposure overhead of 10sec. Zero overhead case is also indicated

EUVL may be first used in the memory manufacturing (e.g., DRAM) due to memory's relaxed defect requirements and its faster scaling pace than CMOS logic devices. While EUVL has made significant progress in its system development, there are still many difficult challenges to overcome. At 13.5 nm wavelength, EUV offers a tempting alternative to its longer wavelength cousin in DUV spectrum, 193 nm. Production of sub-wavelength RMS roughness mirrors to meet imaging targets, near theoretical levels of multilayer reflectivity, continuous improvement in source power and optic lifetime are just some of the successes of EUVL. However, for insertion into high volume manufacturing (HVM), many more challenges need to be overcome. Of these, source power is the most prominent and the laser

[2] The use of a full collector would be too expensive for evaluation purpose, usually partial collector based calculated value translates well to the full-collector power, assuming thermal considerations hold true.

industry has achieved significant improvement to meet the HVM requirement. Others serious issues are mask defects [26], resist performance [64], and cost of ownership [63]. Table 3 outlines some key specifications in EUVL for 2013 HVM [18] applications.

3.1. EUV Source Power Requirement

The light source comprises source and collector modules (SoCoMo) that are responsible for bringing in-band[3] photons to the intermediate focus (IF) plane where the source is mated to the scanner. From the vantage point of end consumers (i.e., IC manufacturers), the source parameters drive wafer throughput, CD control, and cost. To meet these requirements, apart from bandwidth and power, the light source specifications also include étendue, dose stability, operating cost, and component lifetime. There are two main technologies currently in contention for EUV: discharge produced plasma (DPP) and laser produced plasma (LPP). From an initial set of fuel choices of xenon [68], lithium, and tin, for high-volume manufacturing, only tin is currently viewed as most viable for providing the necessary power to meet throughput requirements [61]. Xenon has the benefit of low debris given its inert and gaseous nature; however, it has the lowest conversion efficiency (CE) of the three choices. Lithium has one of the highest theoretical conversion efficiency but a practical system capable of operating at that limit has so far evaded researchers [65, 66]. Therefore, liquid tin with its reasonably high CE has remained the fuel of choice. Over the years of EUV source development, CE% has also improved, thereby allowing source suppliers to meet the growing power requirement challenge. With each node that EUV is extended, the power requirements go up such that from early 2000s when EUV was first expected to hit commercialization, the requirement of 100W at IF has now gone up to about 200W (Table 4) and possibly even 400W by the time it can be realized for high volume manufacturing.

Wafer overhead includes time to change wafer, alignment, etc. Exposure overhead includes stepping time and acceleration and deceleration time for each step. Therefore, the total throughput, measured as wafer per hour is dependent on number of fields exposed, overhead terms, total input power, resist sensitivity, and finally, transmission through the optical column (IU, PO, and reticle). A representative wafer throughput versus power scenario is shown in Figure 5. A power requirement of 180W at IF plane is necessary for a throughput of 100WPH[4] assuming the following: 115 fields of size 20 mm x 25 mm exposed with resist sensitivity of 10mJ/cm^2, exposure and wafer overhead of 15sec, and average IU & PO mirror reflectivity of 63%. Each of these assumptions can take on a range of values; therefore, it is too premature to predict the precise power requirements at this stage. For instance, if the resist industry fails to meet LWR and resolution requirements at sensitivity lower than 15mJ/cm^2, then the required power may jump up to 270W.

[3] Source bandwidth, $\Delta\lambda/\lambda \leq 2\%$.

[4] All calculations are based on 300mm (diameter) wafers.

3.2. EUV Mask Defect

There are three critical areas where defects are a concern in EUVL: mask blank inspection, patterned mask inspection and patterned wafer inspection. Defect inspection for both blank and patterned masks has been a growing challenge for EUV [38, 58]. The International Technology Roadmap for Semiconductor (ITRS) has defined specifications for all aspects of EUV mask defects such as size and opacity for various technology nodes. Defects are seemingly endemic to every aspect of multilayer in EUV reticles. They could be bumps or pits on the substrate; and these defects can propagate up the multilayer stack and manifest as phase defects. These defects are known as buried defects. Surface defects are also a cause of concern; however, they are more easily identifiable and can be readily cleaned. How detrimental is a defect is determined largely by its proximity effect to a main feature [22].

The two major problems in this area are mask inspection and defect printability. Actinic Inspection Tools (AIT) exist and they are capable of detecting buried defects as small as 1nm in height, however, AIT is a research tool and as such is not capable of supporting high volume of a mask pilot line, let alone HVM [23]. Several defect studies conducted on AIT have given us insights on defect printability, especially through focus.

The EUV reticle specifications include substrate/blank defects, substrate reflectivity, sidewall angle, absorber line-width roughness, and mask flatness. The blank defect density is based on phase defects that can occur when fabricating the multilayer reflective coating and substrate defects (pits and bumps). Determining substrate defect specs is complicated by the fact that it has both a phase defect component driven by depth (or height) of the pit (or bump) and a size characteristic. Worst case performance induced by a defect determines the specification. In this case, a step of 3.4 nm is sufficient to induce a 180degree phase shift in the reflected wavefront. On the other hand, a 90 degree defect (1.7 nm height) can also cause a CD defect if it is about twice the specification size [26].

EUV mask flatness has a very tight specification, since non-flatness causes a registration error associated with the off-axis illumination of the masks[5]. The specification for flatness is such that it contributes an equal amount to the wafer registration error as mask patterning does. Ongoing research may allow this specification to be loosened in the future. This is based on the concept of measuring a mask's non-flatness and then correcting the position of the patterns such that the printed pattern will be in the correct location even if the mask does not meet the flatness specification.

EUV mask blanks must start and remain free of defects. In DUV lithography a thin non-absorptive organic film is used as a pellicle to defend against ambient particles from depositing on the patterned surface of the reticle. Though EUV pellicle solution is proposed [72], around 30-40% of light absorbed by the pellicle makes the solution less than ideal. It is desirable to find pellicle-free solutions to keep patterned masks defect free. These different NGL mask requirements can be expected to exacerbate, rather than relieve, the high costs of masks.

For successful insertion of EUVL into HVM of logic devices at the 22nm half-pitch node by 2013 (which might not be practical though), the development and readiness of a commercial EUV mask inspection tool infrastructure is a critical gap that remains to be

[5] Over 142x142 mm2 area of the mask, the expected P-V is <30nm for 22nm node, as per SEMI P37 spec developed by SEMATECH Inc.

addressed [45]. Mask pilot line activities will accelerate as pre-production tools are deployed at multiple sites worldwide over the next few years.

3.3. EUV Mask Inspection

Mask inspection is essential to qualify reticles for use in production environment [23]. Printable defects need to be identified, evaluated, and hopefully repaired. Defects can cause amplitude change or phase change in reflected wavefront. Particles on the surface of the multilayer are generally amplitude defects while defects caused by pits or bumps in the glass substrate manifest as phase defects. Amplitude defects can be detected by commercial tools that use DUV wavelength inspection. However, in order to detect impact of phase defects on aerial image, actinic tools must be utilized [34]. Defect inspection is carried out at several steps in the EUVL reticle manufacturing process. The sequence of fabrication steps is highlighted in Figure 6 with critical inspection steps indicated.

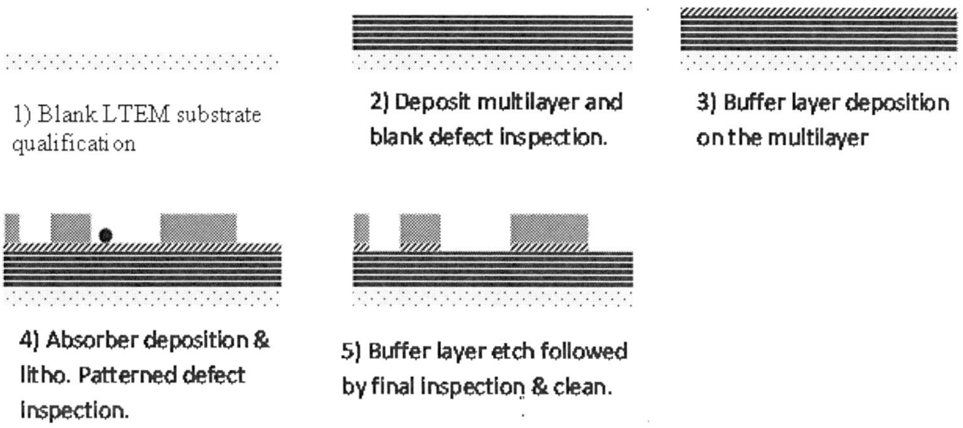

Figure 6. EUV reticle fabrication process flow. Defect inspection points are inserted along with defect removal and cleaning steps before the mask is shipped to the factory

Preparation of defect free glass substrates has been an ongoing research topic but it is difficult to produce completely defect free glass substrates for HVM applications. Pits are more prevalent and account for 75% of the total defects detected [19]. Inability to repair multilayer defects is one reason for tight specification on mask blank defects.

4. EUV Source Development

In this section we will cover the basic types of EUV light sources and current technical challenges of EUV source technology. EUV light can be generated by excitation of very hot plasma. The spectral emission properties of plasma are known from the well known Plank relation between wavelength and temperature.

$$\lambda_{max} = 2898\mu m \cdot Kelvin/T \quad (1.2)$$

For λ=13.5 nm, the required plasma temperature is about 220 000 *K* which can be reached via excitation through either pulsed electrical discharge or by pulsed laser discharge [16].

The gas discharge produced plasma (GDPP) method is being pursued jointly by Phillps/Xtreme in Europe using tin as a fuel for plasma [69]. Laser produced plasma (LPP) is favored by Cymer in USA and Gigaphoton in Japan, also using tin. Both methods have their supporters and detractors and both have shown a gradually increasing power output at increasing burst duration. Current development of EUV sources is focused almost exclusively on tin (Sn) to generate plasma. Sn fueled discharge sources benefit from high conversion efficiency. For GDPP, this implies having a small pinch size and for LPP sources less debris for given power at intermediate focus (IF). Alpha tool EUV sources are at 3-7 W power at IF plane. These sources are sufficient for research purpose but fall short in terms of power and cost efficiency when considering high volume production environment. Downtime related to fast use of consumables such as electrodes is a significant obstacle. The requirements are high uptime (reliability), low cost of operation, high power, and dose stability. Ongoing work at several source manufacturers aims to solve these issues, albeit with very divergent method.

4.1. Gas Discharge Produced Plasma

A tin GDPP source consists of two rotating electrode wheels that are covered with a thin tin layer. A high power laser ablates tin on one of the electrodes forming a vapor plume that propagates towards the second electrode thereby creating a conductive channel between the two electrodes. The rotating electrodes continually regenerate the tin surface; most of the tin (Sn) is recycled back into the system (see Figure 7). Unlike its xenon based GDPP predecessor, no electrode erosion helps keep operating cost low. Secondly, by collecting and recycling Sn, there is less debris to worry about. The Philips/Xtreme approach uses two laser pulses fired in quick succession followed by a long pause to generate maximum EUV light. The researchers have been able to show that increasing the input laser power directly correlates to increasing the in-band EUV radiation. The diameter of the pinch needs to be less than 1mm to meet the source étendue requirement. To produce 180W at IF, thermal load needs to be managed, which is a topic of ongoing investigation.

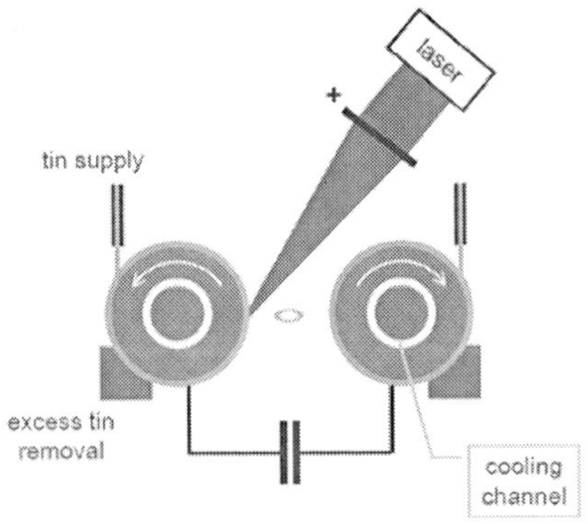

Figure 7. GDPP-Sn based EUV light source (Reprinted from [Ref 34])

Other main components of a GDPP system are an efficient debris mitigation foil and grazing incidence collector. The collector is composed of a series of concentric shells. Each shell has a front-end ellipsoid portion that is mated with a back-end hyperbolic portion to focus EUV radiation at the IF plane. The grazing incidence collector shells have a built-in thermal management and thick EUV reflective metal coating. The debris mitigation system is a foil trap with buffer gas that prevents most high energy Sn ions from impinging upon the collector surface and eroding the shells. The foil trap and collector need to be designed for long lifetime. The high-cost collectors are expected to have a lifetime of one year or more in high volume production environment. Currently the GDPP technology is still in development mode; and even though remarkable progress has been made in recent years, much more improvement in power needs to be made.

4.2. Laser Produced Plasma

The LPP based system in development by Cymer uses a high power CO_2 laser to create highly ionized plasma from tin (Sn) droplets [67]. The laser beam is focused onto a droplet of Sn to generate plasma through a central opening in the collector mirror. The normal incidence collector (NIC) mirror captures the portion of radiation emitted in the backward direction in a 5-sr collection angle and focuses it on the IF plane. Figure 8 shows the system in detail. Even a laser is used to generate EUV source power, the emitted EUV light is spatially incoherent.

Along with in-band EUV and significant out-of-band radiation, these high power sources also generate a lot of debris in the form of high-energy ions, neutral atoms, and clusters of target material [60]. This can, over time, reduce the reflectivity of graded multilayer coated NIC. Current estimates put the lifetime of consumable optics in the LPP source at around one year of operation at HVM.

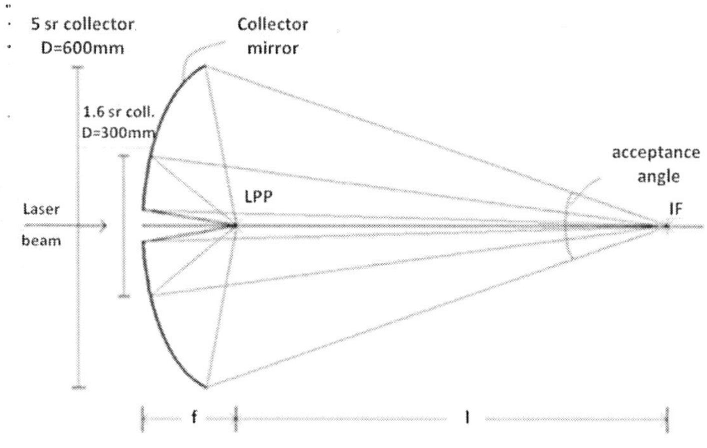

Figure 8. Schematic diagram of an LPP EUV source from Cymer (Reprinted from [Ref 35])

Power improvement is achieved by continuous increase in laser output power. A pre-pulse expands the Sn droplet and the main pulse then fully eviscerates it. This helps improve the output energy per pulse and reduces Sn debris that inevitably collects around the expensive NIC. Improvement in power also comes from operating the laser and droplet generation systems at higher frequencies, currently at 50kHz. At the middle of 2010, the laser industry reported LPP source generating 90-100W raw power at IF. Excluding the raw power loss due to dose control and spectral filtering, EUV exposure power is expected to reach 40W in 2010, and >60W in 2011.

5. EUV Resist Development

Resist development has consistently been highlighted as one of the most critical issues gating HVM deployment of EUVL. Despite the progress in recent years, resist goals remain challenging for 2013 (see Table 3). Resist challenges in EUV revolve around pushing the limits of LER and exposure dose at the tightest pitches of interest. The key lithographic performance indicators - *resolution*, *LWR*, and *sensitivity* – known as RLS, are interdependent. One parameter cannot be improved without degrading performance of the others [20]. To meet EUVL requirements, the RLS triangle (indicated pictorially in Figure 9) needs to be reduced simultaneously. The RLS challenge is not unique to EUV but due to the unavailability of high power source, it takes on a more ominous tone. For instance, line-width roughness, or LWR, is a statistical outcome of the number of photons used to expose a given feature (E_{size}). It is inversely related to sensitivity. Increase in exposure dose improves LWR and gives higher resolution, but resist sensitivity suffers. Several experiments [17, 28, 29] show conclusively that RLS is a limiting factor in resist performance, which is also confirmed by simulations [20].

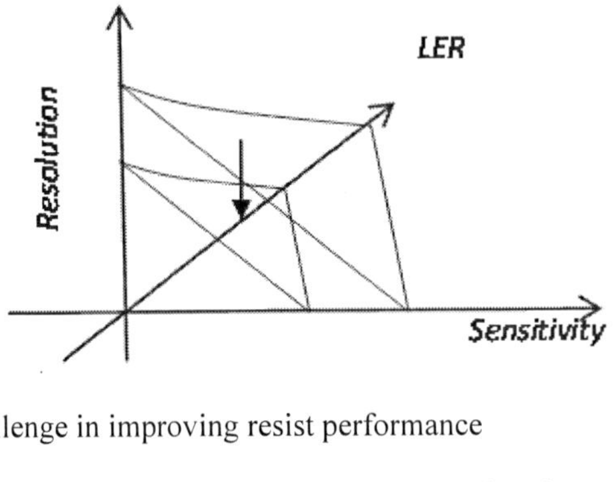

Figure 9. RLS tradeoff challenge in improving resist performance

The impact of shot-noise limiting resist performance has been experimentally observed [17, 21]. Based on the Poisson statistics, a more accurate relationship between LWR and dose has been derived:

$$LER = 4.9/\sqrt{Dose} \tag{1.3}$$

If roughness between two edges of a line is uncorrelated, the values of line edge roughness (LER) and line width roughness (LWR) are related by [20]

$$LWR \cong \sqrt{2}LER \tag{1.4}$$

Thus, $LWR \sim 6.9/\sqrt{Dose}$. Several dozen resists have been evaluated on the 0.3 NA MET and the LER quantified against the dose. Results are plotted in Figure 10 and the solid line in the figure shows LER dependence on dose.

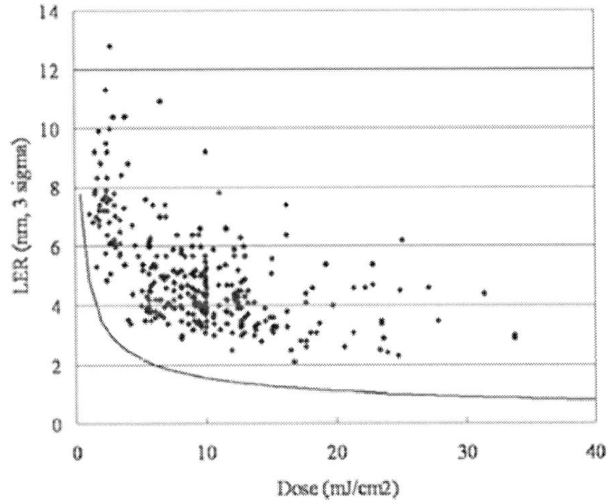

Figure 10. LER due to non-mask sources plotted against dose (reprinted from [Ref 40])

6. EUV Imaging with Multilayer Mirrors

Fundamentally, imaging optics with 13.5 nm wavelength is no different than 193 nm dry or immersion lithography since both use photons for patterning resist. Historically, the development of multi-layer coating on aspheric mirrors was an important milestone to mark technical feasibility of soft x-ray imaging. Along with figure control, multilayer deposition techniques also demand high reflectivity. Low stress (to minimize the reticle bow), thermal stability (to withstand high power EUV light source) drive the interface engineering of EUV multilayers [70]. In this section we outline some of the significant challenges with multilayer technology and present state of art.

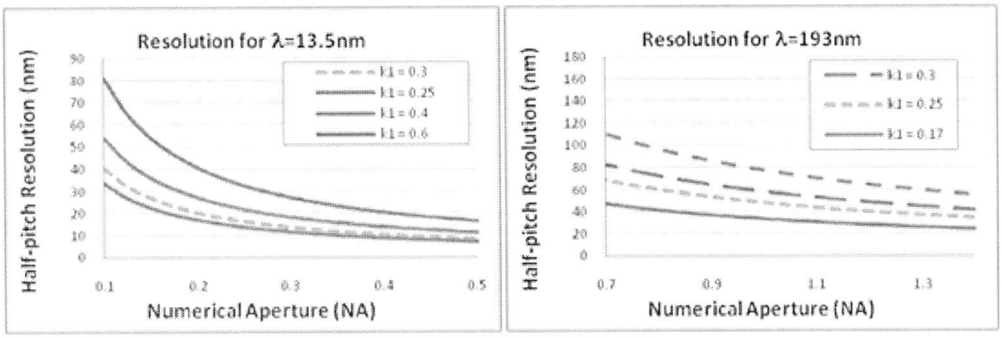

Figure 11. (a) DUV resolution as a function of NA and process metric, k_1; (b) similar chart for EUV shows resolution requirements sub 20nm HP can be met with much simpler process (higher k_1)

6.1. EUV Mirror Structure

All EUV mirrors have four key components. The optics consist of a low thermal expansion material (LTEM) substrate on which alternating layers of molybdenum and silicon (Mo/Si) are deposited. The substrate coefficient of thermal expansion (CTE) is less than 5ppb/K because expansion of the substrate results in registration errors of the image projected on the wafer [71]. The substrate flatness spec according to SEMI P37 needs to be less than 30 nm. Use of interface layer such as B_4C is useful in improving the reflectivity of the mirrors at high operating temperatures. Finally, a capping layer such Ru is utilized to prevent oxidation of the top surface, thereby maintaining the mirror reflectivity. The multilayer stack is deposited on aspheric highly polished low thermal expansion material substrate. Multilayer mirrors typically reflect 60-70% of incoming radiation and the absorbed energy needs to be dissipated. Even so, there is a rise in temperature as the EUV source power increases to meet HVM throughput requirements. Development of Si/Mo multilayers has created significant new technologies, such as atomic level control of deposited layers and development of aspheric highly polished mirrors. The reflectivity of EUV light (at near normal incidence) on Mo/Si multilayers is around 67% at 13.5 nm wavelength. Thus, for a 6-mirror projection optics (PO), the total photon flux from reticle to wafer is less than 7.5%. More importantly, addition of two more mirrors for an 8-mirror PO system decreases the total photon flux such that the source power requirement nearly doubles. To meet the continually increasing high resolution requirement, EUV projection system NA needs to be increased from current designs of around 0.25 to >0.4. At 0.4 NA, we have a choice of either using a PO that has a central obscuration or using an 8-mirror system. For NA≥0.5, 8-mirror system is necessary. Figure shows the resolution trending with numerical aperture. To meet sub-20 nm HP with k_1 of 0.6 we need NA>0.4. Alternatively, phase-shifting mask or other resolution enhancement techniques need to be used for EUV lithography (thereby effectively reducing the k_1 factor and increasing the process complexity). Each option has its detractors and the ultimate choice will be dependent on economics of high volume manufacture and technology readiness.

Compared to DUV optics, EUV mirrors have a much tighter optical spec. This is because the quality of the final finish of the multilayer mirrors determines the ultimate imaging capability of the system. Optical aberrations (usually characterized in the form of Zernike polynomial coefficients) and flare need to be minimized.

6.2. Flare Impact on EUV Imaging

Flare is caused by the scattering of light from roughness of the optics. Light from bright areas of the mask is scattered by the optical components and redirected across the image field including areas meant to be dark. Such internal scattering is mitigated by polishing the optics but is never eliminated [39]. As a result of flare, maximum intensity in the bright areas is reduced and the dark areas become less dark. The resulting loss in image contrast coupled with lower image log slope will reduce the process window [54].

Non-idealities in the performance of imaging optics such as aberrations and flare become more important for low k_1 processes. EUV multilayer non-idealities are bracketed in three major components. These are low, middle, and high spatial frequency roughness. Low spatial frequency roughness is figure error and is characterized by the first 37 Zernike polynomial coefficients. High spatial frequency roughness leads to loss of light outside the pupil collection angle of the projection system [40, 41, 42]. Multilayer mirror surface roughness mainly responsible for flare is in the mid-spatial frequency range of 1/mm to 1/um. In general, flare relates to the total fraction of power scattered away from the specular direction, known as Total Integrated Scattering (TIS). In a smooth surface approximation (SSA), under the assumption of a Gaussian height distribution of roughness, TIS is given as [57]:

$$TIS = \exp\left[\left(\frac{4\pi\sigma\cos(\theta)}{\lambda}\right)^2\right] - 1 \tag{1.5}$$

Here σ is related to rms roughness, or phase difference introduced by surface height variations (instead of partial coherence with the same symbol to be discussed in other section). θ is the incident angle and λ is the exposure wavelength. In the small roughness regime, $(4\pi\sigma\cos(\theta)/\lambda)^2 \ll 1)$, TIS scales linearly with the inverse of the square of wavelength [56].

$$TIS \sim \frac{1}{\lambda^2} \tag{1.6}$$

Due to the short wavelength of radiation, flare in EUVL is difficult to control. Apart from loss in contrast and process margin, flare contributes to CD error as a result of mask pattern density variation across the reticle [55]. Lithographic measurements are most widely used to determine flare, the most popular method being "Kirk test". In Kirk test (see Figure 12), usually a 2 um line is exposed with a fixed blade opening in the scanner [44]. The dose at which the line partially "disappears" is taken to decide the flare percentage.

$$Flare\%|_{2um} = \frac{E0}{Dose_{Clear}} \times 100\% - Mask_{Refl}\% \tag{1.7}$$

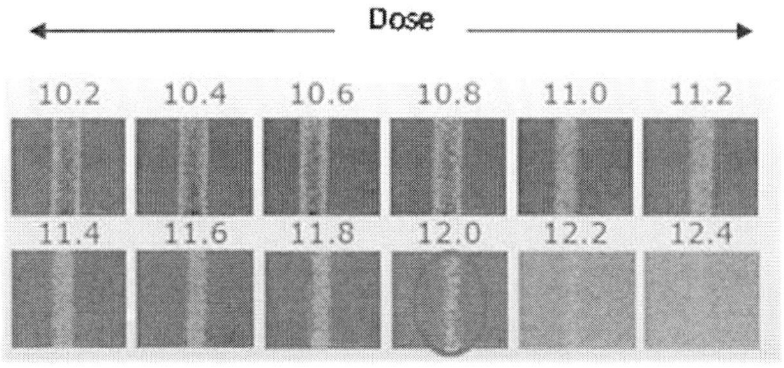

Figure 12. Lithographic measurement of flare using "Kirk" test. Increasing dose eventually results in sufficiently exposing the "dark" pattern. Dose to clear in this case is 12mJ/cm^2. This is on a tool with approximately 7.5% flare ([Ref 46])

E0 is the dose to clear resist for an open frame exposure. The term $Mask_{Refl}$ represents the reflectivity of the absorber stack in the EUV reticle. Figure shows a series of resist SEM (scanning electron microscope) images with an increasing bright-field dose test used to determine the intrinsic flare of the optics. A 50% disappearance of the line is a qualitative indication of flare. The impact of flare on aerial image intensity is calculated using [55]:

$$I(x,y) = (1 - TIS)I_o(x,y) + I_{flare}(x,y) \tag{1.8}$$

The term $I_{flare}(x,y)$ is calculated by convolution of the point spread function (PSF) of the optical system with I_o (the intensity profile without flare):

$$I_{flare}(x,y) = PSF \otimes I_o \tag{1.9}$$

The PSF of the imaging optics can be directly calculated from the power spectral density (based on roughness of the mirrors) or, directly (lithographically) using Kirk test described above. The aerial image of 20/40 L/S pattern with 20% and zero flare is shown in Figure 13. Since the impact of flare is in the order of several millimeters, flare correction to minimize CD variation and improve process window through mask design becomes important [46, 51, 53]. In conjunction with continual improvement in MSFR of projection optics, post design changes in layout via dummification and CD biasing are important techniques utilized by lithography engineers to overcome this critical non-ideality in imaging.

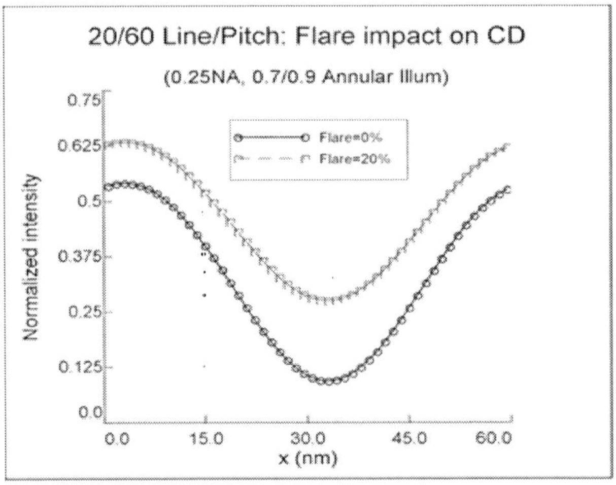

Figure 13. Aerial image of a 20/60 line/pitch pattern indicating the impact of flare on CD, contrast, and image slope

7. EUV Phase-Shifting Masks

EUV is expected to come into high-volume manufacturing for 16 nm or lower half-pitch node. This implies that for an optical system with an NA of 0.25, the process factor, k_1, is of the order of 0.3 or below. This is an extremely aggressive regime for EUV to be inserted. Even with increase in projection NA to 0.4, there will be some need for reticle enhancement techniques at the 16 nm and lower HP nodes. Phase-shifting masks in EUV lithography is one such approach [31].

For many generations, phase-shifting masks (PSM) has been a mainstay of DUV lithography [32, 33]. Optical principles of phase-shifting masks will be discussed later. However, unlike the longer wavelength and transmissive mode of DUV wave propagation, EUV has a much more challenging PSM problem at a shorter wavelength and reflective mode.

In the *alternating* phase-shifting mode (*Alt-PSM*), phase change of 0 and 2π is used with TaN/TiN/Cr as absorbers. Destructive interference of light amongst nearby phase-shifted regions can be used to generate arbitrary patterns in theory. One method of achieving phase-shift in EUV reticles is to etch the multilayers. In this case, the desired shift in phase attributed to the wavefront reflected from etched multilayer can be achieved with multilayer etch-step given by [27]:

$$t_{AltPSM} = \frac{\Delta\phi}{2\pi}\frac{\lambda_{EUV}\cos\theta}{2\Delta n} \quad (1.10)$$

Here $\Delta\phi$ is the desired phase-shift, θ is the angle of incidence of EUV radiation (typically 6deg), Δn is the difference in the refractive index between the multilayer film and vacuum, and λ_{EUV} is the actinic wavelength of 13.5 nm. Since the Mo/Si multilayer stack refractive index is 3% higher than in vacuum, the necessary multilayer step is in the order of 110 nm for 180deg phase shift in the reflected wavefront.

While a relatively high contrast compared to other PSM methods can be achieved, the AltPSM approach is already difficult to execute in practice for DUV lithography; and with a multilayer reticle, it is even more challenging for EUV lithography. On the other hand, by combining a weak resolution enhancement technique such as *attenuated* phase-shifting mask (*Att-PSM*) with optimized illumination, we can greatly improve the process window of low k_1 patterns. Compared to binary masks, attenuated PSM can improve image quality by increasing contrast and depth-of-focus. A phase step in the substrate before deposition of Mo/Si multilayers becomes the starting point for AttPSM mask fabrication. The phase step can created either by directly etching into the substrate, or more accurately by patterning a thin layer on the substrate. The step height is given by [35]:

$$t_{AttPSM\_EUV} = \frac{\phi}{2\pi}\frac{\lambda_{EUV}}{4\cos(\theta)}(2m+1) \quad (1.11)$$

Here m is an integer number. For $m = 0$, the phase step is 3.4 nm for the incident angle of $\theta = 6$ degrees. While AttPSM does provide sharper images compared to binary reticles, it has an image imbalance problem that worsens with defocus. Also, phase errors are harder to control for EUV AttPSM compared with DUV AttPSM. The etch depth in a DUV lithography PSM quartz is given by

$$t_{DUV} = \frac{\phi}{2\pi}\lambda_{DUV} \quad (1.12)$$

For a phase-shift of 180°, if the phase error is required to be less than 5%, it is easy to see that the required etch control for EUV is $\lambda_{EUV}/40$ or about 0.34 nm[6].

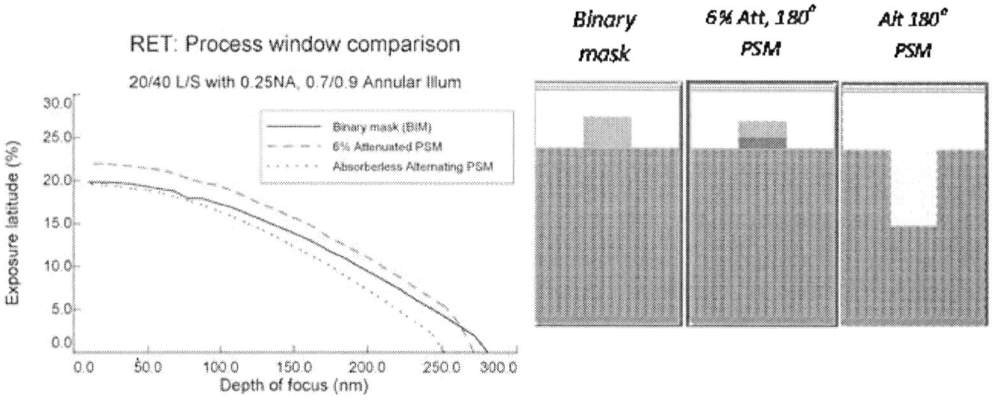

Figure 14. (a) Phase-shifting masks improve process window significantly over binary mask, allowing lower k1 printing while maintaining image fidelity. (b) EUV multilayer design for each of the 3 cases shown here. BIM is 51 nm TaN, AttPSM has 26 nm TaN on 12 nm Mo (for 180deg phase shift), Alt PSM has 110 nm multilayer etched

Other techniques to achieve phase-shifting are related to use of a phase-shifting layer in the absorber stack. By transmitting about 6% of light from the "dark" regions of a layout at 180deg phase shift from "bright" regions, contrast improves over binary mask. The absorber stack consists of attenuator, spacer layer, a phase-shifting layer, and finally an etch-stop or buffer layer. The stack needs low reflectivity at DUV to achieve high contrast in mask inspection. For this reason, an APSM based on a Fabry–Pérot structure can be useful. Alternatively, we may use anti-reflective coating (ARC) to increase DUV contrast. Absorber material such as TaN or SnO, with a large k (extinction coefficient) reduces stack thickness and lower shadowing. In conjunction with a spacer layer such as Al_2O_3 [36], these two layer thicknesses can be tuned to achieve target EUV reflectivity and maximize DUV contrast. Ru is used as a capping layer. Contamination induced loss of reflectivity is a major concern and Ru is one of the leading candidates for oxidation mitigation. It also serves as an etch-stop layer during mask fabrication [31].

In Figure 14(a), we show the relative performance of binary against both types of phase-shift mask approaches. While absorberless alternating phase shift mask has been used to some extent successfully in DUV reticles, in case of reflective masks, this approach is not attractive. The performance gain is lost due to severe shadowing from the deep trenches that need to ion-milled into the multilayer. Secondly, the number of multilayers needs to be increased, otherwise the phase-shifted wavefront is attenuated. Attenuated PSM can be obtained for EUV with the inclusion of a step or etch in the substrate layer or by suitably choice of layers in the absorber stack. The latter approach is shown in the example of Figure 14(b). Mo can be used as a phase-shifter layer and TaN as an attenuator[7]. For simplicity, ARC or spacer layer is not shown in these simulations but needs to be included in a practical implementation.

---

[6] By comparison, for 193 nm, etch control of $\lambda_{DUV}/20$ is about 28x lower at 4.8 nm.

[7] TaN has a much higher extinction coefficient than Mo but both share similar Δn

As a final note on this topic, it remains to be said that while phase-shift masks served a long stint in DUV lithography in improving process margin for given k_1, it is unclear whether this technology will play a similarly important role in EUV. Apart from the obvious difficulty of etching multilayers accurately, phase-shifting increases absorber stack height relative to a binary reticle and adds to H-V (horizontal-vertical) CD biasing concerns.

8. Summary of Mask Based EUV Lithography

Even EUVL is moving closer to pilot line introduction, a number of serious concerns remain regarding its high volume feasibility. Source development has made continuous progress in power at IF plane, going from 3-5W in alpha tool deployments to over 90-100W demonstrated by 2010. Mask defects is the other challenge and in the end it could mean a much more expensive EUV system. Characterization of mask defects with wafer printing or actinic imaging shows that not all defects are critical. Efforts are underway to develop high volume actinic blank and patterned inspection tools, led in part by SEMATECH. However, with each shrinking node that EUVL deployment is shifted out, the size of critical defects decreases and so does the capability of existing tools to support defect evaluation.

IC manufacturers may start inserting EUVL at the 16 nm half-pitch node after 2015. To make this possible, resist improvement in the form of sub-2nm LWR at 10mJ/cm2 sensitivity is necessary, along with defect-free logic masks and >180W light source. Whether the semiconductor industry can find solutions to the mask cost and defect challenges remains to be seen and an alternative technology to consider is the maskless EUV lithography.

9. Maskless EUV Lithography

The predicted cost of ownership (CoO) of mask (reticle) based EUV lithography for 22-nm node is shown in Fig. 15. Assuming a high-volume mask life time of 20,000 wafers per mask (WPM), the EUV lithography and mask consume about 60% of the total cost of six major IC fabrication modules. What does this mean for the foundry and ASIC (application-specific IC) industries wherein a large portion of the lots are run at <1000 wafers per mask? It is evident that EUV CoO (especially the mask related costs) based on above economic model will be prohibitive to prevent it from being adopted into medium-to-low volume production.

MEMS (micro-electro-mechanical systems) micromirror-based maskless lithography is a potential future technology to overcome a variety of challenges facing the semiconductor industry. Figure 16 is a conceptual demonstration of an EUV maskless lithography system, in which 10^8 to 10^9 programmable micromirrors are needed to replace the physical masks and achieve satisfactory throughput. Each micromirror consists of about 40 Mo/Si (reflective) multilayers with total thickness of about 0.3 μm [73, 74], and its surface position can be switched to certain analog level by applying a relevant voltage. This enables the "grayscale" patterning [75, 76] as the wave interference will occur when the incident EUV light is reflected back from neighboring micromirrors (biased at different positions with varying optical path difference).

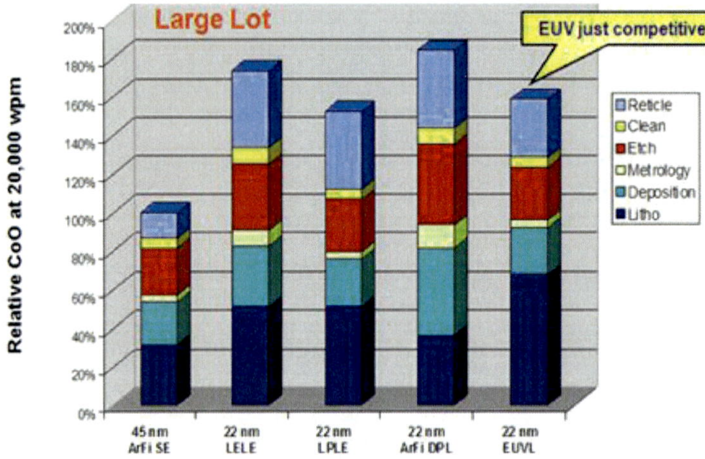

Source: SPIE Litho Asia 2008

Figure 15. Comparison of the cost of ownership of single/double patterning and EUV lithography

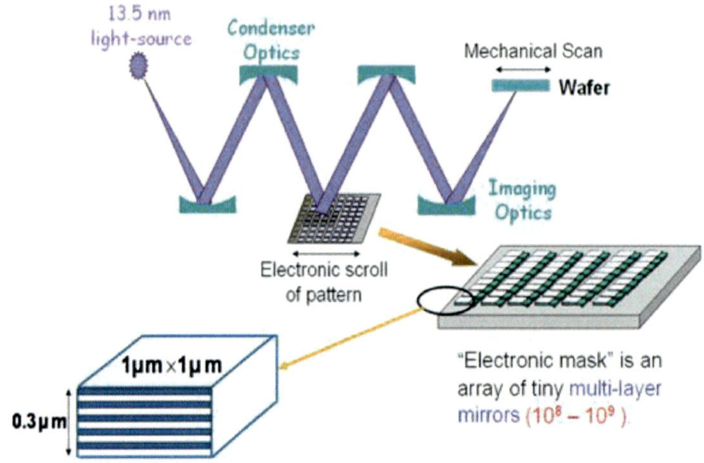

Figure 16. A conceptual demonstration of maskless EUV lithography

In spite of the technical challenges of making 10^8 to 10^9 micromirrors on a small chip (e.g., 2 cm × 2 cm), this method to replace the physical masks by a programmable micromirror array will offer tremendous savings in both mask and scanner costs (due to much smaller printing field of maskless scanner and simpler tool design & manufacturing), and improve our design-for-manufacturing capability by enabling more adaptive process control. Superior imaging performance of micromirror-based maskless lithography using grayscale pattern generation has been demonstrated recently [75, 76]. Maskless lithography can also help to reduce the impact of defects on "masks" by allowing redundant printing and using much higher image demagnification (e.g., 100x). Redundant printing of each pixel with several exposures using different micromirrors will significantly reduce the defect number on the wafer. Commercial defect inspection tools to detect printable mask defects with the size of tens of nanometers are still not available. Thus the high image demagnification required by maskless EUV lithography is of particular interest as the printable defect size will be much larger. This allows the metrology industry to extend the current generation of equipment to inspect much larger printable "mask" defects at the micron scale, consequently saving a large amount of capital expenses and long development time.

The processes to fabricate large-scale and ultra-dense micromirros to meet the requirements of maskless lithography are beyond existing optical MEMS technologies and several dynamics, control and fabrication challenges must be overcome [74]. One of the main technical challenges of fabricating EUV micromirrors comes from the requirement of low-voltage and high-speed device operation in vacuum without air damping. Due to the limited space for each micromirror, complex feedback control circuits are not allowed; thus only optimal electrical (resistive) damping can be considered to minimize the micromirror's settling time.

The stability of micromirrors is another important issue in maskless lithography. The parallel-plate microactuator is well known for suffering from the unstable "pull-in" problem. But parallel-plate microactuators can be fabricated relatively easily and have been widely adopted in many commercial MEMS products. Most of them use a relatively high voltage (which allows a large gap) and incorporate hard-stops. However, device instability is a major concern in maskless lithography, especially in low-voltage actuated EUV micromirrors (operating in vacuum with very small gaps). Furthermore, when the unstable pull-in happens, the low voltage used in micromirror devices will not be able to overcome the stiction force to bring them back up. If micromirrors snap down and get stuck on the substrate, these non-functional mirrors will become defects and degrade the processing yield. Therefore, contact or hard-stop operation is precluded in maskless EUV lithography. More stable devices such as the vertical double-comb tilting and double-flexure piston micromirrors, as shown in Figure 17, have been designed to overcome the instability problem [74].

To be compatible with IC circuitry underneath, micromirrors should be fabricated with low thermal budget (e.g., lower than 420°C). SiGe has been studied as a structural and damping material to meet the low-temperature and resistivity requirements and the processing-related issues [74]. Self-aligned processes using "spacer nanolithography" to define ultra-thin gaps for low-voltage MEMS operation have been developed to fabricate vertical-comb tilting micromirrors and double-flexure piston micromirrors with mirror sizes ranging from 10 to 0.5 micron.

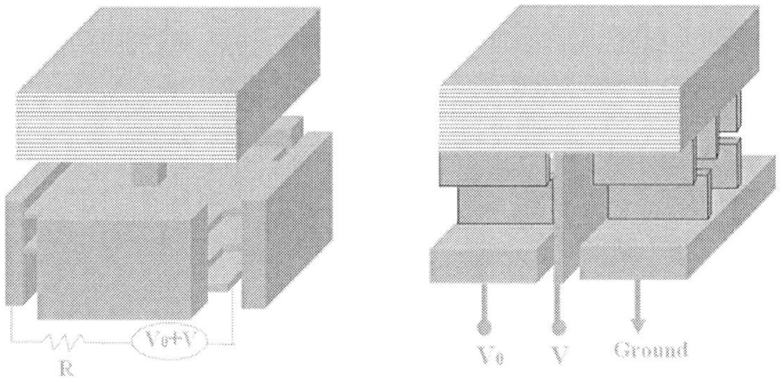

Figure 17. Schematic diagram of a double-flexure piston micromirror (left) and a double-comb tilting micromirror (right)

A multiple-scan imaging theory will be developed for maskless EUV lithography, which includes the effects of both deterministic image blur caused by the wafer scan and stochastic blur due to laser's random timing jitter. More detailed analysis based on partial coherent simulation has shown that the image blur can impact CD control, image slope, line-edge roughness (LER), and process window in maskless lithography [77]. It is also shown that the effects of image blur do not impose serious constraints on lithographic performance in low throughput operation. However, when throughput is high, significant CD shift, lower image slope and higher LER, and process window degradation are observed. The dependence of CD on the wafer's scan speed and the distance between neighboring features will be an important issue of maskless OPC methodology as well.

Assume the wafer is scanned/exposed n_s times at a uniform speed V in x direction, and the stationary/unblurred image corresponding to each scanning is the same: $f(x, y)$. Due to the wafer scanning, each individual exposure will suffer from a deterministic blur effect. Ideally, all these n_s images will overlap perfectly with each other and no further image blur will be observed. However, due to the timing jitters of EUV laser pulses, all the individual images will have random placement errors assumably with the same statistical characteristics. Consequently, the final image as the sum of all individual image profiles will be further blurred. Without loss of generality, we shall ignore y dependence and focus on the 1-D profile $f(x)$ which can be readily extended to the 2-D analysis. Given a rectangle-function laser pulse lasting for a period of T and the time delay of the *ith* pulse as τ_i (relative to the perfect timing determined by the repetition rate/frequency of the laser pulses, see Figure 18), the deterministic blurred image due to wafer scanning is:

$$I_i(x, \tau_i) = \frac{1}{T} \int_{\tau_i}^{T+\tau_i} f(x - Vt)dt. \tag{2-1}$$

Here we have not considered the stochastic blur effect yet since above formula is only for one scan/exposure. Set: $\mu = Vt$, equation (2-1) can be written as:

$$I_i(x, \tau_i) = \frac{1}{VT} \int_{V\tau_i}^{V(T+\tau_i)} f(x - \mu)d\mu. \tag{2-2}$$

We define a "scanning pupil" function as below:

$$p_s(\mu, \tau_i) = \begin{cases} 1/VT \ (V\tau_i < \mu < V\tau_i + VT) \\ 0 \ (\mu < V\tau_i, \ \mu > V\tau_i + VT) \end{cases} \tag{2.3}$$

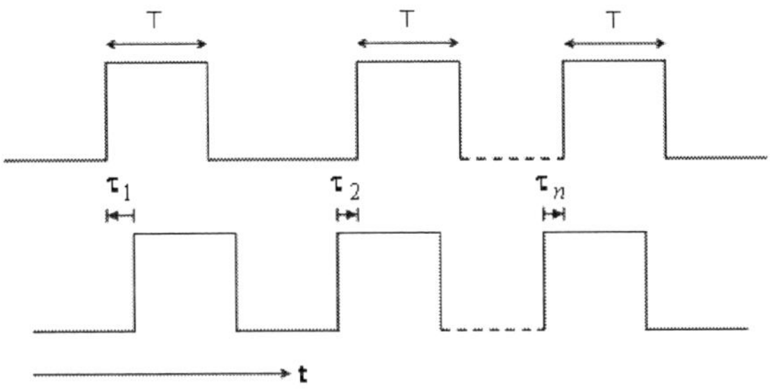

Figure 18. The ideal rectangle EUV laser signals (bottom) and the jittering (top) signals during a multiple-scan exposure

Equation (2-2) can be expressed as a convolution:

$$I_i(x,\tau_i) = \int_{-\infty}^{+\infty} f(x-\mu)\cdot p_s(\mu,\tau_i)d\mu = f \otimes p_s. \tag{2-4}$$

The final image is the sum of all the individual images, i.e., $I_t(x) = \sum_{i=1}^{n_s} I_i(x,\tau_i)$ wherein the time delay τ_i is a random variable described by a probability density function $q(\tau_i)$ that can be experimentally characterized. Strictly speaking, laser pulse intensity amplitude and length T are also random variables, but here we treat them as constant parameters to simplify our analysis. The expected total image intensity is

$$\bar{I}_t(x) = \hat{E}[\sum_{i=1}^{n_s} I_i(x,\tau_i)] = \sum_{i=1}^{n_s} \hat{E}[I_i(x,\tau_i)] = n_s \hat{E}[I_i(x,\tau_i)].$$

Here, the symbol $\hat{E}$ represents the statistical average (or expected function) and we have assumed that each individual image has the same expected profile. It should be kept in mind that the *ith* time delay τ_i will simply shift the position of the corresponding individual (blurred) image without changing its shape, i.e., $I_i(x,\tau_i) = I_{i0}(x - V\cdot\tau_i)$, where $I_{i0}(x)$ is the blurred image with zero delay. Thus we can use the probability density function $q(\tau_i)$ to rewrite the expected total intensity as

$$\bar{I}_t(x) = n_s \hat{E}[I_i(x,\tau_i)] = n_s \int_{-\infty}^{+\infty} I_{i0}(x - V\cdot\tau_i)q(\tau_i)d\tau_i = n_s \int_{-\infty}^{+\infty} I_{i0}(x - V\cdot\tau)q(\tau)d\tau. \tag{2-5}$$

Set: $z = V\cdot\tau$, then above equation becomes

$$\bar{I}_t(x) = n_s \int_{-\infty}^{+\infty} I_{i0}(x-z)q_z(z)dz = n_s \cdot I_{i0} \otimes q_z, \tag{2-6}$$

where we define:

$$I_{i0}(x) = I_i(x, \tau = 0) = f \otimes p_s(\tau = 0) = f \otimes p_{s0}, \; q_z(z) = q(z/V)/V.$$

Combining equation (2-4) and (2-6) yields:

$$\bar{I}_t(x) = n_s \cdot f \otimes p_{s0} \otimes q_z \tag{2-7}$$

Using the convolution theorem, we obtain an important relation in the spectrum domain f_x:

$$\bar{I}_t^F(f_x) = n_s \cdot F(f_x)\, P_{s0}(f_x)\, Q_z(f_x) \tag{2-8}$$

Here, $\bar{I}_t^F(f_x)$, $F(f_x)$, $P_{s0}(f_x)$, $Q_z(f_x)$ are the Fourier transform of $\bar{I}_t$, f, p_{s0}, q_z respectively. The subscript "0" in p_{s0} and P_{s0} represents the zero-delay function ($\tau = 0$).

The simple relation of (2-8) indicates the image blur (in maskless EUV lithography) induced by both wafer scanning and laser jittering can be removed by spectral engineering of EUV wave-front at the Fourier/spectrum plane. For example, if we can introduce an "inverse" filter [80] on the Fourier plane of the image, $M_o(f_x) = \dfrac{1}{P_{s0}(f_x) \cdot Q_z(f_x)}$, it will restore the original image by eliminating both deterministic blur caused by wafer scanning and the statistically-averaged blur caused by the probability density function of the timing jitter. On the other hand, the inverse filtering will suffer from the ill-defined singularity problem if the spectrum of either zero-delay scanning pupil function or the probability density function has zero points within the frequency window of interest. Also, if these two functions are small, the noise can be significantly amplified resulting in a degraded signal-to-noise ratio; thus other type of filters such as Wiener filter [80] may be considered.

Wave-front modulation in general will require an independent control of both phase and amplitude of reflected EUV light at the Fourier plane, which is a non-trivial but achievable engineering task. Further discussion on this topic can be found in other sections that describe EUV phase-shifting mask and multilayer reflective optics. Moreover, it is valuable to study the influence of scan times n_s on the variance of the resist image's position shift. Normally resist CD is measured at the threshold intensity (e.g., 30% of the open-field intensity). To avoid numerically finding the threshold value of the total intensity in a multiple-scan exposure while still being able to gain the physical insight of its statistical characteristics, we assume that the shift of the final image (or the shift of a resist pattern) is the average of all the individual images' shift: $\sum_{i=1}^{n_s}(V\tau_i)/n_s$. Therefore, the variance (symbol: $\hat{\sigma}^2$) of the resist position's shift is given as:

$$\hat{\sigma}^2[\sum_{i=1}^{n_s}(V\tau_i)/n_s] = \frac{1}{n_s^2}\hat{\sigma}^2[\sum_{i=1}^{n_s}V\tau_i] = \frac{1}{n_s}\hat{\sigma}^2[V\tau_i] \quad (2\text{-}9)$$

It is evident that the variance of multiple-scan random displacement is significantly reduced (by a factor of $1/n_s$) from the one-scan case. However, the improved placement accuracy and printing redundancy by increasing scan times directly results in a throughput penalty. Moreover, the impact of image displacement on the process robustness also depends on the proximity effects of surrounding features. Therefore, an optimal n_s to maximize the process yield must be rigorously searched with a stochastic simulation coupled imaging model.

11-16 nm (half pitch) CMOS generation can be the possible insertion point for maskless EUV lithography based on several observations. First, maskless technology is a difficult challenge to lithographic infrastructure which will not be ready in the near future to meet the demand of volume production. Secondly, both micromirror array and data handling issues still need a long development time and the pressure of mask defects and costs will not be severe until the EUV generation (11-16 nm half pitch). Moreover, assuming the pulse-to-pulse timing jitter of a EUV laser to be 10 ns, a scan speed of 0.5 nm/ns (or 50 cm/s) will cause an unacceptable random placement error of 5 nm. In order to increase the wafer's scan speed for higher throughput, we need much tighter control of EUV laser's timing jitters.

To meet a scanning throughput of 1 cm^2/sec (about 5 300-mm wafers/hour), the flash rectangle height (on the wafer) needs to be 1/50 cm based on the 50 cm/s scan speed. If the micromirror chip height is 2 cm, the demagnification will be 100. For the 16-nm CD or 8-nm spot size (assuming a 2×2 micromirror array prints a 16-nm feature), a demagnification of 100 requires the mirror size to be 0.8 μm which is challenging but not impossible. Moreover, for a 20 KHz laser, each flash needs to expose 1/20000 cm^2 area (assuming zero overlap between the exposure areas of neighboring flashes), which requires the flash width of 1/400 cm (flash height is 1/50 cm as calculated before). In other words, a $2\times0.25\,cm^2$ micromirror array is needed. This case study is based on a one-array and one-scan operation, but the actual redundant exposure scheme may need multiple exposures for each pixel with multiple arrays. A typical chip that is manufactured by a foundry can contain about 10 such arrays. Since the field size of EUV maskless lithography is much less than that of a conventional EUV system, much lower cost and smaller footprint of a single writer (using one micromirror chip) will allow us to pack many writers together in a system to meet the high-volume throughput requirement of 100 wafers per hour.

10. Imaging Theory of Projection Lithography

In this section, we shall present the imaging theory of photon-based projection lithography which includes both DUV and EUV lithography. Our goal is to focus on those materials that are "directly" related to imaging optics, rather than the details of the electromagnetism and wave propagation theory. The references [78-81] that we frequently refer to and follow have been listed for the readers to study the topics of their interest.

10.1. Basic Principles of Electromagnetic Field and Geometrical Optics

The basic principles of projection lithography are described by Maxwell's equations [78, 79]:

$$\nabla \times \mathbf{e} = -\frac{\partial \mathbf{b}}{\partial t}$$

$$\nabla \times \mathbf{h} = \frac{\partial \mathbf{d}}{\partial t} + \mathbf{j}$$

$$\nabla \cdot \mathbf{d} = \rho_e$$

$$\nabla \cdot \mathbf{b} = 0$$

$$\frac{\partial \rho_e}{\partial t} + \nabla \cdot \mathbf{j} = 0$$

$$\mathbf{j} = \sigma_z \mathbf{e}$$

$$\mathbf{d} = \varepsilon_z \mathbf{e}$$

$$\mathbf{b} = \mu_z \mathbf{h}$$

Here, $\mathbf{e}$: electric field, $\mathbf{b}$: magnetic induction, $\mathbf{h}$: magnetic field, $\mathbf{d}$: electric displacement, $\mathbf{j}$: electric current density, ρ_e : electric charge density, σ_z : medium conductivity, ε_z : medium permittivity, μ_z : magnetic permeability. As we shall see soon, a rigorous study of EUV multilayer reflective optics needs to start from Maxwell's equations.

Another basic principle of geometrical optics is Snell's Law which states that for any light ray travelling through different parts (such as lens and recording medium) of an imaging system, its $n \cdot \sin\theta$ value remains the same:

$$n_1 \cdot \sin\theta_1 = n_2 \cdot \sin\theta_2 = n_3 \cdot \sin\theta_3 = \ldots = n_i \cdot \sin\theta_i . \qquad (3\text{-}1)$$

n is the refractive index of the medium and θ is the incident angle to the normal axis. As shown in Figure 19, the magnification of an optical imaging system can be described by the ratio of the numerical aperture sustained from the center of objective (mask) plane (

$NA_o = n_o \cdot \sin\theta_o$) to the numerical aperture sustained from the image (wafer) plane ($NA_i = NA = n_i \cdot \sin\theta_i$):

$$M = \frac{NA_o}{NA} = \frac{n_o \cdot \sin\theta_o}{n_i \cdot \sin\theta_i}. \quad (3\text{-}2)$$

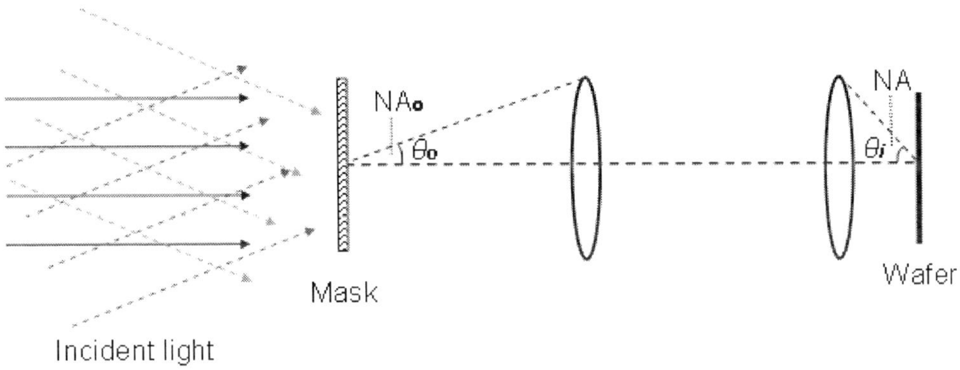

Figure 19. A simplified diagram representing a lithographic imaging system

n_o and n_i are the refractive indexes of the media as indicated in Figure 19. In a scanner system, the minimum resolvable feature size on the mask is 1/*M* times of the minimum resolvable feature on the wafer which is usually described by:

$$\text{Resolution(at wafer scale)} = k_1 \frac{\lambda}{NA}. \quad (3\text{-}3)$$

k_1 is a factor that include all optical and chemical effects except exposure wavelength and NA, thus it is an important number to characterize the complexity of a lithographic process, i.e., a lower k_1 indicates a more complex lithographic process. We are mainly concerned about the resolution capability on the wafer; therefore, NA is usually referred to the sine of the maximum half-angle sustained from the wafer plane (NAi), instead of NAo at the objective (mask) plane as described in some books. The magnification value, *M*, determines the size of mask and is chosen to be 1/4 in many modern projection scanners due to the manufacturing limit. According to equations (3-2) and (3-3), it is impossible to improve the resolution capability of an imaging system by decreasing its magnification (i.e., higher demagnification) without changing the numerical aperture.

10.2. Light Propagation and Spatial Coherence

There are two types of coherence involved in a light wave field: temporal coherence and spatial coherence. Temporal coherence describes the relation between light waves at different times but at the same point in space, while spatial coherence looks at the relation between light waves at different points in space at the same time. If a light wave is an ideal monochromatic (harmonic) wave, its length will be infinity and the wave is temporally coherent. In a lithographic system, spatial coherence evolution of a wave-front during its propagation is a more important issue. Unless specified otherwise, coherence of a wave field in this chapter means "spatial coherence". It should be kept in mind that unlike a "laser" (e.g., 193 nm excimer laser with good spatial coherence) that is often mentioned in the lithographic literature, the wave field on the surface of a scanner's light source is spatially incoherent. It should also be distinguished from the spatially coherent light beams that are used in interference lithography to create interferometric patterns.

The differential equation governing the light wave propagation can be derived from Maxwell's equations. Both electric and magnetic fields are vectors and in general the vector diffraction theory is needed to accurately describe the wave diffraction phenomena. However, the scalar theory treats the wave disturbance at any position and time as a scalar function, which significantly simplifies the analysis and is a good approximation when the diffracting structures are large compared with the length of wave. The basic scalar wave equation that describes the time-independent complex disturbance function U is the Helmholtz equation [80]

$$(\nabla^2 + k^2)U = 0\,, \tag{3-4}$$

where ∇^2 is the Laplacian operator and k is the wave number ($2\pi/\lambda$, λ is wavelength). Calculation of the complex disturbance $U(P)$ at a point P in space can be accomplished using the well-known Green's theorem that can be found in most of mathematical physics textbooks and will not be repeated here. Based on the Helmholtz equation, it can be proved that a plane wave with its frequency higher than $1/\lambda$ (called *evanescent wave*) will decay rapidly over a short distance thus is not recoverable by a conventional imaging system. Mathematical expression of the Huygens-Fresnel principle according to the first Rayleigh-Sommerfeld solution [80] is given below:

$$U(P) = \frac{1}{i\lambda}\iint_S U(Q)\frac{\exp(ikR)}{R}\cos(\gamma)ds\,, \tag{3-5}$$

where Q is an arbitrary point on the secondary wave surface S, R is the distance between point Q and the target location P, and γ is the angle between the outward normal $\vec{n}$ at point Q and the vector $\vec{R}$ joining P to Q. The effects of propagation angle $\cos(\gamma)$ and imaginary number $1/(i\lambda)$ are often ignored as a first-order approximation, which means we only consider the optical phase delay and amplitude decay over a distance R (required by the energy conservation of secondary spherical wave durng its propagation).

According to the Huygens-Fresnel principle, every point of a wave-front may be considered as a secondary disturbance source which causes spherical wavelets to propagate downstream to form new wave-front later, i.e., the wave-front at a later time may be constructed using the envelope of these wavelets. The intensity of light at any point Q is defined to be the time-average (denoted as $<\cdot>$) of the light amplitude (absolute value), i.e., $I(Q) = <u(Q,t)u^*(Q,t)>$ where * is the notation for the conjugate of a complex function.

Using an example as shown in Figure 20, two pinholes Q_1 and Q_2 are considered as the centers of secondary disturbances and the complex disturbance at *P* of the downstream wavefront may be calculated by

$$u(P,t) = k_1 u(Q_1, t - R_1/c) + k_2 u(Q_2, t - R_2/c) = k_1 u_1(t - R_1/c) + k_2 u_2(t - R_2/c),$$

where R_1/c and R_2/c are the time delays needed for light to travel from Q_1 and Q_2 to *P* respectively. The optical phase delay (exponential term) in the Huygens-Fresnel principle is equivalent to the introduction of the time delays in above relation. Moreover, according to the Huygens-Fresnel principle, the factors k_1 (different from k1 in equation 3-3) and k_2 (different from k2 in equation 3-45) are inversely proportional to R_1 and R_2, and also depending on the pinhole size and the incident angle of wave propagation at Q_1 and Q_2. Using the Kirchhoff's diffraction formula [79], k_1 and k_2 are imaginary numbers and the light intensity at *P* is

$$I(P,t) = <u(P,t)\,u^*(P,t)> = k_1 k_1^* <u_1(t - R_1/c)u_1^*(t - R_1/c)> +$$
$$k_1 k_2^* <u_1(t - R_1/c)u_2^*(t - R_2/c)> + k_2 k_1^* <u_2(t - R_2/c)u_1^*(t - R_1/c)> + k_2 k_2^* <u_2(t - R_2/c)u_2^*(t - R_2/c)>.$$

Since

$$<u_1(t - R_1/c)u_1^*(t - R_1/c)> = <u_1(t)u_1^*(t)> = I(Q_1)$$

and

$$<u_2(t - R_2/c)u_2^*(t - R_2/c)> = I(Q_2),$$

light intensity at *P* can be written as

$$I(P,t) = |k_1|^2\, I(Q_1) + |k_2|^2\, I(Q_2) + 2|k_1 k_2|\,\mathrm{Re}[\Gamma(Q_1, Q_2, \frac{R_2 - R_1}{c})].$$

Here, we define mutual coherence function as

$$\Gamma(Q_1, Q_2, \tau) = <u(Q_1, t + \tau)u^*(Q_2, t)>. \quad (3\text{-}6)$$

If $Q_1 = Q_2$, the mutual coherence function will be the auto-coherence which becomes local light intensity if we further assume $\tau = 0$. To describe the spatial coherence of light field between two arbitrary points, e.g., Q_1 and Q_2, a concept of (spatial) mutual intensity is introduced below,

$$J(Q_1,Q_2)=\Gamma(Q_1,Q_2,0)=<u(Q_1,t)u^*(Q_2,t)>. \tag{3-7}$$

If light from Q_1 and Q_2 is incoherent and do not interfere, their mutual intensity will be zero. If $Q_1 = Q_2 = Q$, their (coherent) mutual intensity is equal to the light intensity at point Q.

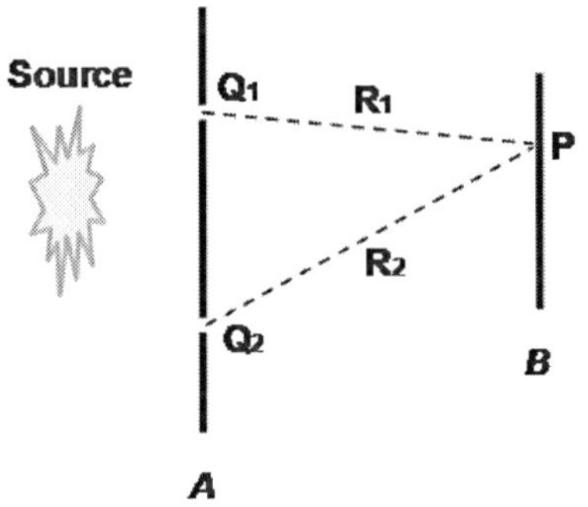

Figure 20. Light propagation through two pin holes

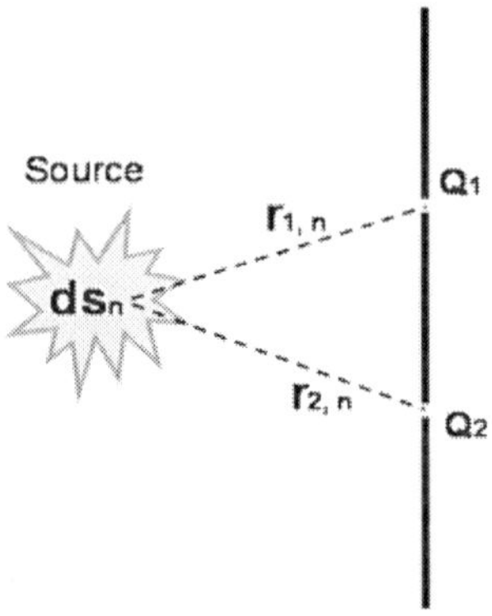

Figure 21. Propagation of mutual intensity of light from the light source to a surface

Based on a similar reasoning using the Huygens-Fresnel principle, wave vibrations at Q_1 and Q_2 may be traced back to the light source which is divided into many small elements as shown in Figure 21. For example, the nth vibration $u_n(Q_1,t)$ is caused by the upstream vibration "v_n" emitted from light source at the time of $t-r_{1,n}/c$. The field at any point of the wave front (e.g., Q_1 or Q_2) is the superposition of small vibrations caused by all the elements on the light source,

$$u(Q_1,t)=\sum_n u_n(Q_1,t), \tag{3-8}$$

$$u(Q_2,t) = \sum_n u_n(Q_2,t) . \tag{3-9}$$

Here, $u_n(Q_1,t)$ and $u_n(Q_2,t)$ are defined to be the disturbance elements at Q_1 and Q_2 respectively, both of which are caused by *the same spatial point* of the light source. In general, these two vibrations can be traced back to the same spatial point of the source, but not necessarily to the same (previous) time depending on their distances $r_{1,n}$ and $r_{2,n}$ as shown in Figure 21. The mutual intensity is then calculated as

$$J(Q_1,Q_2) = < u(Q_1,t)u^*(Q_2,t) > = \sum_n < u_n(Q_1,t)u_n^*(Q_2,t) > + \sum\sum_{i \neq j} < u_i(Q_1,t)u_j^*(Q_2,t) > . \tag{3-10}$$

It should be emphasized that any two points on the source are mutually incoherent (i.e., their phase and amplitude are both random variables) such that the second term in equation (3-10) is zero. However, when wave-front propagates, it will gradually gain certain spatial correlation (between different points of the downstream wave-front) due to the first term in this equation. Therefore, the spatial coherence of a scanner's illumination system does not originate from the coherence of light source (which is zero due to mutually incoherent source assumed), but from the propagation of wave itself obeying the Huygens-Fresnel principle. Next, we shall apply a single variable "v_n" (light disturbance from the nth element of a light source) to replace the downstream wave variables $u_n(Q_1,t)$ and $u_n(Q_2,t)$, both of which are emitted from the same source element at different times but reaching Q_1 and Q_2 simultaneously (at time *t*). Using a single variable "v_n" to replace $u_n(Q_1,t)$ and $u_n(Q_2,t)$ is based on a hidden assumption that the coherent length of light is so long (e.g., for 193-nm laser, the coherence length is on the order of 0.15 m) that both light vibrations $u_n(Q_1,t)$ and $u_n(Q_2,t)$ are induced by the same temporal wave train (a function of time) emitted from the same source element (but in general at different times depending on their optical path). In principle, if two vibrations are induced by different temporal wave trains (even from the same spatial point of the light source), the same variable can not be used for them as temporal wave trains emitted at different times do not have any correlation. Therefore, equations (3-8) and (3-9) can be re-written as

$$u(Q_1,t) = \sum_n u_n(Q_1,t) = \sum_n v_n(t - r_{1,n}/\mathrm{c}) / r_{1,n} ,$$

$$u(Q_2,t) = \sum_n u_n(Q_2,t) = \sum_n v_n(t - r_{2,n}/\mathrm{c}) / r_{2,n} .$$

Here, the effect of wave propagation angle is ignored while the effect of wave decay over a distance is included. For the purpose of simplicity, the "*k*" factors we previously used are not explicitly shown in above two equations. The mutual intensity then becomes

$$J(Q_1,Q_2)=\sum_n < u_n(Q_1,t)u_n^*(Q_2,t) > = \sum_n < v_n(t-r_{1,n}/c)\, v_n^*(t-r_{2,n}/c) > /(r_{1,n}\, r_{2,n}). \quad (3\text{-}11)$$

Since the coherence length of light wave trains is assumed to be long, complex amplitudes $v_n(t-r_{1,n}/c)$ and $v_n(t-r_{2,n}/c)$ emitted from the same wave train have a constant optical path difference $r_{1,n}-r_{2,n}$ at any time. Consequently, it is straightforward to prove that

$$< v_n(t-r_{1,n}/c)v_n^*(t-r_{2,n}/c) > = < v_n(t)v_n^*[t-(\frac{r_{2,n}-r_{1,n}}{c})] > =< v_n(t)v_n^*(t)\exp[i\varpi(\frac{r_{1,n}-r_{2,n}}{c})] >$$

$$=< v_n(t)v_n^*(t) > \exp[i\varpi(\frac{r_{1,n}-r_{2,n}}{c})] = \exp[i\varpi(\frac{r_{1,n}-r_{2,n}}{c})]\, I(s_n)ds_n.$$

Substitute this result into equation (3-11), the mutual intensity can be given as

$$J(Q_1,Q_2)=\sum_n \exp[i\varpi(\frac{r_{1,n}-r_{2,n}}{c})]\, I(s_n)/(r_{1,n}\, r_{2,n})\, ds_n = \sum_n \frac{\exp[i\bar{k}(r_{1,n}-r_{2,n})]\, I(s_n)}{r_{1,n}\cdot r_{2,n}} ds_n \cdot$$

Using an integral to represent above sum of discrete disturbance elements, we have

$$J(Q_1,Q_2)=\iint_S I(S)\frac{e^{i\bar{k}(r_1-r_2)}}{r_1\, r_2} ds \quad (3\text{-}12)$$

This simple integral result to describe how mutual intensity (or spatial coherence) propagates from an incoherent light source is called Van Cittert-Zernike theorem, which is one of the most important theorems of modern optics.

Moreover, it is a general question of interest to study how mutual intensity evolves when light waves propagates from an upstream surface **A** to a downstream surface **B** as shown in Figure 22. A simple method to illuminate how mutual intensity propagates is to similarly start with the Huygens-Fresnel principle in the integral form. Again, for the purpose of simplicity, we shall ignore the effects of wave propagation angle and imaginary coefficient that are included in the Kirchhoff's approximation and assume long quasi-monochromatic wave trains.

By definition, the mutual intensity is $J(P_1,P_2) =< u(P_1,t)u^*(P_2,t) >$. The wave disturbances on surface **B** may be related to the upstream wave disturbances using the Huygens-Fresnel principle. In particular, the complex amplitude of two arbitrary wave disturbances can be written as

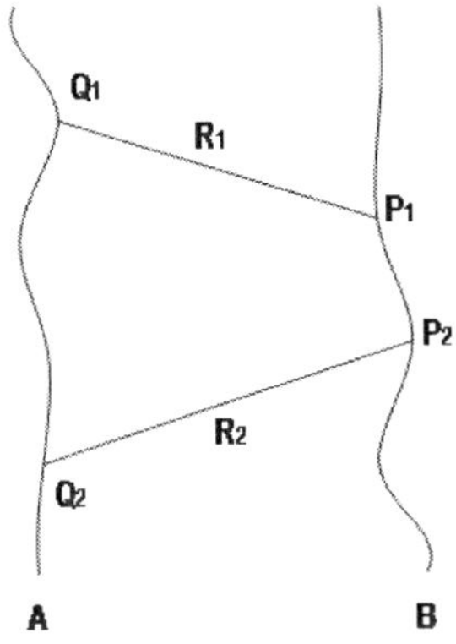

Figure 22. Propagation of mutual intensity of light from surface A to surface B

$$u(P_1,t)=\iint_A \frac{u(Q_1,t-R_1/c)}{R_1}\,ds\,,$$

$$u(P_2,t)=\iint_A \frac{u(Q_2,t-R_2/c)}{R_2}\,ds'.$$

where $R_1 = R_1(Q_1,P_1)$, $R_2 = R_2(Q_2,P_2)$ are distance variables from P_1 and P_2 to the upstream surface **A** respectively. Substituting $u(P_1,t)$ and $u(P_2,t)$ into $J(P_1,P_2)=<u(P_1,t)u^*(P_2,t)>$ and interchanging orders of integration and averaging, we have

$$\begin{aligned} J(P_1,P_2) &= \iint_A \iint_A \frac{<u(Q_1,t-R_1/c)\,u^*(Q_2,t-R_2/c)>}{R_1R_2}\,ds\,ds' \\ &= \iint_A \iint_A \frac{<u(Q_1,t)\,u^*[Q_2,t-(R_2-R_1)/c]>}{R_1R_2}\,ds\,ds' = \iint_A \iint_A \Gamma[Q_1,Q_2,\frac{R_2-R_1}{c}]\frac{dsds'}{R_1R_2}. \end{aligned} \qquad (3\text{-}13)$$

Similar to the reasoning used to derive the Van Cittert-Zernike theorem, for a wave train with a long coherence length, we assume $u[Q_2,t-(R_2-R_1)/c]$ and $u(Q_1,t)$ are emitted from the same wave train and they have a constant optical path difference $R_1 - R_2$ at any time. Thus

$$\begin{aligned} &\Gamma[Q_1,Q_2,\frac{R_2-R_1}{c}] =< u(Q_1,t)\,u^*[Q_2,t-(R_2-R_1)/c]> \\ &=< u(Q_1,t)\,u^*(Q_2,t)\exp[-i\varpi(R_2-R_1)/c]>=< u(Q_1,t)\,u^*(Q_2,t)>\exp[-i\varpi(R_2-R_1)/c] \\ &= J[Q_1,Q_2]\exp[i\varpi(R_1-R_2)/c]. \end{aligned}$$

Substituting this result into equation (3-13) yields the basic law for mutual intensity to propagate from one to the other surface

$$J(P_1, P_2) = \iint_A \iint_A \frac{J(Q_1, Q_2)}{R_1 R_2} \exp(i\varpi \frac{R_1 - R_2}{c}) ds ds'. \tag{3-14}$$

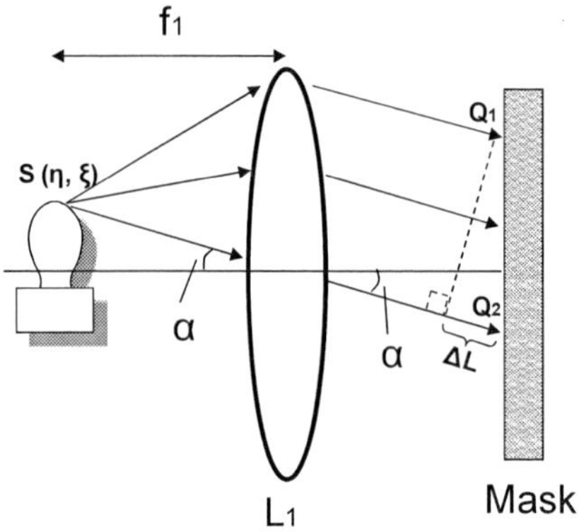

Figure 23. Köhler's illumination method

Next, we shall examine the spatial coherence of incident light wave on the mask surface in a modern lithographic system. Illumination systems in a scanner implement Köhler's method as shown in Figure 23, in which the light source is located at the focal plane of a condenser lens (L1) and light rays emitted from any point on the source will travel through this condenser lens and become parallel light rays before they hit the mask. As a result, the energy emitted from any point source is uniformly distributed across the mask, and the total intensity of light incident on the mask is always uniform regardless of the non-uniformity of source brightness.

From equation (3-12), we can see that the difference between a free-space and through-lens mutual intensity propagation is mainly on how the distance variable $1/(r_1 r_2)$ and the exponential term are treated. Again, this needs a physical interpretation of the Huygens-Fresnel principle. First, the distance variable $1/(r_1 r_2)$ affects the amplitude (or intensity) of light wave propagating to a downstream surface. As we analyzed, light rays (emitted from any point source) travelling through a condenser lens will become parallel light rays with uniformly-distributed intensity regardless what their corresponding "distances" are. Thus we expect distance variable $1/(r_1 r_2)$ does not play an explicit role in calculating the downstream amplitude on the mask when a condenser lens is used at an upstream location. Another factor that will impact the downstream wave-front is the optical path difference expressed as the exponential term, which is only related to the optical phase difference between two points. Thus, for Köhler's illumination, the formula (3-12) to calculate the mutual intensity should be slightly modified by removing the distance variable $1/(r_1 r_2)$ and replacing the exponential term with an optical phase difference (e.g., $e^{i(\Phi_1 - \Phi_2)}$) between two arbitrary points on the mask,

$$J(Q_1, Q_2) = \iint_S A_0\, e^{i(\Phi_1 - \Phi_2)}\, ds\,. \tag{3-15}$$

Here we replace the source light intensity distribution with a constant A_0 by assuming the source has uniform light brightness. The optical phase can be calculated by a simple method as described below. We draw a dashed line which is perpendicular to the propagation direction of the (parallel) light rays behind the lens. According to the Fermat's principle and through-lens focusing behaviour of parallel light rays, those points at which the light rays are intersected by the dashed line have the same phase. Using points $Q_1(x_1, y_1)$ and $Q_2(x_2, y_2)$ on the mask illuminated by light rays emitted from a source point (η, ξ) as an example, the optical path difference between these two points is ΔL as indicated in Figure 23,

$$\Phi_1 - \Phi_2 = \frac{2\pi}{\lambda}\Delta L = \frac{2\pi}{\lambda}[\sin\varphi \cdot (x_1 - x_2) + \sin\alpha \cdot (y_1 - y_2)]\,.$$

When φ and α (φ is the incident angle on the η plane pointing to the reader, perpendicular to the mask plane) are small, above phase difference can be approximated as

$$\Phi_1 - \Phi_2 \approx \frac{2\pi}{\lambda}[\frac{\eta}{f_1}(x_1 - x_2) + \frac{\xi}{f_1}(y_1 - y_2)]\,.$$

Substituting this approximation into equation (3-15) yields

$$J(Q_1, Q_2) \approx \iint_S A_0 \cdot \exp\{\frac{2\pi i}{\lambda}[\frac{\eta}{f_1}(x_1 - x_2) + \frac{\xi}{f_1}(y_1 - y_2)]\}\, ds \propto \frac{2J_1(v)}{v}, \tag{3-16}$$

where $v = \frac{2\pi}{\lambda}\frac{\xi_0}{f_1} \cdot \sqrt{(x_1 - x_2)^2 + (y_1 - y_2)^2}$ and $J_1(v)$ is the Bessel function of the first kind. In above calculation, we have assumed a circular uniform light source whose semi-aperture angle is shown in Figure 23 ($\sin\alpha_0 \approx \xi_0 / f_1$ is the sine of the maximum half-angle sustained by the source). The mutual intensity reaches the maximum value when Q_1 and Q_2 coincide, and it vanishes when the distance between Q_1 and Q_2 is equal to $0.61\lambda/\sin\alpha_0$. A further increase in the distance reintroduce certain correlation but it decays gradually over distance. Physical insight can be gained from this small-angle approximation; for example, when the sustained angle of the light source approaches zero (i.e., a point source), the mutual intensity decreases very slowly from its maximum value (when v=0) which corresponds to a coherent illumination. Namely, spherical light rays emitted from an ideal point source (at the focal plane) become parallel after they pass through the condenser lens and the light disturbance at every point of the mask is coherent with the disturbance at other points. On the other hand, when the size of light source increases, the radius ($0.61\lambda/\sin\alpha_0$) of the (circular) coherent area can significantly drop to the wavelength scale.

It is apparent the expression of (3-16) is a space-invariant relation which can be rewritten as:

$$J(x_1-x_2,y_1-y_2)=\iint_S A_0\cdot\exp\{\frac{2\pi i}{\lambda}[\frac{\eta}{f_1}(x_1-x_2)+\frac{\xi}{f_1}(y_1-y_2)]\}ds$$
$$=\int_{-\infty}^{+\infty}\int S(\eta,\xi)\cdot\exp\{\frac{2\pi i}{\lambda f_1}[\eta(x_1-x_2)+\xi(y_1-y_2)]\}d\eta\, d\xi\,. \tag{3-17}$$

Here, $S(\eta,\xi)$ is the light intensity distribution of the source which is equal to A_0 at every point of the source and zero outside the source. A space-invariant function is dependent only on the difference in coordinates (distances) between two points but not their specific values. Therefore above analysis can be extended to make a general conclusion that *under the small-angle approximation, the mutual intensity (as a function of the space-invariant variables) of the downstream wave-front is proportional to the Fourier transform of the light intensity distribution of the source.* In the extreme case of a point source or an infinitely large source, the mutual intensity distribution on the mask will be a constant (coherent) or a Dirac delta function (incoherent).

When α and φ are large (e.g., close to $\pi/2$), above small-angle approximation does not hold well and we must use a more strict relation:

$$\Phi_1-\Phi_2=\frac{2\pi}{\lambda}\Delta L=\frac{2\pi}{\lambda}[\frac{\eta}{\sqrt{\eta^2+f_1^2}}(x_1-x_2)+\frac{\xi}{\sqrt{\xi^2+f_1^2}}(y_1-y_2)]\,.$$

Substituting this relation into equation (3-15) yields

$$J(Q_1,Q_2)=\iint_S A_0\cdot\exp\{\frac{2\pi i}{\lambda}[\frac{\eta}{\sqrt{\eta^2+f_1^2}}(x_1-x_2)+\frac{\xi}{\sqrt{\xi^2+f_1^2}}(y_1-y_2)]\}ds\,.$$

We derived equation (3-14) under the assumption that the medium between the source and the image points is homogeneous. For a heterogeneous optical imaging system usually consisting of the object, lens, free space, and recording medium, the simple terms corresponding to the optical phase delay and amplitude decay over a distance (in equation (3-14)) must be replaced with a more general propagation function (the mutual intensity formula under Köhler's illumination is one particular example). In general, we define a mapping function $H(S,Q)$ as the complex disturbance field at Q caused by a point source S which emits light rays to pass through an imaging system. To answer the question why we can apply a mapping function $H(S,Q)$ to a partially coherent imaging system, we shall give a short explanation here and readers may wish to consult the textbook [80] by Goodman for a detailed interpretation. If we go back to the equation governing the mutual intensity propagation (3-14), the term $\frac{1}{R_1 R_2}\exp(i\varpi\frac{R_1 - R_2}{c})$ is related with the fixed time/phase delay and amplitude decay during the wave propagation and it does not contain any statistical/time-average information embedded in the other term: $J(Q_1, Q_2)$. Here we introduce a general mapping function $H(S,Q)$ for a more complicated imaging system based on its physical role similar to the term in equation (3-14): $\frac{1}{R_1 R_2}\exp(i\varpi\frac{R_1 - R_2}{c})$. This is a time-independent term that only depends on the intrinsic structure of an imaging system regardless of coherent or incoherent illumination is used. It should not bring the confusion to the reader that the mapping function $H(S,Q)$ is valid only for a coherent imaging system. Thus for an extended light source, the mutual intensity can be rewritten as an integral

$$J(Q_1, Q_2) = \iint_S I(S)H(S,Q_1)H^*(S,Q_2)\,ds = \iint_S U(S,Q_1)U^*(S,Q_2)\,ds\,. \quad (3\text{-}18)$$

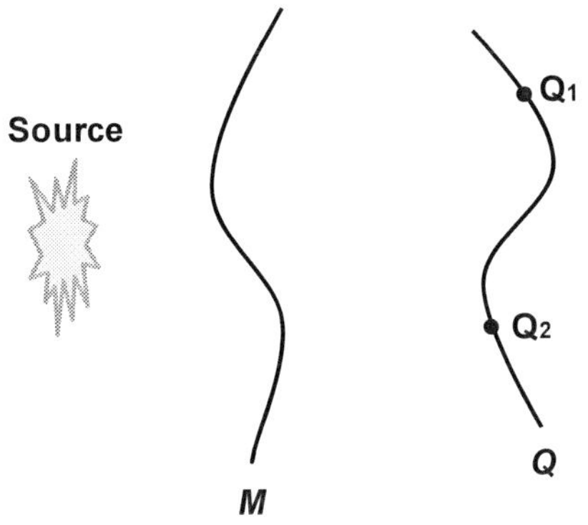

Figure 24. Propagation of mutual intensity in a heterogeneous medium

Here, the relation $U(S,Q) = \sqrt{I(S)}H(S,Q)$ is used to simplify above integral. Moreover, based on the Huygens-Fresnel principle, we consider both fields $U(S,Q_1)$ and $U(S,Q_2)$ as the summation of upstream disturbances from a midway surface M (e.g., the mask surface as we shall discuss in detail later) between source S and surface Q as shown in Figure 24.

$$U(S,Q_1) = \iint_M U(S,M')H(M',Q_1)\,ds',$$

$$U(S,Q_2) = \iint_M U(S,M'')H(M'',Q_2)\,ds''.$$

Using these two expressions, we have

$$U(S,Q_1)U^*(S,Q_2) = \iint_M \iint_M U(S,M')U^*(S,M'')H(M',Q_1)H^*(M'',Q_2)ds'\,ds''.$$

Thus the mutual intensity (equation (3-18)) can be written as

$$J(Q_1,Q_2) = \iint_M \iint_M J(M',M'')H(M',Q_1)H^*(M'',Q_2)\,ds'\,ds''. \quad (3\text{-}19)$$

Here, $J(M',M'')$ is the mutual intensity of the light incident on the midway surface M and $H(M',Q)$ is the propagation function that relates the light complex amplitude at a arbitrary point Q to the upstream disturbance at point M'. Apparently, equation (3-19) is a generalized form of the homogeneous equation (3-14) and can be used to analyze the mutual intensity of a heterogeneous optical system. In particular, when Q_1 and Q_2 coincide, the mutual intensity becomes the light intensity:

$$I(Q) = J(Q,Q) = \iint_M \iint_M J(M',M'')H(M',Q)H^*(M'',Q)\,ds'\,ds''. \quad (3\text{-}20)$$

In a scanner, surface M represents the object plane such as the mask; and the correlation between an image point on the mask and the induced image on the wafer can be characterized by the so-called object transmission function $G(x_M, y_M)$ and transfer function $T(x_M, y_M; x, y)$. Mathematically, the propagation function is expressed as

$$H(M,Q) = H(x_M, y_M; x, y) = G(x_M, y_M)T(x_M, y_M; x, y).$$

The propagation function $H(M,Q)$ contains the mask transmission information and describes how the image on a mask is projected onto the wafer; while the critical function used to examine the imaging performance of a scanner (excluding the mask) is the transfer function $T(x_M, y_M; x, y)$. Thus the image intensity is

$$I(x,y) = \iint_M \iint_M J(x_{M'}, y_{M'}; x_{M''}, y_{M''})\, G(x_{M'}, y_{M'})T(x_{M'}, y_{M'}; x, y)\cdot G^*(x_{M''}, y_{M''})T^*(x_{M''}, y_{M''}; x, y)dx_{M'}\,dy_{M'}\,dx_{M''}\,dy_{M''}. \quad (3\text{-}21)$$

According to the mutual intensity formula (3-17) under Köhler's illumination, $J(x_{M'}, y_{M'}; x_{M''}, y_{M''})$ is dependent only on the difference in coordinates (distances) between two illuminated points but not their specific values. This characteristic exists in the transfer function of a space-invariant (i.e., isoplanatic) linear imaging system. Thus we have

$$J(x_{M'}, y_{M'}; x_{M''}, y_{M''}) = J(x_{M''} - x_{M'}, y_{M''} - y_{M'}) = J(\Delta x_M, \Delta y_M),$$
$$T(x_{M'}, y_{M'}; x, y) = T(x - x_{M'}, y - y_{M'}) = T(\Delta x_{CM}, \Delta y_{CM}),$$
$$T^*(x_{M'}, y_{M'}; x, y) = T^*(x - x_{M'}, y - y_{M'}) = T^*(\Delta x_{CM}, \Delta y_{CM}).$$

In modern optics, the widely adopted technique to analyze the image performance is the Fourier transform and spectral analysis (or frequency filtering). For example, the lens is considered as a low-pass filter that collects certain range of frequency and performs inverse Fourier transform to form the image. In the frequency domain, it is convenient to model the size effect of light source using the parameter of "partial coherence", which is defined to be the ratio of the sine of the half-aperture angle of the source (sustained at the condenser lens) to the entrance pupil's numerical aperture (sustained at the mask plane). We assume that the reader has been familiar with Fourier transform and no details about this mathematical tool will be described here. The Fourier integrals of some related functions are listed below:

$$J(\Delta x_M, \Delta y_M) = \iint_{-\infty}^{+\infty} J_F(f_x, f_y) e^{-i2\pi(f_x \Delta x_M + f_y \Delta y_M)} df_x df_y ,$$

$$T(\Delta x_{CM}, \Delta y_{CM}) = \iint_{-\infty}^{+\infty} T_F(f_x, f_y) e^{-i2\pi(f_x \Delta x_{CM} + f_y \Delta y_{CM})} df_x df_y ,$$

$$G(x, y) = \iint_{-\infty}^{+\infty} G_F(f_x, f_y) e^{-i2\pi(f_x x + f_y y)} df_x df_y .$$

The reader should be aware that there are several conventions to define the Fourier transform. For example, one convention uses a positive sign ($+i$) in the exponential term while the other uses a negative sign ($-i$). While these two conventions are equivalent and produce the same physical results, we use the negative sign when performing the inverse Fourier transform unless specified otherwise. Using above integrals and the shift theorem of Fourier transform, it can be proved that [79]:

$$\begin{aligned} I(x, y) = &\int_{-\infty}^{+\infty} \ldots \int J_F(f_x, f_y) T_F(f_x + f_x', f_y + f_y') \, T_F^*(f_x + f_x'', f_y + f_y'') \cdot \\ &G_F(f_x', f_y') \, G_F^*(f_x'', f_y'') \cdot \exp\{-i2\pi[(f_x' - f_x'')x + (f_y' - f_y'')y]\} df_x df_y df_x' df_y' df_x'' df_y'' . \end{aligned} \quad (3\text{-}22)$$

Interchanging the integration order of equation (3-22) becomes

$$I(x, y) =$$
$$\iint_{-\infty}^{+\infty} J_F(f_x, f_y) \left| \iint_{-\infty}^{+\infty} T_F(f_x + f_x', f_y + f_y') \cdot G_F(f_x', f_y') \cdot \exp[-i2\pi(f_x' x + f_y' y)] df_x' df_y' \right|^2 df_x df_y .$$

(3-23)

In order to separate the contributions of the mask (mask object transmission function: $G(x_M, y_M)$) and the imaging system (image transfer function: $T(x_M, y_M; x, y)$), we introduce a function called *transmission cross-coefficient* (TCC) such that equation (3-22) can be written as

$$I(x,y) = \iiiint_{-\infty}^{+\infty} TCC(f_x^{'}, f_y^{'}; f_x^{''}, f_y^{''}) \cdot G_F(f_x^{'}, f_y^{'})\, G_F^{*}(f_x^{''}, f_y^{''}) \cdot \exp\{-i2\pi[(f_x^{'} - f_x^{''})x + (f_y^{'} - f_y^{''})y]\} df_x^{'} df_y^{'} df_x^{''} df_y^{''}, \quad (3\text{-}24)$$

where $TCC(f_x^{'}, f_y^{'}; f_x^{''}, f_y^{''})$ is defined as

$$TCC(f_x^{'}, f_y^{'}; f_x^{''}, f_y^{''}) = \iint_{-\infty}^{+\infty} J_F(f_x, f_y) T_F(f_x + f_x^{'}, f_y + f_y^{'})\, T_F^{*}(f_x + f_x^{''}, f_y + f_y^{''}) df_x df_y. \quad (3\text{-}25)$$

The image is interpreted as the summation of wave components (in 2-D frequency domain) of mask image spectrum weighted by the TCC function. TCC is a integral of the product of three functions: $J_F(f_x, f_y)$, $T_F(f_x + f, f_y + f_y^{'})$, and $T_F^{*}(f_x + f_x^{''}, f_y + f_y^{''})$. It is nonzero only if these three functions have some overlap in (f_x, f_y) space. TCC is independent on the mask transmission function and only related with the mutual intensity of light incident on the mask (i.e., the effective source) and the image transfer function of the optical system, both in the frequency domain. Precalculation of the TCC function can significantly speed up image calculation for applications (e.g., optical proximity correction, or OPC) where the mask pattern needs to be tuned while the imaging system (including the source and pupil shapes) remains the same.

The Fourier transform of the mutual intensity on the mask, $J_F(f_x, f_y)$, is called *the effective source*. The reason to call $J_F(f_x, f_y)$ "the effective source" becomes apparent if we understand that the mutual intensity of the light incident on the mask (under the small-angle approximation) is proportional to the Fourier transform of the light intensity distribution of the source $S(\eta, \xi)$ (see our discussion about equation (3-17)). Therefore, $J_F(f_x, f_y)$ is proportional to the double Fourier transforms of $S(\eta, \xi)$, i.e., simply a function that is rotated by 180° from the original distribution of the source light intensity (if source image magnification M_s is ignored). The effect of 180° rotation can always be removed by choosing a relevant coordinate system. For example, the "effective source" of an imaging system illuminated by a uniform circular light source is:

$$J_F(f_x, f_y) = \begin{cases} \text{constant } J_0 (\text{ if } \sqrt{f_x^2 + f_y^2} < \dfrac{\sin\alpha_0}{\lambda}) \\ 0 \text{ (otherwise)} \end{cases}. \quad (3\text{-}26)$$

On the other hand, the light emitted from a point source (its intensity distribution is Dirac delta function) will become parallel light rays (coherent light) when striking the mask. As a result, the Fourier transform of the mutual intensity on the mask due to a point source (the effective source) is also a Dirac delta function: $J_F(f_x, f_y) = \delta(f_x - f_{x0}, f_y - f_{y0})$. In this case, the TCC can be simplified to

$$TCC(f_x^{'}, f_y^{'}; f_x^{''}, f_y^{''}) = T_F(f_{x0} + f_x^{'}, f_{y0} + f_y^{'})\, T_F^*(f_{x0} + f_x^{''}, f_{y0} + f_y^{''}) \ .$$

Substitute above relation to equation (3-24), we have

$$I(x, y) = \iiiint_{-\infty}^{+\infty} T_F(f_{x0} + f_x^{'}, f_{y0} + f_y^{'})\, T_F^*(f_{x0} + f_x^{''}, f_{y0} + f_y^{''}) \cdot G_F(f_x^{'}, f_y^{'})\, G_F^*(f_x^{''}, f_y^{''})$$

$$\exp\{-i2\pi[(f_x^{'} - f_x^{''})x + (f_y^{'} - f_y^{''})y]\} df_x^{'} df_y^{'} df_x^{''} df_y^{''}$$

$$= \left| \iint_{-\infty}^{+\infty} T_F(f_{x0} + f_x^{'}, f_{y0} + f_y^{'})\, G_F(f_x^{'}, f_y^{'}) \exp[-i2\pi(f_x^{'}x + f_y^{'}y)] df_x^{'} df_y^{'} \right|^2$$

$$= \left| \iint_{-\infty}^{+\infty} T_F(f_x^{'}, f_y^{'})\, G_F(f_x^{'} - f_{x0}, f_y^{'} - f_{y0}) \exp[-i2\pi(f_x^{'}x + f_y^{'}y)] df_x^{'} df_y^{'} \right|^2 = |U(x, y)|^2 \ .$$

The image field can be expressed with the mask transmission function and the system transfer function as

$$U(x, y) = \iint_{-\infty}^{+\infty} T_F(f_x^{'}, f_y^{'})\, G_F(f_x^{'} - f_{x0}, f_y^{'} - f_{y0}) \exp[-i2\pi(f_x^{'}x + f_y^{'}y)] df_x^{'} df_y^{'} \ . \quad (3\text{-}27)$$

Above relation indicates that the Fourier transform of $U(x, y)$ is:

$$T_F(f_x^{'}, f_y^{'})\, G_F(f_x^{'} - f_{x0}, f_y^{'} - f_{y0}) .$$

In the other extreme case of incoherent illumination on the mask, its mutual intensity (on the mask) can be represented by a Dirac delta function such that its Fourier transform (the effective source) is a constant value. Nevertheless, an easier approach to calculate the light intensity in the incoherent case is directly substituting the mutual intensity on the mask

$$J(x_{M'}, y_{M'}; x_{M''}, y_{M''}) = \delta(x_{M''} - x_{M'}) \cdot \delta(y_{M''} - y_{M'})$$

into equation (3-21) and applying the properties of the delta function to simplify it to below form:

$$I(x,y) = \iint_M \iint_M \delta(x_{M''} - x_{M'})\delta(y_{M''} - y_{M'})G(x_{M'}, y_{M'})T(x_{M'}, y_{M'}; x, y) \cdot$$

$$G^*(x_{M''}, y_{M''})T^*(x_{M''}, y_{M''}; x, y)dx_{M'}dy_{M'}dx_{M''}dy_{M''} = \iint_M G^*(x_{M'}, y_{M'})T^*(x_{M'}, y_{M'}; x, y) \cdot$$

$$G(x_{M'}, y_{M'})T(x_{M'}, y_{M'}; x, y)dx_{M'}dy_{M'} = \iint_M |G(x_{M'}, y_{M'})|^2 \cdot |T(x_{M'}, y_{M'}; x, y)|^2 dx_{M'}dy_{M'}$$

$$= \iint_M |G(x_{M'}, y_{M'})|^2 \cdot |T(x - x_{M'}, y - y_{M'})|^2 dx_{M'}dy_{M'}$$

According to the convolution theorem, the Fourier transform of $I(x,y)$ under an incoherent illumination is the product of the Fourier transforms of $|G(x,y)|^2$ and $|T(x,y)|^2$.

In above analysis, we have described how to calculate the light intensity of three types of imaging schemes: coherent, partially coherent and incoherent illuminations. In the coherent case, the spectrum of light (complex) amplitude is determined by the spectra of mask transmission function and system transfer function. In the incoherent case, the spectra of image light intensity (instead of complex amplitude) and mask light intensity are related by the spectrum of $|T(x,y)|^2$. Between these two extreme schemes, the fundamental functions under a partially coherent illumination are the mutual intensity and transmission cross-coefficient.

To find out the transfer function of an optical imaging system wherein a lens system is placed between the mask and wafer, we shall introduce some preliminary knowledge about the lens and interpret its physical meaning. A lens is composed of a transparent optical material (such as fused silica for 193 nm lithography) in which the optical wavelength becomes shorter than that in air. A lens is said to be "thin" if a light ray entering at coordinates (normal to the incident direction) on one face exits at about the same coordinates on the opposite face of the lens. As a consequence, a thin lens simply introduces optical phase delay to the incident wave-front by an amount proportional to the local lens thickness. The finite extent of the lens can be described by a pupil function $P_u(x_L, y_L)$ which is equal to 1 inside the lens aperture and zero outside. Using the paraxial approximation, the complex amplitude modulation due to the lens can be expressed as: $P_u(x_L, y_L)\exp[-i\frac{k}{2f}(x_L^2 + y_L^2)]$, where f is the lens' focal length. As shown in Figure 25, the distance R_1 can be approximated by retaining only the first two terms of its binomial expansion, $R_1 \approx z_1[1 + \frac{1}{2}(\frac{x_M - x_L}{z_1})^2 + \frac{1}{2}(\frac{y_M - y_L}{z_1})^2]$. The system's transfer function $T(x_M, y_M; x, y)$ is actually the through-lens impulse response (or *point-spread function*) on the wafer caused by a point on the mask (object) image. Therefore, we shall start with a spherical wave $\frac{1}{i\lambda R_1}\exp(ikR_1)$ diverging from a point on the mask $P_M(x_M, y_M)$. The binomial approximation to that spherical wave is:

$$\frac{e^{ikz_1}}{i\lambda z_1}\exp\{i\frac{k}{2z_1}[(x_M-x_L)^2+(y_M-y_L)^2]\}.$$

Taking the complex amplitude modulation due to the lens into account, the field distribution after passage through the lens becomes

$$U_L(x_M,y_M;x_L,y_L)=\frac{e^{ikz_1}}{i\lambda z_1}\exp\{i\frac{k}{2z_1}[(x_M-x_L)^2+(y_M-y_L)^2]\}P_u(x_L,y_L)\exp[-i\frac{k}{2f}(x_L^2+y_L^2)]. \quad (3\text{-}28)$$

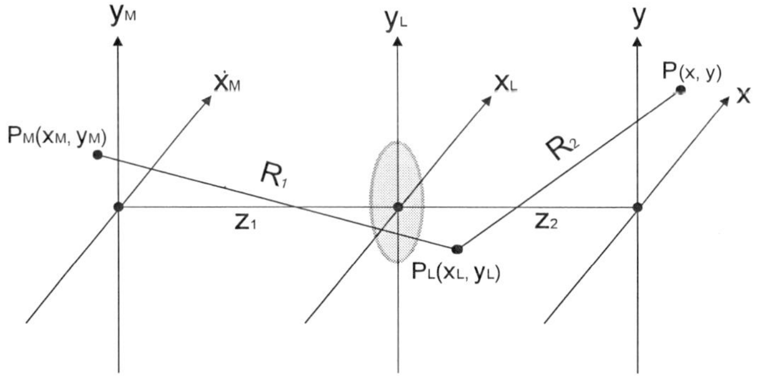

Figure 25. Propagation of light through a lens (the middle circle)

Using the Huygens-Fresnel principle and the binomial approximation, the transfer function (impulse response/point-spread function) can be re-written as a 2-D integral at coordinates shown in Figure 25:

$$T(x_M,y_M;x,y)=\frac{e^{ikz_2}}{i\lambda z_2}\int\int_{-\infty}^{+\infty}U_L(x_M,y_M;x_L,y_L)\exp\{i\frac{k}{2z_2}[(x-x_L)^2+(y-y_L)^2]\}dx_L\,dy_L. \quad (3\text{-}29)$$

Here, the wave-front behind the lens $U_L(x_M,y_M;x_L,y_L)$ is similarly considered as the secondary wave sources and its each point (x_L,y_L) emits a spherical wave which is described by the binomial approximation. Substituting the field distribution (3-28) to equation (3-29) yields

$$T(x_M,y_M;x,y)=\frac{e^{ik(z_1+z_2)}}{-\lambda^2 z_1 z_2}\int\int_{-\infty}^{+\infty}\exp\{i\frac{k}{2z_1}[(x_M-x_L)^2+(y_M-y_L)^2+i\frac{k}{2z_2}[(x-x_L)^2+(y-y_L)^2]\}$$

$$P_u(x_L,y_L)\exp[-i\frac{k}{2f}(x_L^2+y_L^2)]dx_L\,dy_L=\frac{e^{ik(z_1+z_2)}}{-\lambda^2 z_1 z_2}\cdot\exp[\frac{ik}{2z_2}(x^2+y^2)+\frac{ik}{2z_1}(x_M^2+y_M^2)]\cdot$$

$$\int\int_{-\infty}^{+\infty}P_u(x_L,y_L)\cdot\exp[i\frac{k}{2}(x_L^2+y_L^2)(\frac{1}{z_1}+\frac{1}{z_2}-\frac{1}{f})]\exp\{-ik[x_L(\frac{x_M}{z_1}+\frac{x}{z_2})+y_L(\frac{y_M}{z_1}+\frac{y}{z_2})]\}dx_L\,dy_L.$$

However, above equation is tedious unless further simplifications are adopted. From the lens law of geometrical optics, the term $\frac{1}{z_1}+\frac{1}{z_2}-\frac{1}{f}=0$. Moreover, Goodman [80] points out that the impulse response of the system at particular object coordinates should extend over only a small region surrounding the exact image point corresponding to that particular object point, thus we have

$$\exp[\frac{ik}{2z_2}(x^2+y^2)+\frac{ik}{2z_1}(x_M^2+y_M^2)]\approx\exp[\frac{ik}{2z_2}(x^2+y^2)+\frac{ik}{2z_1}\frac{x^2+y^2}{m^2}]\approx\text{constant},$$

where $m=-\frac{z_2}{z_1}$ is the magnification of the optical system. Dropping $\frac{e^{ik(z_1+z_2)}}{-\lambda^2 z_1 z_2}$ and the constant term mentioned above, the simplified form of the impulse response is

$$\begin{aligned}&T(x_M,y_M;x,y)\\&=\int\int_{-\infty}^{+\infty}P_u(x_L,y_L)\exp\{-ik[x_L(\frac{x_M}{z_1}+\frac{x}{z_2})+y_L(\frac{y_M}{z_1}+\frac{y}{z_2})]\}dx_L\,dy_L\\&=\int\int_{-\infty}^{+\infty}P_u(x_L,y_L)\exp\{-\frac{ik}{z_2}[(x-mx_M)x_L+(y-my_M)y_L]\}dx_L\,dy_L\,.\end{aligned}$$

Set new variables: $\widetilde{x}_M=mx_M$, $\widetilde{y}_M=my_M$, the impulse response reduces to

$$T(\widetilde{x}_M,\widetilde{y}_M;x,y)=\int\int_{-\infty}^{+\infty}P_u(x_L,y_L)\exp\{-\frac{ik}{z_2}[(x-\widetilde{x}_M)x_L+(y-\widetilde{y}_M)y_L]\}dx_L\,dy_L\,.$$

This is a space-invariant function, thus we have

$$T(x-\widetilde{x}_M,y-\widetilde{y}_M)=\int\int_{-\infty}^{+\infty}P_u(x_L,y_L)\exp\{-\frac{ik}{z_2}[(x-\widetilde{x}_M)x_L+(y-\widetilde{y}_M)y_L]\}dx_L\,dy_L\,,$$

or

$$T(x-\widetilde{x}_M,y-\widetilde{y}_M)=T(\Delta\widetilde{x}_{CM},\Delta\widetilde{y}_{CM})=\int\int_{-\infty}^{+\infty}P_u(x_L,y_L)\exp\{-\frac{ik}{z_2}[\Delta\widetilde{x}_{CM}\cdot x_L+\Delta\widetilde{y}_{CM}\cdot y_L]\}dx_L\,dy_L\,.$$

(3-30)

Thus for a linear space-invariant imaging system, its image transfer function (or impulse response/point-spread function) is the Fourier transform of the pupil function.

10.3. Spectral Analysis of Image Formation

As we discussed before, the mask object pattern can be represented by a 2-D function $G(x, y)$ which is the complex amplitude distribution of wave field right behind the mask. According to the Fourier analysis, this 2-D function can be equivalently expressed as a Fourier integral

$$G(x,y) = \int\int_{-\infty}^{+\infty} G_F(f_x, f_y) e^{-i2\pi(f_x x + f_y y)} df_x df_y \tag{3-31}$$

For certain functions, above integral may not exist and therefore our analysis will not be complete without mentioning some "existence conditions". In the normal mathematical sense, the mask object function should be absolutely integrable over the infinite space and only have finite number of discontinuities and maxima/minima in any finite region. However, the Dirac delta function is one exception and it requires a generalized Fourier transform as discussed in reference [80]. Above integral expression carries not only mathematical convenience to image calculation, but also profound physical meanings. For example, $e^{-i2\pi(f_x x + f_y y)}$ corresponds to a plane wave with spatial frequency (f_x, f_y) and $G_F(f_x, f_y)$ stands for its amplitude. Namely, in general a 2-D function can be decomposed into infinite number of elementary plane waves. It should be emphasized that both the space and frequency domains of the elementary plane waves in above relation are defined in the infinite region ($-\infty, +\infty$). For example, assuming $G(x, y)$ as a simple circle function whose value is constant within the pupil while zero outside, $G(x, y)$ can be composed of a series of plane waves with different spatial frequencies (propagating at different angles) and by definition each plane wave spreads throughout the infinite space ($-\infty, +\infty$). This physical view is helpful to understand why the first lens (L2, behind the mask, see Figure 26(1)) will collect all the incident plane waves with propagation angles between -90° and +90°, instead of being limited by NA (see Figure 26(2)). But there is other limit on the maximum value of collected spatial frequency due to the phenomenon of evanescent waves. Those plane waves with their frequencies satisfying $f_x^2 + f_y^2 > 1/\lambda^2 = (\sin 90°)/\lambda^2$ will decay over a short distance (from the mask) and can not propagate to the first lens (L2). The phenomenon of decaying evanescent waves is an intrinsic characteristic of any wave field that obeys the Helmholtz equation (3-4) and has nothing to do with the low-pass filtering effect of the lens. In other words, the wave system (that is described by the Helmholtz equation) itself is a low-pass filter that only allows certain range of frequencies to propagate away from the object (mask). Nevertheless, the upper limit of its spatial frequencies allowed to pass is always higher than that of a lens since the NA of a lens (in air) is always smaller than one.

In the lithographic literature, we frequently see one illustrative picture which replaces the lens L2 and L3 by a single lens and demonstrates the low-pass filter effect of the lens by stating that the 0, +1, and -1 orders of a periodic pattern (different-frequency components) are collected by the lens while +2 and -2 orders are missed due to the size limit of the lens aperture. This simplification can often lead to misinterpretation of the real physical process of optical low-pass filtering. For example, one might think that the "beam width of plane wave components" are limited only within the mask area. The line connecting the mask center to the edge of the lens forms an angle θ_2 as shown in Figure 26(3) and this angle might then be used to determine whether the corresponding order of wave component will be collected by the lens or not (e.g., if $\sin\theta_2 > NA$, it will be missed).

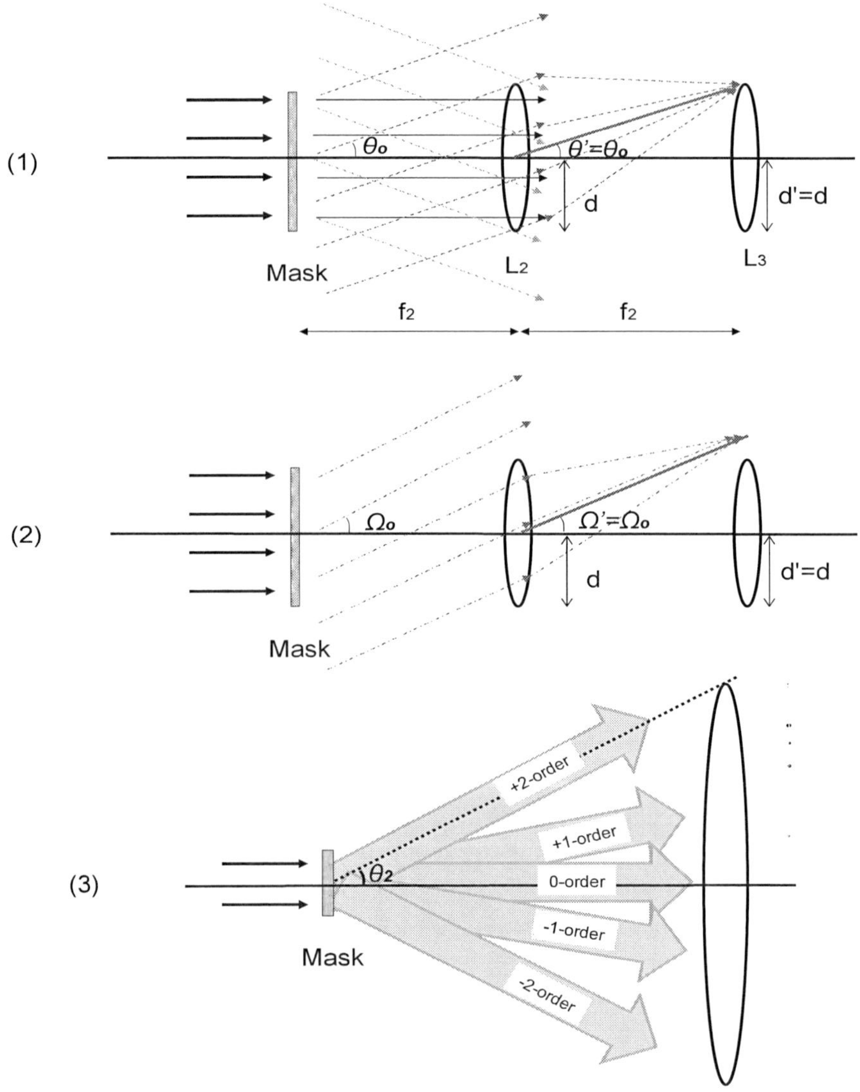

Figure 26. A schematic illumination of the optical low-pass filtering

However, the reader can question this kind of reasoning by simply arguing that about half energy of the higher order (e.g., +2) will be collected as the "beam width of this plane wave component" is limited within the mask area and only the energy flow above the dotted line (as shown in Figure 26(3)) is missed. Consequently, a partial inclusion of higher-order frequency components beyond the limit of NA (of the lens system) sometimes will be unavoidable, causing a fundamental issue in the imaging theory. More persuasive questioning comes directly from the expression of an arbitrary mask image distribution (e.g., circle function) by the Fourier integral (3-31) which clearly defines all its plane wave components in the infinite space ($-\infty, +\infty$), both in X and Y. Therefore, we can see that the width of any plane wave component behind the mask spreads out extensively and is not limited only within the mask area. As a result, all the plane wave components except those evanescent waves are partially collected by the lens L2 even the sine of the incident angle of some components can be larger than NA, as demonstrated in Figure 26(2). Of course, the light component energy collected by the lens is dependent on the "collection area" (normal to the propagation direction) sustained by the lens. As the incident angle θ approaches +90° (or -90°), both the amplitude of the plane wave component (represented by its Fourier transform) and the collection area (and the corresponding energy collected by the lens L2) become smaller; therefore its contribution to image formation gradually decreases.

The reader may ask why NA of an imaging system is determined by the semi-aperture angle sustained by L2 if it actually does not act as a low-pass filter. With the help of Figures 26(1) and 26(2), it is apparent that the pupil of the lens L3 (instead of L2) is where the frequency cut-off occurs. Since the lens L3 is located at the focal plane of L2, an elementary plane wave will focus at the L3 plane after it passes through the lens L2. If the focus point of an elementary wave is out of the pupil of the lens L3 (i.e., $\sin\Omega' > \frac{d'}{\sqrt{d'^2+f_2^2}} \approx \frac{d'}{f_2}$), this elementary wave will not be collected by the lens L3. The incident angle (Ω_0) of this elementary plane wave is equal to the angle (Ω') subtended by the focus point according to the geometrical optics. Thus it is equivalent to say when the incident angle of an elementary plane wave satisfies $\sin\Omega_0 > \frac{d'}{\sqrt{d'^2+f_2^2}} = \frac{d}{\sqrt{d^2+f_2^2}} \approx \frac{d}{f_2}$, this plane wave will be missed by the imaging system.

To demonstrate the fundamental mechanism of spectrum decomposition and recombination in an imaging process, let us start from the coherent illumination as described by equation (3-27). Without loss of generality, we shall change the integral variable f_x' to f_x as below

$$\begin{aligned} U(x,y) &= \iint_{-\infty}^{+\infty} T_F(f_x, f_y)\, G_F(f_x - f_{x0}, f_y - f_{y0}) \exp[-i2\pi(f_x x + f_y y)] df_x df_y \\ &= \mathbf{F}^{-1}[T_F(f_x, f_y)\, G_F(f_x - f_{x0}, f_y - f_{y0})], \end{aligned} \tag{3-32}$$

where $\mathbf{F}^{-1}$ represents the inverse Fourier transform. $T_F(f_x, f_y)$ is the Fourier transform of the impulse response, thus it is simply the pupil function $P_u(f_x, f_y)$ of the imaging system in the frequency domain (see our previous discussion). Thus equation (3-32) can be rewritten as:

$$U(x,y) = \mathbf{F}^{-1}[P_u(f_x, f_y)\, G_F(f_x - f_{x0}, f_y - f_{y0})] \tag{3-33}$$

Assuming the point source is located at the center of the source ($f_{x0} = 0, f_{y0} = 0$), above relation can be further simplified to:

$$U_0(x, y) = \mathbf{F}^{-1}[P_u(f_x, f_y)\, G_F(f_x, f_y)] \qquad (3\text{-}34)$$

The diffraction process related to the first lens behind the mask (L2) is to perform the Fourier transform of the mask (object) pattern to obtain the spectrum of the mask transmission function $G_F(f_x, f_y)$. As we analyzed using the dry lithography (in air) as an example, excluding those high frequencies satisfying $f_x^2 + f_y^2 > 1/\lambda^2$, the available spectrum for the second lens (L3, a low-pass filter) to collect is wider than the passing range (NA/λ) of L3. The pupil function of L3 only allows those frequencies lower than NA/λ to enter the lens, namely,

$$P_u(f_x, f_y) = \begin{cases} 1\ (\text{if} \sqrt{f_x^2 + f_y^2} < NA/\lambda) \\ 0\ (\text{otherwise}) \end{cases}. \qquad (3\text{-}35)$$

In the coordinates rotated by 180°, the focusing behavior of the lens L3 is to perform the inverse Fourier transform of collected frequency components. As a result, those higher frequencies corresponding to higher resolution and better fidelity are lost during two low-pass filtering processes: the evanescent wave decay right behind the mask and the frequency cut-off at the pupil plane of the lens L3. Even the size of the lens is infinitely large, we still can not collect the full spectrum of the mask image due to the evanescent wave phenomenon.

The analysis of a point-source caused coherent imaging process is particularly important to understand the numerical calculation of the final image when the mask is illuminated by an extended incoherent source (i.e., by a partially coherent wave-front striking the mask). Besides the TCC approach mentioned before, sometimes a different method (called *Abbe's method*) is applied directly to calculate the integral (3-23). In In Abbe's method, a finite number of Dirac delta functions are distributed at the discrete points on the source to simulate the continuous distribution of the source light intensity. Namely, the effective source can be discretized as:

$$J_F(f_x, f_y) = \sum_n A_n\, \delta(f_x - f_{xn}, f_y - f_{yn}).$$

When the point source is not located at the center, the corresponding plane wave incident on the mask will tilt at an angle which is equal to the angle sustained by the point source (i.e., α as indicated in Figure 23). In the frequency domain, it will only cause a shift of the mask spectrum by the amount of $\sin\alpha/\lambda$. Therefore, according to equations (3-23) and (3-27), the final image on the wafer is the sum of a finite number of elementary coherent intensity distributions arising from the discretized point sources described by shifted delta-functions in the frequency domain. The relation between the spectrum shift and a parameter called "partial coherence" will be discussed in detail later.

Using the mask as an example, it is straightforward to understand why the elementary intensity distributions can be added together to form the final image. The spatial coherence of incident wave-front on a mask usually is not zero because the complex amplitude of any elementary wave striking a point on the mask (at any time) has certain correlation with an elementary wave arising from the same source element while reaching a different mask point simultaneously. On the other hand, the complex amplitude of any elementary wave striking a point on the mask is incoherent with other elementary waves arising from different source elements while reaching the same mask point simultaneously. Therefore the total light intensity on the mask is simply the sum of the elementary light intensity distributions caused by different point sources. Whether we can add the elementary light intensity distributions (arising from the points of an upstream surface) to calculate the total light intensity at a downstream plane only depends on the spatial coherence of the upstream surface we choose. If the upstream surface (e.g., a light source) is completely incoherent, the answer is yes regardless of whether the mask or wafer plane will be the downstream surface of interest. However, if we choose a partially coherent plane (e.g., the mask) as the upstream surface, then the answer is no.

10.4. Oblique Illumination and Partial Coherence

In calculation of final aerial image, it is necessary to consider the effect of incident angle of light on the mask as it will shift the center of its diffraction pattern by the same amount of angle before it is collected by the lens L3. From geometrical optics point of view, the incident angle of parallel light rays (behind the condenser lens L1, see Figure 23) is determined by the sustained angle of the source point (located at the focal plane of the condenser lens). Thus an explicitly defined parameter that can be related to the source size and used in the frequency domain to calculate the aerial image will help to understand the involved physics. This parameter is often called *partial coherence* defined as

$$\sigma = \frac{\rho'}{d'} \approx \frac{\sin\alpha'}{\sin\theta_i}. \tag{3-36}$$

Here, ρ' is the radius of the source image (assumed to be a circular shape) located at the exit pupil plane of the objective lens, and d' is the radius of the exit pupil as shown in Figure 27. It is a straightforward practice left to the reader to prove that the image plane of source is right at the exit pupil based on the rules of geometrical optics. The magnification of source image is equal to $\frac{f_2}{f_1} = |m_s| = M_s$ which in general can be different than the magnification M of the mask-to-wafer imaging system; thus the radius of the source image at the exit pupil is $\rho' = M_s\rho = \frac{f_2}{f_1}\rho$ and the partial coherence can be written as

$$\sigma = \frac{\rho'}{d'} \approx \frac{(f_2/f_1)\rho}{M f_2 \sin\theta_i} \approx \frac{(f_2/f_1) f_1 \sin\alpha_0}{M f_2 \cdot \sin\theta_o / M} = \frac{\sin\alpha_0}{\sin\theta_o} = \frac{\sin\alpha_0}{NA_o} \qquad (3\text{-}37)$$

The sine ($\sin\alpha'$) of the semi-aperture angle of the source image seen at the wafer plane is *1/M* times the sine ($\sin\alpha_0$) of the semi-aperture angle seen at the condenser plane:

$$\sin\alpha' = \frac{\rho'}{M f_2} = \frac{M_s \rho}{M f_2} = \frac{(f_2/f_1)\rho}{M f_2} = \frac{1}{M}\frac{\rho}{f_1} \approx \frac{1}{M}\sin\alpha_0 .$$

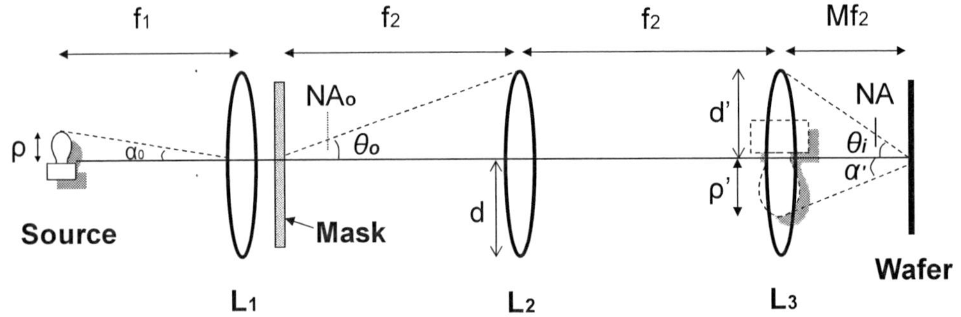

Figure 27. Representation of a lithographic imaging system

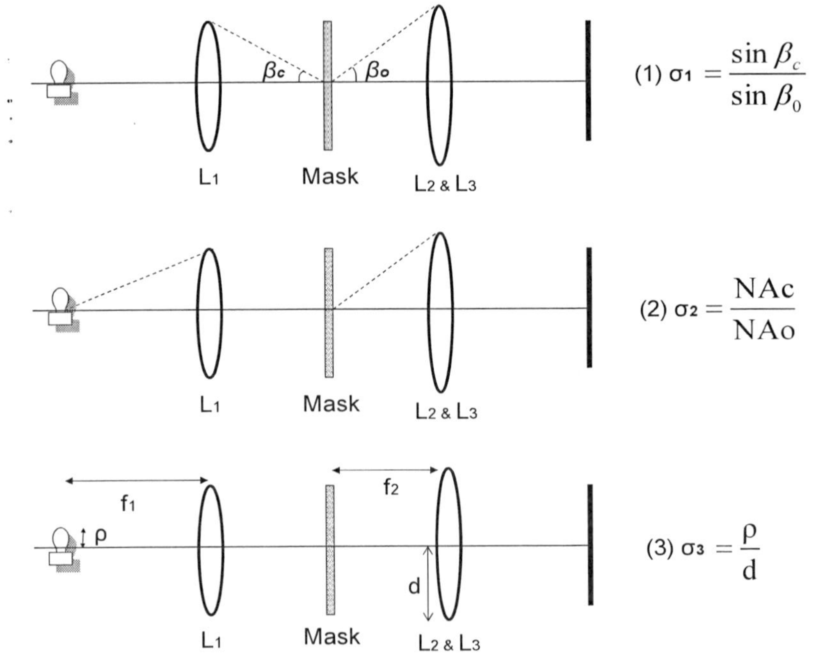

Figure 28. Three definitions of partial coherence found in the literature

In the lithography literature, there are different definitions of partial coherence as pointed out in Figure 28, some of which unfortunately cause confusion to the readers. For example, here we quote three definitions of partial coherence found in references [81-83]. The first definition shown in Figure 28 involves the sine of the sustained half-aperture angle β_c (at the mask plane) of the condenser lens which has nothing to do with the source size. The second definition does not clarify what NAc (condenser lens NA) means and the guessed answer is that it might be the numerical aperture of condenser lens seen at the source or mask plane, which again is not related with the source size. From geometrical optics, the radius (*d*) of the entrance pupil of the objective lens (which is a virtual lens combining L2 and L3 together, see Figures 26 and 27) is equal to that of exit pupil, i.e., *d=d'* (or NA=NAo/*M*). However, the radius of source is different than the radius of the source image at the exit pupil ($\rho' = \frac{f_2}{f_1}\rho$).

Apparently, $\frac{\rho}{d} = \frac{\rho}{d'} = \frac{f_1}{f_2}\frac{\rho'}{d'} = \frac{1}{M_s}\sigma$ and the third definition is also incorrect.

The physical meaning of partial coherence can be best understood using equation (3-37). In theoretical analysis, it is often convenient to define the normalized frequency as

$$\tilde{f}_x = \frac{f_x}{NA_o / \lambda}, \tilde{f}_y = \frac{f_y}{NA_o / \lambda}.$$

In the normalized frequency domain, the Fourier transform of the mutual intensity on the mask (i.e., the effective source, see equation (3-26)) due to a uniform circular light source becomes:

$$\tilde{J}_F(\tilde{f}_x, \tilde{f}_y) = \begin{cases} \text{constant } \tilde{J}_0 (\text{ if} \sqrt{\tilde{f}_x^{\,2} + \tilde{f}_y^{\,2}} < \frac{\sin\alpha_0 / \lambda}{NA_o / \lambda} = \sigma) \\ 0 \text{ (otherwise)} \end{cases}. \tag{3-38}$$

We can see that *the partial coherence* σ *is the radius of the circular "effective source" in the normalized frequency domain.* As we analyzed before, light rays emitted from every point of the source become parallel after passage through the condenser lens (L1); and the incident angle of each set of parallel light rays is equal to the angle that source point subtends on the condenser lens. The incident angle α of a plane wave will only cause a shift of the mask spectrum by the amount of $\sin\alpha / \lambda$. The semi-aperture angle α_0 (see Figure 23) that the source subtends on the condenser lens is actually the maximum incident angle of the parallel light rays striking the mask. This angle determines the maximum spatial frequency shift ($\frac{\sin\alpha_0}{\lambda} \approx \frac{\rho / f_1}{\lambda}$, see Figure 27) of the diffraction pattern in the frequency domain. If the incident angle of a plane wave on the mask is out of the range of collectable angles, the (shifted) zeroth order of its mask diffraction pattern will not be collected by the lens L3. Mathematically, if the partial coherence of a circular source satisfies

$\sigma > 1$, (i.e., $\sin\alpha_0 > NA_o$), the elementary plane wave caused by a point source at the source boundary will tilt at an angle α_0 (before striking the mask) and the center (the location of the zeroth order) of the induced mask diffraction pattern will shift by the same amount of angle, thus missing the pupil of an imaging system. The zeroth order of a diffraction pattern does not carry any feature signal and only adds a uniform light distribution which causes a degradation of the image contrast.

A widely used technique called *off-axis illumination* will specify certain value of partial coherence σ when setting up the scanner's illumination conditions. In this manner, the center of a diffraction pattern can be shifted such that the higher frequencies that used to be out of the collectable range will enter the window of the lens pupil for a collection. According to the Fourier analysis, being able to collect higher frequencies of the object pattern is beneficial to the resolution improvement of an imaging system. But in general, the optimization of partial coherence is quite complicated and numerical simulations are needed to determine its impact on lithographic performance of different types of features. For example, there is always a trade-off between increasing σ (and NA) to resolve the features with higher spatial frequencies (i.e., dense lines/spaces with a smaller pitch) and reducing σ (and NA as well) to improve the process window (e.g., depth of focus) of more isolated features (with a larger pitch).

It is often mentioned that when the range of incident angles striking the mask extends from -90° to +90° (an infinitely large source with $\sigma \to +\infty$), the illumination will be incoherent. This conclusion comes directly from the small-angle approximation that the spatial coherence of incident light on the mask can be expressed as the Fourier transform of the source light intensity distribution such that the coherent area of an arbitrary point on the mask is inversely proportional to the size of light source. It can be proved that every point on the mask is incoherent with all other points (i.e., the mutual intensity between any two points is the Dirac delta function) when the size of a uniform circular source is infinitely large. Nevertheless, the concept of partial coherence is different than the spatial coherence (or the mutual intensity of the light on the mask) which should be calculated from a modified form of Van Cittert-Zernike theorem (i.e., equation (3-15)). Unless specified otherwise, the spatial coherence of a lithographic system normally means the coherence distribution of the wave-front incident on the mask; while the partial coherence is a single-value parameter that specifies the size of a light source which is completely incoherent. The reader may notice that the original definition $\sigma = \frac{\rho'}{d'}$ allows σ to approach infinity when the source size is large, but the small-angle approximation $\sigma = \frac{\sin\alpha_0}{NA_o} \le \frac{1}{NA_o} = \frac{1}{M \cdot NA}$ yields a finite value of partial coherence for an imaging system with a non-zero NA. Practically, the partial coherence of a scanner system does not exceed one.

10.5. Resolution Enhancement Techniques

Theoretically, the resolution limit of an optical imaging system depends on how the optical signals are recorded in the medium. For example, if the recording medium is sensitive enough to respond to multiple thresholds of light intensity in significantly different manners, it may record more features than a less sensitive medium does. Thus, any physical criterion to define the resolution of an imaging system is based on the current capability of the recording/detecting system. The well-known Rayleigh criterion is $0.61\frac{\lambda}{NA}$ for an optical system with a circular aperture. Rayleigh criterion is usually derived under a coherent illumination by examining how close two intensity peaks can be placed together without merging into a new peak. In a lithographic system, the mask object is usually illuminated by partially coherent light and the imaging formation on the wafer is more complicated than in a coherent system. As a result, deriving a general criterion of the resolution of a lithographic system will be difficult by just examining the light intensity profile in the physical space. More important parameters to characterize the performance of a lithographic system are *minimum half-pitch* (defined as half of the spatial period of a pattern) and *critical dimension (CD)* or minimum dimension. Minimum half-pitch determines the density of integrated circuits while critical dimension (CD) is directly related to transistors' gate length or speed of a circuit. Sometimes critical dimension (CD) also stands for minimum half-pitch, but the reader should be aware that they are two different concepts.

From Fourier point of view, it is convenient to decompose an optical pattern into a series of plane waves with different spatial frequency (or period). The minimum half-pitch that can be resolved by an optical system is determined by the highest spatial frequency (or the smallest spatial period) that can be collected the system. To give a general analysis that is independent on specific exposure wavelength and numerical aperture of a scanner, we shall use the dimensionless normalized variables defined as:

$$\widetilde{x} = \frac{x}{\lambda / NA},\quad \widetilde{y} = \frac{y}{\lambda / NA},\quad \widetilde{f}_x = \frac{f_x}{NA / \lambda},\quad \widetilde{f}_y = \frac{f_y}{NA / \lambda}.$$

Without loss of generality, we have ignored the effect of magnification M (i.e., $NA = NA_o$) since it does not have an impact on the resolution capability of an imaging system. According to equation (3-24), the Fourier transform of the light intensity distribution on the wafer is:

$$\begin{aligned}
I_F(\widetilde{f}_x, \widetilde{f}_y) &= \iint_{-\infty}^{+\infty} I(\widetilde{x}, \widetilde{y}) e^{i2\pi(\widetilde{f}_x \widetilde{x} + \widetilde{f}_y \widetilde{y})} d\widetilde{x}\, d\widetilde{y} = \int_{-\infty}^{+\infty}\!\!\ldots\!\int TCC(\widetilde{f}_x', \widetilde{f}_y'; \widetilde{f}_x'', \widetilde{f}_y'') G_F(\widetilde{f}_x', \widetilde{f}_y')\, G_F^*(\widetilde{f}_x'', \widetilde{f}_y'') \cdot \\
&\exp\{i2\pi[(\widetilde{f}_x - \widetilde{f}_x' + \widetilde{f}_x'')\widetilde{x} + (\widetilde{f}_y - \widetilde{f}_y' + \widetilde{f}_y'')\widetilde{y}]\}\, d\widetilde{x}\, d\widetilde{y}\, d\widetilde{f}_x' d\widetilde{f}_y' d\widetilde{f}_x'' d\widetilde{f}_y'' \\
&= \iiiint_{-\infty}^{+\infty} TCC(\widetilde{f}_x', \widetilde{f}_y'; \widetilde{f}_x'', \widetilde{f}_y'') G_F(\widetilde{f}_x', \widetilde{f}_y')\, G_F^*(\widetilde{f}_x'', \widetilde{f}_y'') \delta(\widetilde{f}_x - \widetilde{f}_x' + \widetilde{f}_x'') \delta(\widetilde{f}_y - \widetilde{f}_y' + \widetilde{f}_y'') d\widetilde{f}_x' d\widetilde{f}_y' d\widetilde{f}_x'' d\widetilde{f}_y'' \\
&= \iint_{-\infty}^{+\infty} TCC(\widetilde{f}_x + \widetilde{f}_x'', \widetilde{f}_y + \widetilde{f}_y''; \widetilde{f}_x'', \widetilde{f}_y'') G_F(\widetilde{f}_x + \widetilde{f}_x'', \widetilde{f}_y + \widetilde{f}_y'') G_F^*(\widetilde{f}_x'', \widetilde{f}_y'') d\widetilde{f}_x'' d\widetilde{f}_y''.
\end{aligned} \tag{3-39}$$

From equation (3-25), the transmission cross-coefficient function can be calculated as:

$$TCC(\tilde{f}_x + \tilde{f}_x^{"}, \tilde{f}_y + \tilde{f}_y^{"}; \tilde{f}_x^{"}, \tilde{f}_y^{"}) = \iint_{-\infty}^{+\infty} J_F(\tilde{a}, \tilde{b})\, T_F(\tilde{a} + \tilde{f}_x + \tilde{f}_x^{"}, \tilde{b} + \tilde{f}_y + \tilde{f}_y^{"}) \cdot$$
$$T_F^*(\tilde{a} + \tilde{f}_x^{"}, \tilde{b} + \tilde{f}_y^{"})\, d\tilde{a}\, d\tilde{b}. \tag{3-40}$$

In our preceding discussion, it has been shown the image transfer function T (or impulse response) of a linear space-invariant imaging system is the Fourier transform of the pupil function. Therefore, the Fourier transform of the image transform function, T_F, is simply the pupil function P_u which can be defined in the normalized coordinates (with the effects of images magnification and rotation removed) as:

$$P_u(\tilde{f}_x, \tilde{f}_y) = \begin{cases} 1 \text{ (if} \sqrt{\tilde{f}_x^{\,2} + \tilde{f}_y^{\,2}} < 1) \\ 0 \text{ (otherwise)} \end{cases} \tag{3-41}$$

For TCC function to be non-zero, it is apparent that the effective source $J_F(\tilde{a}, \tilde{b})$ and the displaced pupil functions should overlap. For example, an imaging system with a circular aperture and light source will have a pupil function with a radius of 1 and an effective source with a radius of σ in the normalized frequency coordinates. From Figure 29, we can see that $T_F(\tilde{a} + \tilde{f}_x + \tilde{f}_x^{"}, \tilde{b} + \tilde{f}_y + \tilde{f}_y^{"})$ and $T_F^*(\tilde{a} + f_x^{"}, \tilde{b} + f_y^{"})$ can overlap only if $\sqrt{\tilde{f}_x^{\,2} + \tilde{f}_y^{\,2}} < 2$. But this is only a necessary (not enough) condition to obtain a non-zero TCC function.

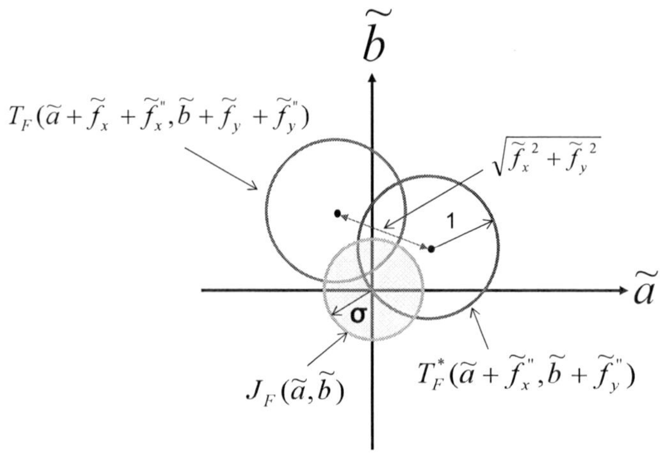

Figure 29. A schematic description of TCC integral

The maximum value 2 of the normalized frequency corresponds to a spatial frequency of $2\frac{NA}{\lambda}$, or a spatial period of $\frac{1}{2}\frac{\lambda}{NA}$ and a half-pitch of $\frac{1}{4}\frac{\lambda}{NA}$. This is the minimum pitch that can be resolved by a lithographic system. As we mentioned before, $\sqrt{\widetilde{f}_x^{\,2}+\widetilde{f}_y^{\,2}}<2$ is a necessary but not enough condition to obtain a non-zero TCC function; therefore, other conditions based on equation (3-40) are required and this means a scanner can not achieve the ideal resolution of $\frac{1}{4}\frac{\lambda}{NA}$.

The reader should keep it in mind that the Dirac delta function is not a normal mathematical function obeying the well-know multiplication properties such as the associative and commutative laws. For example,

$$(a).\ (\frac{1}{x}x)\delta(x)=\delta(x)\neq\frac{1}{x}[x\delta(x)]=\frac{0}{x}=0,$$

$$(b).\ \int_{-\infty}^{+\infty}[f(x)\delta(x-x_0)]\delta(x-x_1)dx=\int_{-\infty}^{+\infty}[f(x_0)\delta(x-x_0)]\delta(x-x_1)dx=f(x_0)\int_{-\infty}^{+\infty}\delta(x-x_0)\delta(x-x_1)dx$$

$$=f(x_0)\delta(x_0-x_1)=f(x_0)\delta(x_1-x_0)\neq f(x_1)\delta(x_1-x_0)=\begin{cases}f(x_1)\cdot\infty & (x_0=x_1)\\ 0 & (x_0\neq x_1),\end{cases}$$

$$(c).\ \int_{-\infty}^{+\infty}[f(x)\delta(x-x_1)]\delta(x-x_0)dx=\int_{-\infty}^{+\infty}[f(x_1)\delta(x-x_1)]\delta(x-x_0)dx=f(x_1)\int_{-\infty}^{+\infty}\delta(x-x_1)\delta(x-x_0)dx$$

$$=f(x_1)\delta(x_1-x_0)=f(x_1)\delta(x_0-x_1)\neq f(x_0)\delta(x_0-x_1)=\int_{-\infty}^{+\infty}[f(x)\delta(x-x_0)]\delta(x-x_1)dx.$$

To demonstrate the spectrum analysis of an aerial image, we approximate a 1-D periodic binary mask transmission function as

$$G(\widetilde{x})=\frac{1}{2}[1+\cos(2\pi\widetilde{f}_0\widetilde{x})],\ (\widetilde{f}_0=1/\widetilde{p}),$$

where $\widetilde{p}$ is the spatial period and there is no change of the sign of the transmission function due to the constant term 1/2. The Fourier transform of this 1-D mask transmission function is

$$G_F(\widetilde{f})=\frac{1}{2}\delta(\widetilde{f})+\frac{1}{4}[\delta(\widetilde{f}-\widetilde{f}_0)+\delta(\widetilde{f}+\widetilde{f}_0)]. \qquad (3\text{-}42)$$

Substituting equation (3-42) into equation (3-39) yields

$$I_F(\widetilde{f})=\int_{-\infty}^{+\infty}TCC(\widetilde{f}+\widetilde{f}^{''};\widetilde{f}^{''})\{\frac{1}{2}\delta(\widetilde{f}+\widetilde{f}^{''})+\frac{1}{4}[\delta(\widetilde{f}+\widetilde{f}^{''}-\widetilde{f}_0)+\delta(\widetilde{f}+\widetilde{f}^{''}+\widetilde{f}_0)]\}\cdot$$
$$\{\frac{1}{2}\delta(\widetilde{f}^{''})+\frac{1}{4}[\delta(\widetilde{f}^{''}-\widetilde{f}_0)+\delta(\widetilde{f}^{''}+\widetilde{f}_0)]\}d\widetilde{f}^{''}.$$

Based on the multiplication rules of Dirac delta function described above, the background (no signal modulation) light intensity can be obtained when $\widetilde{f} = 0$,

$$I_F(\widetilde{f}=0) = \int_{-\infty}^{+\infty} TCC(\widetilde{f}^{"};\widetilde{f}^{"})\{\frac{1}{2}\delta(\widetilde{f}^{"}) + \frac{1}{4}[\delta(\widetilde{f}^{"}-\widetilde{f}_0)+\delta(\widetilde{f}^{"}+\widetilde{f}_0)]\}\cdot$$
$$\{\frac{1}{2}\delta(\widetilde{f}^{"}) + \frac{1}{4}[\delta(\widetilde{f}^{"}-\widetilde{f}_0)+\delta(\widetilde{f}^{"}+\widetilde{f}_0)]\}d\widetilde{f}^{"} = \frac{1}{4}TCC(0;0) + \frac{1}{16}TCC(\widetilde{f}_0;\widetilde{f}_0) + \frac{1}{16}TCC(-\widetilde{f}_0;-\widetilde{f}_0).$$

Similarly, the spectrum component that has the same spatial period as the mask pattern is

$$I_F(\widetilde{f}=\pm\widetilde{f}_0) = \int_{-\infty}^{+\infty} TCC(\pm\widetilde{f}_0+\widetilde{f}^{"};\widetilde{f}^{"})\{\frac{1}{2}\delta(\pm\widetilde{f}_0+\widetilde{f}^{"}) + \frac{1}{4}[\delta(\pm\widetilde{f}_0+\widetilde{f}^{"}-\widetilde{f}_0)+\delta(\pm\widetilde{f}_0+\widetilde{f}^{"}+\widetilde{f}_0)]\}\cdot$$
$$\{\frac{1}{2}\delta(\widetilde{f}^{"}) + \frac{1}{4}[\delta(\widetilde{f}^{"}-\widetilde{f}_0)+\delta(\widetilde{f}^{"}+\widetilde{f}_0)]\}d\widetilde{f}^{"} = \frac{1}{8}TCC(\widetilde{f}_0;0) + \frac{1}{8}TCC(0;-\widetilde{f}_0).$$

Assuming a uniform circular light source, the transmission cross-coefficient is symmetric such that $I_F(\widetilde{f}=\pm\widetilde{f}_0) = \frac{1}{4}TCC(\widetilde{f}_0;0)$. If the partial coherence of the light source is σ, the "effective source" described by equation (3-38) can be used to obtain $TCC(\widetilde{f}_0;0)$ as below

$$TCC(\widetilde{f}_0;0) = \iint_{-\infty}^{+\infty} J_F(\widetilde{a},\widetilde{b})\, T_F(\widetilde{a}+\widetilde{f}_0,\widetilde{b})\, T_F^*(\widetilde{a},\widetilde{b})\, d\widetilde{a}\, d\widetilde{b}.$$

$TCC(\widetilde{f}_0;0)$ is determined by the overlapping area of three circles: the "effective source" circle of radius σ centered at the origin $J_F(\widetilde{a},\widetilde{b})$, a unit "pupil" circle centered at center $T_F^*(\widetilde{a},\widetilde{b})$, and a unit "pupil" circle centered at $(-\widetilde{f}_0,0)$ $T_F(\widetilde{a}+\widetilde{f}_0,\widetilde{b})$. The reader may refer to our previous discussion about why the spectrum of transfer function becomes the pupil function (in the frequency domain).

When σ<1, the overlapping area is determined by the two circles: $J_F(\widetilde{a},\widetilde{b})$ and $T_F(\widetilde{a}+\widetilde{f}_0,\widetilde{b})$. We can see that the maximum value of $\widetilde{f}_0$ to obtain a non-zero TCC (a necessary condition to resolve certain frequency component) is $1+\sigma$ as shown in Figure 30(left). Since $\widetilde{f}_0$ is the normalized spatial frequency of the mask pattern, the minimum half-pitch that can be resolved by a lithographic system is $\frac{1}{2}\frac{1}{(1+\sigma)}\frac{\lambda}{NA}$.

When σ>1, the "effective source" circle is larger than pupil circle, thus the overlapping area is determined by the two circles: $T_F(\widetilde{a}+\widetilde{f}_0,\widetilde{b})$ and $T_F^*(\widetilde{a},\widetilde{b})$. Under this condition, the maximum value of $\widetilde{f}_0$ to obtain a non-zero TCC is 2 as shown in Figure 30(right); and consequently the minimum half-pitch that can be resolved by a lithographic system is $\frac{1}{4}\frac{\lambda}{NA}$. However, when the effective source is larger than the pupil window, some light components will not be collected by the lens because their incident angles are higher than the maximum collectable angle that is defined by the system's numerical aperture (see our discussion about the partial coherence). Practically, the partial coherence σ of a modern scanner will not exceed 1.

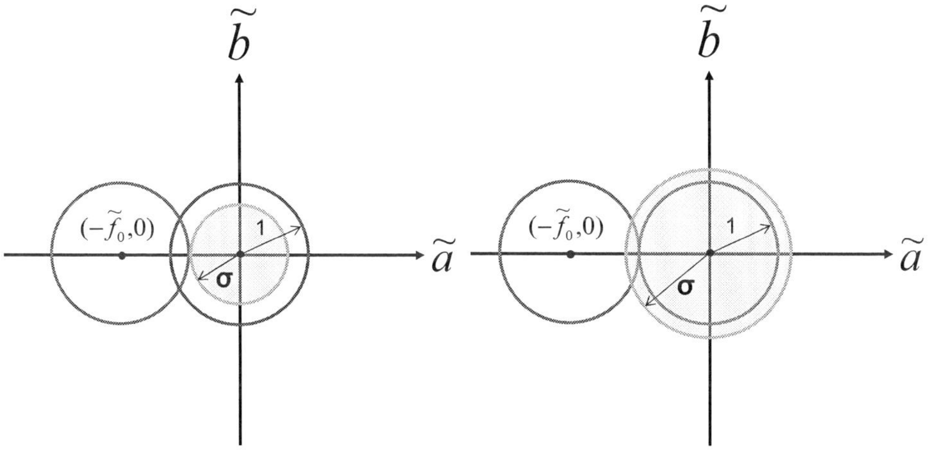

Figure 30. Optical resolution explained by TCC function: σ<1 (left), σ>1 (right)

The resolution enhancement due to the off-axis illumination can also be explained schematically using the example with σ<1. When the light source is moved away from the axis, the "effective source" in the (normalized) frequency domain will be shifted accordingly. As a result, the maximum value of $\tilde{f}_0$ to obtain a non-zero TCC can reach 2 as shown in Figure 31(left) and the minimum resolvable half pitch will be reduced from the on-axis value of $\frac{1}{2}\frac{1}{(1+\sigma)}\frac{\lambda}{NA}$ to $\frac{1}{4}\frac{\lambda}{NA}$.

Another widely-used resolution enhanced technique is the phase shifting mask which has two types: the alternating and attenuated phase-shifting masks. Unlike the discussed binary mask whose mask transmission function does not change sign due to a constant term, the alternating mask transmission function can be approximated as

$$G(\tilde{x}) = \cos(\pi \tilde{f}_0 \tilde{x}), (\tilde{f}_0 = 1/\tilde{p}),$$

with its Fourier transform as

$$G_F(\tilde{f}) = \frac{1}{2}[\delta(\tilde{f} - \frac{\tilde{f}_0}{2}) + \delta(\tilde{f} + \frac{\tilde{f}_0}{2})]. \quad (3\text{-}43)$$

It is a straightforward practice to show that the wafer image spectrum when $\tilde{f} = \pm\tilde{f}_0$ is

$$I_F(\tilde{f} = \pm\tilde{f}_0) = \frac{1}{4}TCC(\tilde{f}_0/2; -\tilde{f}_0/2).$$

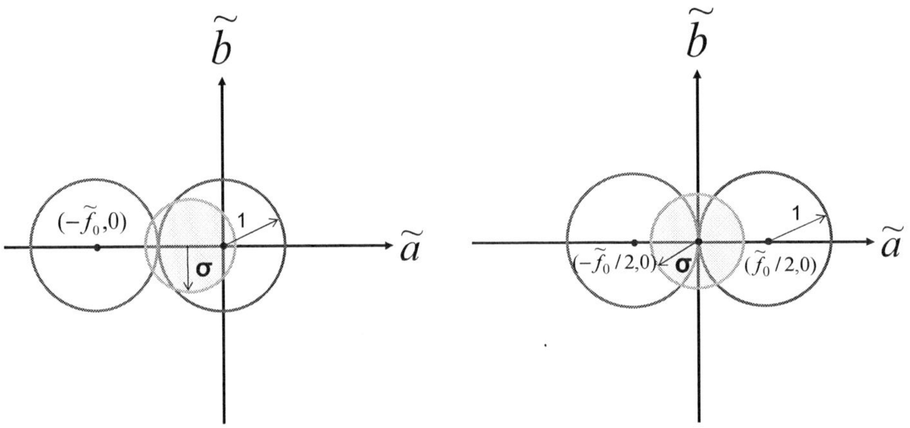

Figure 31. Off-axis illumination (left) and phase-shifting mask (right) explained by TCC function

When the effective source is located at the origin with a radius σ<1, the overlapping area is determined by the two circles: $T_F(\tilde{a} + \tilde{f}_0/2, \tilde{b})$ and $T_F^*(\tilde{a} - \tilde{f}_0/2, \tilde{b})$. The maximum value of $\tilde{f}_0$ to obtain a non-zero TCC is 2 as shown in Figure 31(right) and the minimum resolvable half pitch will be reduced from the binary value of $\frac{1}{2}\frac{1}{(1+\sigma)}\frac{\lambda}{NA}$ to $\frac{1}{4}\frac{\lambda}{NA}$.

The attenuated phase-shifting mask allows certain amount of background transmission that is 180° out of phase. Mathematically, its spectrum is a slight modification of equation (3-42) as below

$$G_F(\tilde{f}) = \frac{1+T_B}{2}\delta(\tilde{f}) + \frac{1-T_B}{\pi}[\delta(\tilde{f} - \tilde{f}_0) + \delta(\tilde{f} + \tilde{f}_0)]. \quad (3\text{-}44)$$

The wafer image spectrum when $\tilde{f} = \pm\tilde{f}_0$ is

$$I_F(\tilde{f} = \pm\tilde{f}_0) = \frac{1-T_B^2}{2\pi}[TCC(\tilde{f}_0;0) + TCC(0;-\tilde{f}_0)].$$

To characterize the imaging performance, we often define a modulation contrast parameter in the (normalized) frequency domain: $M_c = \frac{I_F(+\tilde{f}_0) + I_F(-\tilde{f}_0)}{I_F(0)}$. This parameter determines the ratio of the desired modulation signal intensity to the undesired background intensity and apparently a higher value is better. Compared with the binary on-axis case discussed before, it can be proved that M_c will be significantly improved by setting

T_B at a value that results in about 6% transmission of background light intensity. And this is why the 6% attenuated phase-shifting mask is widely adopted by the semiconductor industry. As a summary, the mask transmission functions (1-D) of three types of masks (binary, alternating and attenuated phase-shifting masks) are shown in Figure 32 for a comparison.

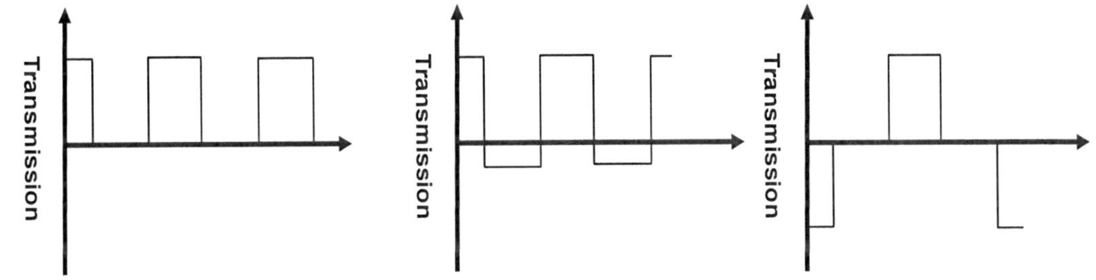

Figure 32. Comparison of the transmission profiles of three types of masks: binary mask (left), attenuated phase-shifting mask (middle), and alternating phase-shifting mask (right)

10.6. DUV Immersion Lithography

Using an immersion fluid to improve the resolution capability of an imaging system is not a new concept. For example, in an optical microscope, oil can be used on top of an objective to increase its numerical aperture NA and resolution. From Fourier (spectrum) point of view, the resolution improvement of an immersion imaging system is a result of being able to collect higher spatial frequencies of diffraction light by introducing an immersion fluid between the objective and lens. Since the fluid is placed directly on top of the objective, it is straight forward to use Snell's Law (equation (3-1)) to verify that higher frequencies (corresponding to larger incident angles) of the objective image indeed are collected (by the upstream parts) into the imaging system. Here, the upstream parts of an optical imaging system stand for the optical elements that light rays travel through first, while the downstream parts represent those elements that light rays travel through later. The goal of lithographic resolution improvement is to collect spatial frequencies as high as possible in the recording medium (resist) which is the last downstream element. On the other hand, the highest-frequency components that can be collected by an optical element (behind the mask) is always equal to or lower than the highest-frequency components propagating from the upstream part as no new frequencies besides those coming from the mask are generated in a passive signal processing system such as a scanner.

Assuming all the lens systems remain the same and only water is placed between the downstream parts (e.g., between the last lens and wafer) which does not affect the frequencies (higher spatial frequencies corresponding to larger incident angles) of light that are collected by the upstream parts of a scanner, an immediate question about the scanner's resolution arises: how can water improve the resolution capability (in a resist) from Fourier's point of view? This question can also be described from geometrical optics point of view. According to Snell's Law, $n \cdot \sin\theta$ value of any light ray travelling through different parts of an imaging system remains constant. If the refractive index of a resist is not increased, the maximum incident angle of a light ray in the resist is determined only by the $n \cdot \sin\theta$ value of the light ray in the upstream parts which remains the same regardless whether water is placed between the downstream parts or not. The resolution capability of a scanner is directly determined by the maximum incident angle of light inside the resist. Therefore, it is apparent that without changing the design of the lens system, the resolution capability of a scanner can not be improved by only placing water between the last lens and wafer.

In general, it is a necessary and complex task to re-design the imaging system (especially the last lens) to collect higher-frequency components (with higher incident angles) and allow them to be incorporated to form images by using new lens material, immersion fluid and resist with higher refractive indexes. The immersion fluid will be useful for resolution improvement only when higher-angle light rays can be collected by the upstream parts of an imaging system. To understand this, we need to introduce the concept of total reflection. If the refractive index is lower on the other side of the boundary, an optical phenomenon called total reflection can occur when a light ray hits the medium boundary at an angle larger than a critical angle with respect to the normal to the surface. Under this condition, no light can be transmitted and all of the light is reflected back. The critical angle is the incident angle above which the total reflection occurs. Using the last lens (which has a refractive index of n_L) in a 193 nm scanner as an example, we assume the whole imaging system (including the upstream parts and the last lens) have been modified such that some light rays inside the last lens have $n_L \cdot \sin\theta$ values between 1 and 1.35. If no water is placed between the last lens and wafer, those light rays corresponding to $1 < n_L \cdot \sin\theta < 1.35$ will be reflected back to the lens and unable to travel into the air due to the total reflection at the critical angle θ_c (which satisfies the relation of $n_L \cdot \sin\theta_c = 1$). As a consequence, the critical angle inside the last lens defines the maximum incident angle of light that can travel from the lens into the air. Therefore, those light rays with higher frequencies (or with larger incident angles satisfying $1 < n_L \cdot \sin\theta < 1.35$) will be wasted as they are not collected to form the image. However, if the gap between the last lens and wafer is filled with water (refractive index: n_w), the total reflection will occur only when $n_L \cdot \sin\theta_c > n_w \cdot 1 \approx 1.44$. Apparently, those light rays in the last lens with $1 < n_L \cdot \sin\theta < 1.35$ can travel into the water and contribute to forming the image on the wafer, which helps to improve the resolution of an optical imaging system according to the Fourier analysis.

Currently the highest NA that can be achieved using water as an immersion fluid in a scanner system is 1.35. Even higher NA was pursued by the lithographic industry in the past, while little progress was made due to many technical challenges, e.g., being unable to find a functional lens material with higher refractive index than currently available in 193 nm systems. In a scanner, the tunable optical parameters/elements include: NA, source shape, partial coherence σ range (with certain upper and lower limits) and polarization modes.

Another advantage of immersion lithography besides the higher resolution is the improved depth of focus. As shown in Figure 33, ideally the wafer surface should be placed at the focal plane (located at point *O*) such that a perfect spherical wave coming out of the exit pupil will converge to a point on the wafer. The Fourier expression of the aerial image applies only to the image at the focal plane (i.e., the best focus). However, in the manufacturing environment, there always exists certain level of inaccuracy when placing the wafer surface to the target (focal) plane. Therefore it is important to know the impact of a small distance away from the best focus (defocus Z_0) on the image and critical-dimension (CD) control.

For this purpose, we introduce a concept of depth of focus (DOF) to characterize the maximum defocus that a lithographic process can tolerate to produce images of suitable quality. For a production worthy process, it is often required that DOF should be at least 0.15 μm based on the current capability of positioning control in a scanner system. From Figure 33, we can see that when the wafer surface is placed at the best focus, the optical phase difference (OPD) between different light rays converging to point *O* is zero resulting in a constructive interference (which is also required by the Fermat's principle in geometrical optics).

When the wafer surface is displaced by a distance Z_0 from its focal plane, the image point of interest moves from *O* to *W* correspondingly and the optical path of different light rays reaching point *W* will not be the same any more. In the extreme case, let us examine the optical path difference between the vertical light ray and the light ray incident at an angle θ which corresponds to the maximum sustained half-angle of exit pupil. Mathematically, this OPD can be expressed as

$$OPD = \frac{2\pi}{\lambda}(R_0 + Z_0 - L) = \frac{2\pi}{\lambda}(\overline{WO} - \overline{WN}) \approx \frac{2\pi}{\lambda} Z_0(1-\cos\theta) = \frac{2\pi}{\lambda} Z_0 \cdot 2\sin^2\frac{\theta}{2}.$$

A widely accepted criterion defines DOF as the defocus (displacement Z_0) when above OPD reaches certain value (e.g., $k_2\pi$ where k_2 is a constant) to produce significant destructive interference. Thus we have DOF as

$$DOF = \frac{k_2\lambda}{4\sin^2\frac{\theta}{2}} = \frac{k_3\lambda}{\sin^2\frac{\theta}{2}}. \tag{3-45}$$

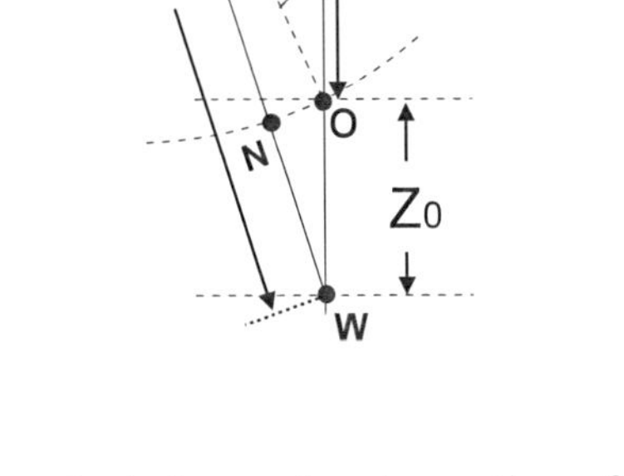

Figure 33. The defocus effect on a spherical wave focusing on the wafer plane

Here, λ is the light wavelength in medium and θ is the maximum incident angle of light rays coming out of the exit pupil. In the dry lithography, this wavelength becomes the wavelength in air: λ_0, θ is replaced by θ_i (see Fig. 19), and $4\sin^2(\theta_i/2)$ can be approximated as $\mathrm{NA}_{\mathrm{dry}}^2 (\mathrm{NA}_{\mathrm{dry}} = \sin\theta_i)$ such that equation (3-45) becomes a dry formula frequently seen in the literature:

$$DOF_{dry} = \frac{k_2 \lambda_0}{NA_{dry}^2}.$$

It is evident that the resolution improvement by increasing NA is at the price of reduced depth of focus and process window. It should be kept in mind that the dry formula of DOF is not accurate enough to predict the immersion DOF for which the general form (3-45) should be used. For a fixed resolution ($k_1 \frac{\lambda_0}{\sin\theta_i} = k_1 \frac{\lambda}{\sin\theta}$), the ratio of immersion DOF to dry DOF is:

$$\frac{k_3\lambda}{\sin^2\frac{\theta}{2}} \Bigg/ \frac{k_3\lambda_0}{\sin^2\frac{\theta_i}{2}} \approx \frac{\sin^2\frac{\theta_i}{2}}{\sin^2\frac{\theta}{2}} \frac{\lambda}{\lambda_0} \approx \frac{\sin\theta_i}{\sin\theta} = n_w > 1$$

10.7. Vector Imaging Theory and Polarization

The previous treatment of diffraction is based on the scalar imaging theory wherein the disturbance at any position and time is represented by a scalar function. The scalar imaging theory is a good approximation to describe wave propagation phenomena when NA is not high such that the perpendicular (transverse electric, or TE) component/polarization of the electric field vector is much larger than the parallel (transverse magnetic, or TM) component/polarization. When NA is high (e.g., in immersion lithography), the effect of the TM polarization on the imaging performance can not be ignored and a rigorous treatment of the diffraction problem using the vector imaging theory is necessary.

The scalar Huygens-Fresnel principle can be slightly modified to get the vectorial form for a high-NA imaging system as below:

$$\vec{E}(P) = \iint_S \vec{U}_s(Q) \frac{\exp(ikR)}{R} ds.$$

The electric field vector is denoted by its Cartesian components: $\vec{E}(P) = [E_x(P), E_y(P), E_z(P)]$. Let us consider two light rays interfere to form an image on the wafer after passage through the lens. With the TE polarization, the components $E_{1\perp}$ and $E_{2\perp}$ have the same vibration direction and their scalar sum is the total TE component. However, the other components E_1 and E_2 do not align and they must add in a vectorial manner to obtain the total TM component (see Figure 34). It has been demonstrated using a multiple-beam example that the TM polarization is undesired as it results in lower image contrast [84]. Unfortunately, in a high-NA immersion lithographic system, the TE modes is usually reduced more than the TM mode during the wave propagation and the image contrast degration due to TM mode is not negligible.

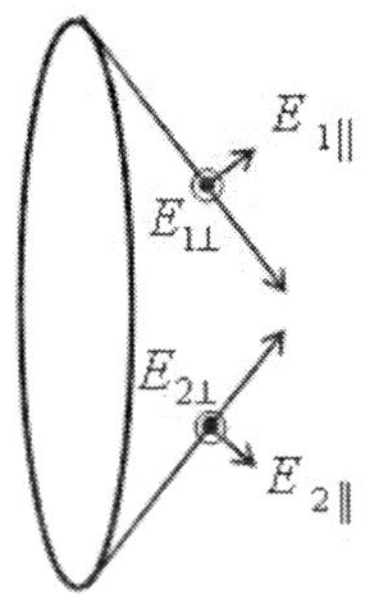

Figure 34. TE and TM polarization of light propagation through a lens

10.8. EUV Multilayer Reflective Optics

As we discussed in the previous EUV sections, the multilayer reflective mirrors are the key component of EUV lithography. In the past century, the theory of x-ray diffraction, scattering and reflection on a multilayer medium using wave interference analysis was well developed [85, 86]. Nevertheless, a rigorous treatment of multilayer reflective optics needs to use Maxwell's equations [78] described in the previous section.

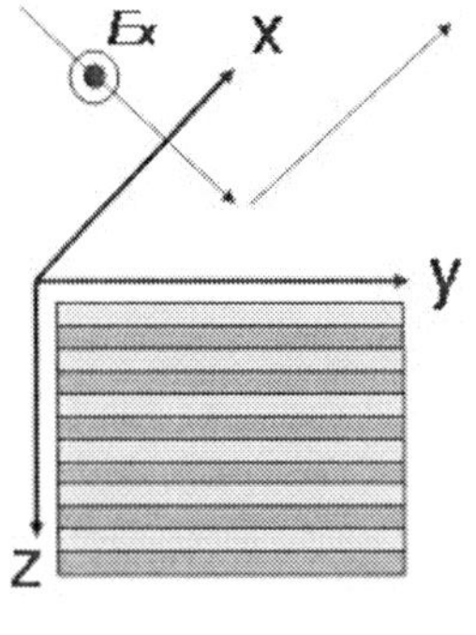

Figure 35. EUV light reflection on a multilayer mirror (not to scale)

Without loss of generality, let us assume the y-z plane is the incident plane of a TE wave which is non-zero only in x direction (i.e., $\vec{E} = [E_x, E_y = 0, E_z = 0]$) and propagates through a stratified (multilayer) medium in z direction as shown in Figure 35. Any polarized plane wave can be decomposed into TE and TE modes, and the theorem relating to TE mode can be readily extended to TM mode because of the duality principle which governs the propagation of a time-harmonic wave (time dependence of $e^{-i\omega t}$ assumed). The duality principle states that Maxwell's equations remain unchanged when $\vec{E}$ and $\vec{H}$ (and simultaneously ε_z and $-\mu_z$) of a time-harmonic wave are interchanged.

For a TE wave, Maxwell's equations become six scalar equations:

$$\frac{\partial H_z}{\partial y}-\frac{\partial H_y}{\partial z}+\frac{i\varepsilon_z\omega}{c}E_x=0,$$
$$\frac{\partial H_x}{\partial z}-\frac{\partial H_z}{\partial x}=0,$$
$$\frac{\partial H_y}{\partial x}-\frac{\partial H_x}{\partial y}=0,$$
$$\frac{i\omega\mu_z}{c}H_x=0,$$
$$\frac{\partial E_x}{\partial z}-\frac{i\omega\mu_z}{c}H_y=0,$$
$$\frac{\partial E_x}{\partial y}+\frac{i\omega\mu_z}{c}H_z=0.$$

With the help of the first equation (in Maxwell's scalar equations), differentiating the fifth equation with respect to z and differentiating the sixth equation with respect to y yields:

$$\frac{\partial^2 E_x}{\partial y^2}+\frac{\partial^2 E_x}{\partial z^2}+n^2k_0^2E_x-\frac{d(\ln\mu_z)}{dz}\frac{\partial E_x}{\partial z}=0, \tag{4-1}$$

where $n=\sqrt{\varepsilon_z\mu_z}, k_0=\frac{2\pi}{\lambda_0}$. To solve equation (4-1), we assume $E_x(y,z)=Y(y)U(z)$ and substitute it into this equation to obtain:

$$\frac{1}{Y}\frac{d^2Y}{dy^2}=-\frac{1}{U}\frac{d^2U}{dz^2}-n^2k_0^2+\frac{d(\ln\mu_z)}{dz}\frac{1}{U}\frac{dU}{dz}. \tag{4-2}$$

The left term is a function of y only while the right term is a function of z only, thus equation (4-2) can be true only if both sides are equal to a constant, e.g., $-(k_0\cdot n_a)^2$. Namely,

$$\frac{1}{Y}\frac{d^2Y}{dy^2}=-(k_0n_a)^2, \tag{4-3}$$

$$\frac{1}{U}\frac{d^2U}{dz^2}+n^2k^2-\frac{d(\ln\mu_z)}{dz}\frac{1}{U}\frac{dU}{dz}=(k_0n_a)^2. \quad (4\text{-}4)$$

The solution of equation (4-3) is: $Y(y) \propto \exp(ik_0 n_a y)$, and consequently $E_x(y,z)$ is:

$$E_x(y,z)=U(z)\exp[i(k_0n_a y-\omega t)]. \quad (4\text{-}5)$$

From Maxwell's equations (the fifth and sixth equations), we can set

$$H_y(y,z)=\Gamma(z)\exp[i(k_0n_a y-\omega t)], \quad (4\text{-}6)$$

$$H_z(y,z)=W(z)\exp[i(k_0n_a y-\omega t)]. \quad (4\text{-}7)$$

Substituting equations (4-5), (4-6) and (4-7) into the first, fifth and sixth Maxwell's equations gives the following differential equations for $U(z)$ and $\Gamma(z)$:

$$\frac{d^2U}{dz^2}-\frac{d(\ln\mu_z)}{dz}\frac{dU}{dz}+k_0^2(n^2-n_a^2)U=0, \quad (4\text{-}8)$$

$$\frac{d^2\Gamma}{dz^2}-\frac{d[\ln(\varepsilon_z-n_a^2/\mu_z)]}{dz}\frac{d\Gamma}{dz}+k_0^2(n^2-n_a^2)\Gamma=0. \quad (4\text{-}9)$$

When the wave is a homogeneous plane wave, its representation form is: $\exp[i(k_0\cdot n\sin\theta\cdot y-\omega t)]$. In this special case, a simple relation exists: $n_a=n\cdot\sin\theta$, where θ is the angle the z-axis makes with the normal to the wave. Therefore, for a TE wave, equations (4-8) and (4-9) can be simplified to:

$$\frac{d^2U}{dz^2}+(k_0^2n^2\cos^2\theta)U=0,$$

$$\frac{d^2\Gamma}{dz^2}+(k_0^2n^2\cos^2\theta)\Gamma=0.$$

The solutions of these two equations (subject to the requirement of Maxwell's equations) are:

$$U(z)=A\cos(k_0nz\cdot\cos\theta)+B\sin(k_0nz\cdot\cos\theta),$$

$$\Gamma(z) = \frac{1}{i}\sqrt{\frac{\varepsilon_z}{\mu_z}}\cos\theta[B\cos(k_0 nz\cdot\cos\theta) - A\sin(k_0 nz\cdot\cos\theta)].$$

Since $U_0 = U(z=0) = A$, $\Gamma_0 = \Gamma(z=0) = \frac{B}{i}\sqrt{\frac{\varepsilon_z}{\mu_z}}\cos\theta$, these solutions can be expressed as a matrix operation:

$$\begin{bmatrix} U(z=0) \\ \Gamma(z=0) \end{bmatrix} = \begin{bmatrix} \cos(k_0 nz\cdot\cos\theta), \frac{-i}{\cos\theta}\sqrt{\frac{\mu_z}{\varepsilon_z}}\sin(k_0 nz\cdot\cos\theta) \\ -i\cos\theta\sqrt{\frac{\varepsilon_z}{\mu_z}}\sin(k_0 nz\cdot\cos\theta), \cos(k_0 nz\cdot\cos\theta) \end{bmatrix} \begin{bmatrix} U(z) \\ \Gamma(z) \end{bmatrix} = \mathbf{M}\begin{bmatrix} U(z) \\ \Gamma(z) \end{bmatrix} \tag{4-10}$$

The characteristic transfer matrix $\mathbf{M}$ is

$$\mathbf{M} = \begin{bmatrix} \cos(k_0 nz\cdot\cos\theta), \frac{-i}{\cos\theta}\sqrt{\frac{\mu_z}{\varepsilon_z}}\sin(k_0 nz\cdot\cos\theta) \\ -i\cos\theta\sqrt{\frac{\varepsilon_z}{\mu_z}}\sin(k_0 nz\cdot\cos\theta), \cos(k_0 nz\cdot\cos\theta) \end{bmatrix}. \tag{4-11}$$

This matrix methodology can be extended from one layer to multilayer film stack:

$$\begin{bmatrix} U(z=0) \\ \Gamma(z=0) \end{bmatrix} = \mathbf{M_T}\begin{bmatrix} U(z_N) \\ \Gamma(z_N) \end{bmatrix}, \tag{4-12}$$

where $\mathbf{M_T} = \mathbf{M_1}(z_1)\cdot\mathbf{M_2}(z_2 - z_1)\cdots\mathbf{M_N}(z_N - z_{N-1})$.

The EUV multilayer consists of a periodic homogeneous (N-pair) layers of alternately low and high refractive indices and thickness, often called quarter-wave films ($n_2 h_2 = n_3 h_3 = \lambda_0/4$) as shown in Figure 36. Let $\mathbf{M} = \begin{bmatrix} m_{11}, m_{12} \\ m_{21}, m_{22} \end{bmatrix}$ be the characteristic matrix of one period ($h = h_2 + h_3$), the N-pair transfer matrix can be expressed as

$$\mathbf{M_T} = \mathbf{M}(h)\cdot\mathbf{M}(h)\cdots\mathbf{M}(h) = [\mathbf{M}(h)]^N.$$

It can be proved that:

$$\mathbf{M}_{\mathrm{T}} = \begin{bmatrix} m_{11}\chi_{N-1}(\overline{m}) - \chi_{N-2}(\overline{m}), m_{12}\chi_{N-1}(\overline{m}) \\ m_{21}\chi_{N-1}(\overline{m}), m_{22}\chi_{N-1}(\overline{m}) - \chi_{N-2}(\overline{m}) \end{bmatrix},$$

where $\overline{m} = (m_{11} + m_{22})/2$ and χ_N is the Chebyshev polynomials of the second kind: $\chi_N(x) = \dfrac{\sin[(N+1)\cos^{-1} x]}{\sqrt{1-x^2}}$.

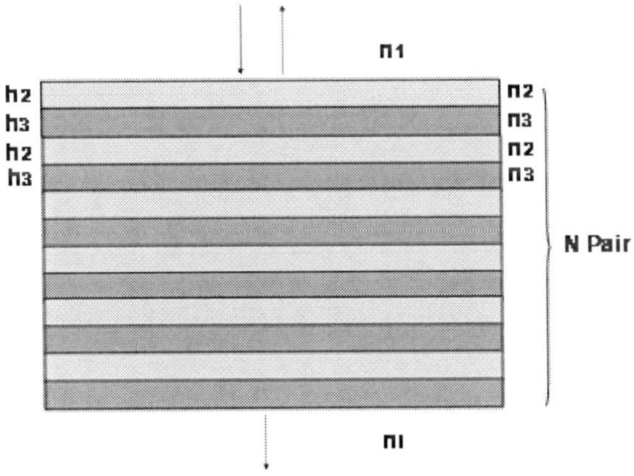

Figure 36. N-pair EUV multilayer mirror with the thickness and refractive index specified in the figure

The characteristic matrix of one period ($n_2h_2 = n_3h_3 = \lambda_0 / 4$) of nonmagnetic material ($\mu_z = 1$) at normal incidence ($\theta = 0$) can be obtained from equations (4-11) and (4-12) as below:

$$\mathbf{M} = \begin{bmatrix} 0, -i\sqrt{\dfrac{1}{\varepsilon_{z,2}}} \\ -i\sqrt{\dfrac{\varepsilon_{z,2}}{1}}, 0 \end{bmatrix} \begin{bmatrix} 0, -i\sqrt{\dfrac{1}{\varepsilon_{z,3}}} \\ -i\sqrt{\dfrac{\varepsilon_{z,3}}{1}}, 0 \end{bmatrix} = \begin{bmatrix} -\dfrac{n_3}{n_2}, 0 \\ 0, -\dfrac{n_2}{n_3} \end{bmatrix}.$$

Thus the N-pair multilayer characteristic matrix is

$$\mathbf{M}_{\mathrm{T}} = \begin{bmatrix} (-\dfrac{n_3}{n_2})^N, 0 \\ 0, (-\dfrac{n_2}{n_3})^N \end{bmatrix} = \begin{bmatrix} m'_{11}, m'_{12} \\ m'_{21}, m'_{22} \end{bmatrix}. \qquad (4\text{-}13)$$

The relation between the multilayer reflectivity and the characteristic matrix must be solved using the boundary conditions of the stratified medium and the relation between electric and magnetic fields: $\vec{H} = \sqrt{\frac{\varepsilon_z}{\mu_z}} \vec{k} \times \vec{E}$ ($\vec{k}$ is the unit vector in the direction of propagation). Let *A*, *R*, *T* denote as the amplitudes of the incident, reflected, and transmitted (electric) waves. The incident functions U_0 and Γ_0, and transmitted functions $U(z_l)$ and $\Gamma(z_l)$ can be expressed as:

$$U_0 = A + R, \Gamma_0 = \sqrt{\frac{\varepsilon_{z,1}}{\mu_{z,1}}} \cos\theta_1 (A - R); U(z_l) = T, \Gamma(z_l) = T\sqrt{\frac{\varepsilon_{z,l}}{\mu_{z,l}}} \cos\theta_l .$$

Substitute these relations into below matrix equation

$$\begin{bmatrix} U_0 \\ \Gamma_0 \end{bmatrix} = \mathbf{M}_\mathbf{T} \begin{bmatrix} U(z_l) \\ \Gamma(z_l) \end{bmatrix} = \begin{bmatrix} m'_{11}, m'_{12} \\ m'_{21}, m'_{22} \end{bmatrix} \begin{bmatrix} U(z_l) \\ \Gamma(z_l) \end{bmatrix},$$

we can solve *A*, *R*, *T* and the reflectivity which is calculated by

$$\Re = \left|\frac{R}{A}\right|^2 = \left[\frac{(m'_{11} + m'_{12}\sqrt{\frac{\varepsilon_{z,l}}{\mu_{z,l}}}\cos\theta_l)\sqrt{\frac{\varepsilon_{z,1}}{\mu_{z,1}}}\cos\theta_1 - (m'_{21} + m'_{22}\sqrt{\frac{\varepsilon_{z,l}}{\mu_{z,l}}}\cos\theta_l)}{(m'_{11} + m'_{12}\sqrt{\frac{\varepsilon_{z,l}}{\mu_{z,l}}}\cos\theta_l)\sqrt{\frac{\varepsilon_{z,1}}{\mu_{z,1}}}\cos\theta_1 + (m'_{21} + m'_{22}\sqrt{\frac{\varepsilon_{z,l}}{\mu_{z,l}}}\cos\theta_l)} \right]^2 .$$

According to equation (4-13), the EUV multilayer reflectivity is:

$$\Re(n_2, n_3, N) = \left[\frac{1 - \frac{n_l}{n_1}(\frac{n_2}{n_3})^{2N}}{1 + \frac{n_l}{n_1}(\frac{n_2}{n_3})^{2N}} \right]^2 = \left[\frac{1 - (\frac{n_2}{n_3})^{2N}}{1 + (\frac{n_2}{n_3})^{2N}} \right]^2, (n_l = n_1 = 1).$$

EUV light is absorbed by the medium that it travels through, hence the refractive index of the medium is a complex number. The complex refractive index of molybdenum/silicon (Mo/Si) is a function of wavelength [86]: $n = 1 - \kappa\lambda^2 (f_1^0 - if_2^0)$, where κ, f_1^0, and f_2^0 are material parameters. Therefore, we can calculate the EUV reflectivity as a function of both wavelength and the pair number of multilayer. The calculation results based on a Mo/Si mirror are shown in Figure 37 and a good agreement with experimental data [86] is observed.

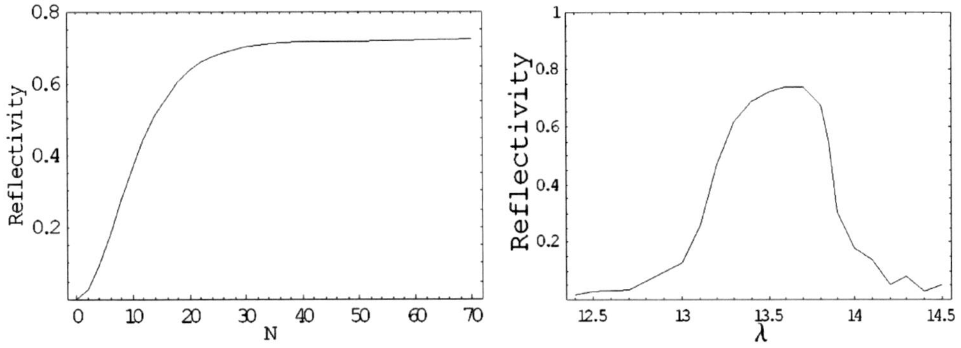

Figure 37. Calculated EUV mirror's reflectivity as a function of the Mo/Si pair number and wavelength

11. Multiple Patterning Technologies

Any lithographic system has a resolution limit beyond which the multiple patterning techniques to reduce the "equivalent k_1" value must be considered. Multiple patterning can be done in an optical or non-optical manner. Optical multiple patterning needs to decompose the dense structure to multiple semi-dense sub-structures (with larger pitch) which can then be printed (and/or etched) separately and recombined together onto the final layer to form the dense structure. Optical multiple patterning requires a tight control of overlay accuracy. Non-optical multiple patterning includes self-aligned spacer double patternng (SADP) that has been applied for NAND flash memory manufacturing.

11.1. Double Patterning

Figure 38 is a cartoon to demonstrate two optical double patterning schemes: LELE (litho-etch-litho-etch) and LFLE (litho-freeze-litho-etch) processes. Two freezing methods: chemical and ion-implantation freezing are indicated in the figure. Detailed discussion about optical double patterning is available in many papers published in recent SPIE Advanced Lithography conferences and will not be repeated here.

The fabrication technique to use sidewall spacers to double the spatial frequency of transistor gates and MEMS/NEMS devices dates back to a long time ago [74, 87-89], while its adoption for large-scale manufacturing (e.g., NAND flash with HP $\leq$ 35 nm) happened only recently. The obstacles to implanting the spacer patterning process from research labs to the high-volume manufacturing fabs are mainly due to significantly increased process complexity and costs, especially when more cost-effective lithographic solutions are available. There are two non-lithographic approaches to overcome the resolution limit of lithography: implementation of a vertical device/gate design [90] and spacer based multiple patterning. In a vertical CMOS device [90], the gates, sources, and drains of PMOS and NMOS cells are designed in the vertical direction (and can even be integrated together to consume less space). The vertical channel/gate length is defined by an epitaxy process while the channel/gate width is defined by the Si shell/cylinder circumference. Epitaxy film thickness can be accurately controlled in the growth process with much less gate CD variation and line width roughness than patterned by lithography. This vertical approach, however, will force a revolution in design and fabrication technologies of integrated circuits; thus is only considered as a long-term solution when all other techniques (e.g., the spacer based multiple patterning) reach their physical limits.

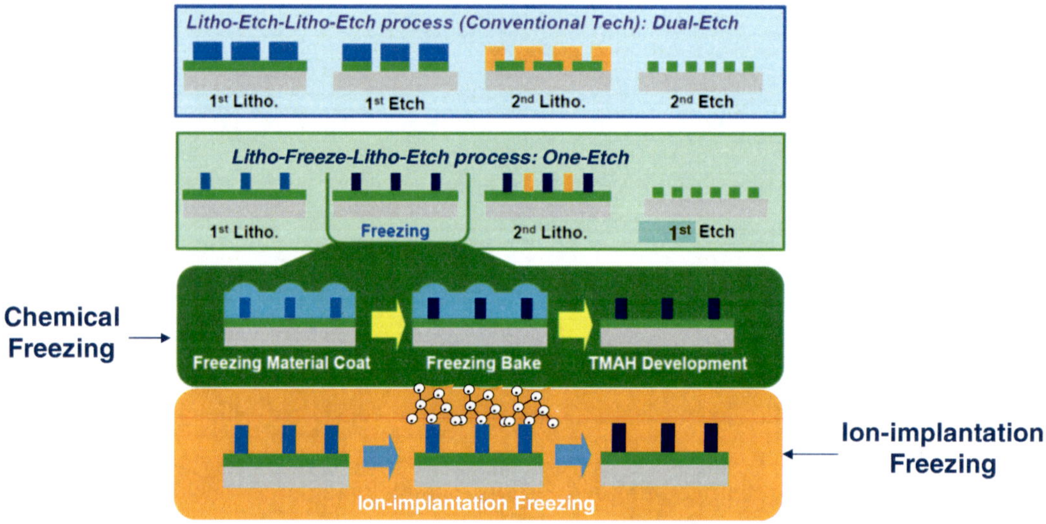

Figure 38. Two possible schemes of optical double patterning (LELE and LFLE) process

The spacer based self-aligned multiple patterning (including, but not limited to double, triple, quadruple, and sextuple patterning techniques) offers a near-term solution to overcome the resolution limit of immersion lithography (about 38 nm HP) even the caused penalty in throughput and cost is nontrivial. Figure 39 shows the SEM top view of the gate layer of a SRAM cell (circled area) that is patterned by self-aligned spacer double pattering (X-direction) and a cropping step (Y-direction). Normally, 3 masks for each critical layer are needed: core mask, cut mask, and peripheral circuit (including line pads) mask. Again, many papers about SADP and its applications in NAND, DRAM, and gridded logic design can be found in recent SPIE Advanced Lithography conferences.

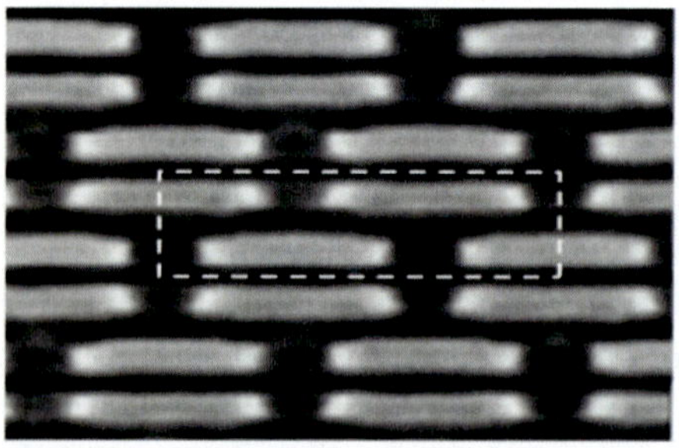

Figure 39. SEM image of the gate layer of a SRAM cell patterned by a SADP technology

11.2. Self-aligned Triple and Quadruple Patterning

193 nm immersion lithography based double patterning can only drive the half pitch down to about 19 nm. EUV, e-beam direct writing, and nano-imprint technologies, all with severe technical or cost barriers, will not be ready for high-volume manufacturing in the near future. Therefore, non-optical approaches such as the self-aligned triple patterning (SATP) and quadruple patterning (SAQP) processes are proposed and developed to pattern 18-10 nm (half-pitch) lines/spaces.

SATP technology was invented by Chen et al. in Applied Materials [91, 92], which enables line/space density tripling (as shown in Figure 40) and many potential innovations in circuit design and EDA methodology (e.g., mask layout decomposition) as well. The key idea is to keep the lithographically defined lines/mandrels (instead of being etched away such as in SADP/SAQP), and have a sacrificial layer formed between the hard-mask layers 1 and 2. A wet etch (or highly selective dry etch) process can release this sacrificial layer as shown in step (8). As a result, the final features are separately defined by the lithographic and deposition processes. The advantages of a SATP process are apparent: simple process integration/less complexity and its capability to include both 1-D dense lines/spaces and 2-D peripheral circuits within one mask. It has been proposed that with SATP, only 1-2 masks are needed to pattern dense lines/spaces with pads and 2-D peripheral circuits on one critical layer. Compared with a SADP/SAQP process (which needs 3 masks for a similar critical layer), a SATP process totally can save 2-6 masks plus the following lithographic and etch steps for one device (assuming it has 2-3 critical layers that need multiple patterning). This can significantly reduce the cost for each wafer, potentially saving hundreds of million dollars for the semiconductor industry. However, patterning smaller lines in steps (2) & (3) is challenging and a major focus of the early-stage SATP research is to develop an etching/trimming process that can pattern small lines without bending/wiggling and necking problems. In Figure 41, we show that 20 nm (half-pitch) lines/spaces have been patterned successfully using a SATP process and 193 nm dry lithography (0.93 NA). With an immersion scanner (1.35 NA), we should be able to extend the optical lithography to about 13 nm half pitch using the SATP technology.

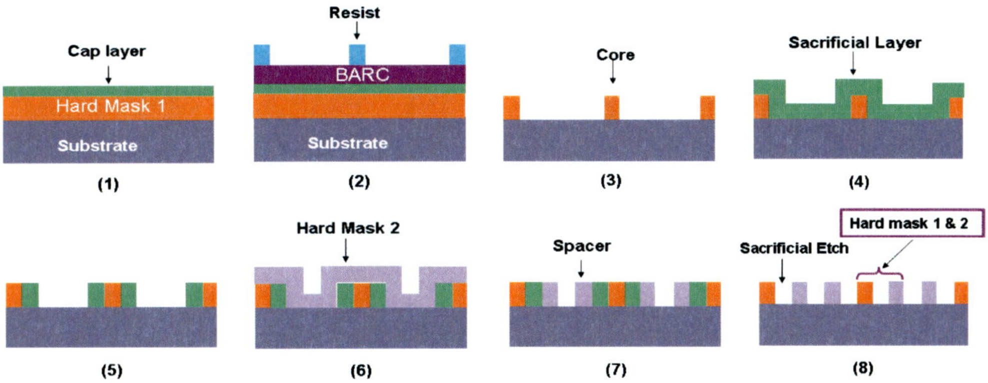

Figure 40. A cartoon of the self-aligned triple patterning process flow [92]

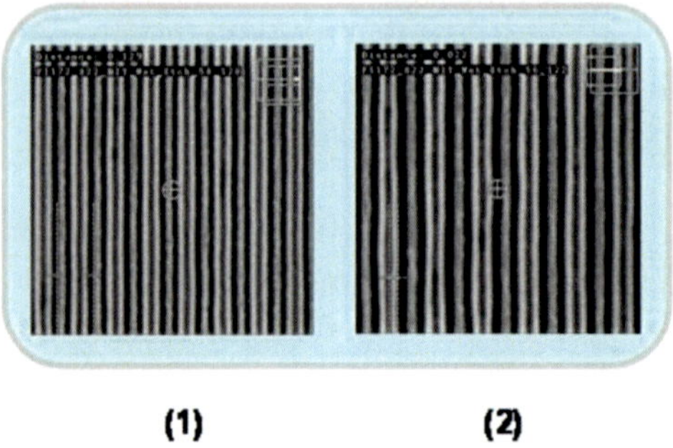

Figure 41. Half-pitch 21 nm (left) and 20 nm (right) lines/spaces patterned by self-aligned triple patterning technology using dry 193 nm lithography [92]

The idea of self-aligned quadruple patterning (SAQP) by doing SADP twice as shown in Figure 42 is straightforward, other simplified approaches of quadruple patterning are possible (e.g., direct deposition of the 2nd spacer on top of the 1st spacer). However, it was pointed out that the simplified process may suffer from the asymmetry of spacer profiles and a consequent CD difference on the final layer unless tall and vertical spacers can be formed [93].

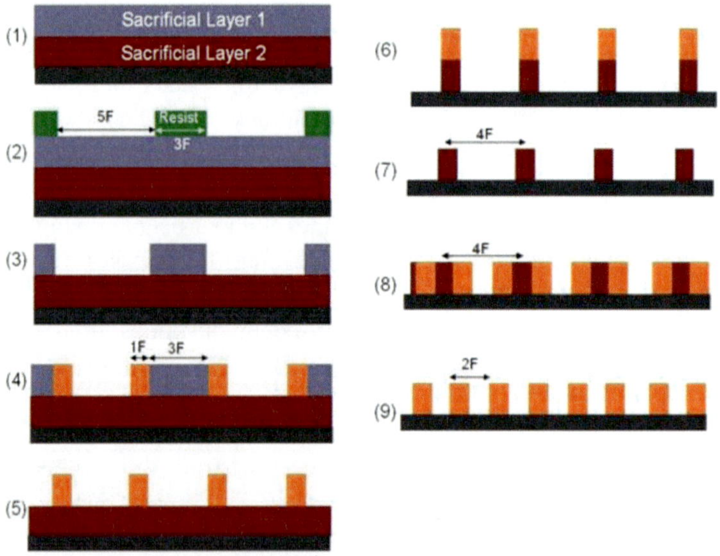

Figure 42. A cartoon of one possible scheme of the self-aligned quadruple patterning process flow, "F" stands for the final half-pitch CD

Whether a SATP or SAQP process will be adopted by the industry depends on the process cost and manufacturability. For example, the triple patterning (SATP) process simplicity and design flexibility make it a promising lithographic candidate with a resolution capability of 18-13 nm (half-pitch) for application in CMOS logic devices. But it may suffer from the LER/CDU challenge of lithography defined small features (with large pitch) because of its reduced process window and required trimming. Also, the sidewall profile of the second spacers may not be as vertical as the first (sacrificial) spacers, and some surface treatment (e.g., CMP) may be needed to reduce the corner rounding/slope of the first spacers before the deposition of the second spacers. It is an interesting research topic to study how circuits can be designed to take advantage of the SATP resolution capability while avoid its worse LER/CDU of lithographically defined features. For even smaller device nodes, quadruple patterning may or may not be needed, depending on whether the high-volume EUVL can be ready in time or not.

SUMMARY

In this chapter, we have presented a general review of CMOS lithography technology, covering a full spectrum of imaging theory, process integration, scaling challenges and potential solutions. Using a systematic approach, we focus on EUV lithography and the fundamental principles of projection lithography system. Key physical concepts that often confuse beginners are explained in detail and some incorrect points frequently seen in the literature are corrected. Several non-optical multiple patterning technologies to overcome the resolution limit of DUV immersion lithography are also discussed.

REFERENCES

[1] Kinoshita, H; Wood, O. "EUV lithography -- an historical perspective," Chapter 1 in *EUV Lithography*, V. Bakshi, ed., pp. 1-54, SPIE Press, *Bellingham*, WA, 2008.

[2] Goldsmith, JEM; Berger, KW; Bozman, DR; et al., "Sub-100-nm imaging with an EUV 10x microstepper," *Proc. SPIE*, 1999, 3676.

[3] Tichenor, D; Kubiak, G; Replogle, W; et al., "EUV engineering test stand," *Proc. SPIE*, 2000, 3997.

[4] Goldberg, K; Naulleau; Denham, P; et al., "At-wavelength alignment and testing of the 0.3-NA MET optic," *J. Vac. Sci. Tech. B*, 2005, 22(6), 2956-2961.

[5] Naulleau, P; Goldberg, K; Cain, J; et al., "EUV microexposures at the ALS using the 0.3-NA MET projection optics," *Proc. SPIE*, 2005, 5751.

[6] Naulleau, P; Anderson, C; Dean; K. et al., "Recent results from the Berkeley 0.3-NA EUV microfield exposures tool," *Proc. SPIE*, 2007, 6517.

[7] Cao, H; Wang, Y; Roberts, J. "EUV resist patterning performance from the Intel microexposure tool (MET)," *Proc. SPIE*, 2005, 5753.

[8] Tanaka, Y; Aoyama, H; Tawarayama; K. et al., "Effects of aberration and flare on lithographic performance of SFET," *Proc. SPIE*, 2008, 6921.

[9] http://www-cxro.lbl.gov/research/euvlithography/next.

[10] Meiling, H; Meijer, H; Banine, V; et al., "First performance results of the ASML α demo tool," *Proc. SPIE*, 2006, 6151.

[11] Lorusso, GF; Hermans, J; Goethals, AM; et al., "Imaging performance of the EUV α demo tool at IMEC," *Proc. SPIE*, 2008, 6921.

[12] Murakami, K; Oshino; T; Kondo, H; et al., "Development of EUV lithography tools at Nikon," *Proc. SPIE*, 2008, 7140.

[13] Wagner, C; Harned; N; Kuerz, P; et al., "EUV into production with ASML's NXE platform," *Proc. SPIE*, 2010, 7636.

[14] Lowisch, M; Kuerz; P; Mann, HJ; et al., "Optics for EUV production," *Proc. SPIE,* 2010, 7636.
[15] Murakami, K; Oshino, T; Kondo, H. "Development progress of optics for extreme ultraviolet lithography at Nikon," *Jour. of Microlithography, Microfabrication, and Microsystems,* 2009, 8(4).
[16] Editor: Bakshi, V. *"EUV Lithography,"* 150.
[17] Brainard, RL; Trefonas; P; Lammers, JH; et al., "Shot noise, LER, and quantum efficiency in EUV photoresists," *Proc. SPIE* vol, 2004, 5374.
[18] Sematech Inc. *International Tech. Roadmap for Semiconductors*. 2009.
[19] Kinoshita, H; Yoshizumi; T; Osugi, M. "Study on critical dimension of printable phase defects using an EUV microscope," *Microelectronic Eng.*, 2009, 86(4-6), 505-8.
[20] Bainard, R. "Photoresist for Extreme Ultraviolet Lithography," *EUV lithography*, pp 384-437, Wiley, SPIE, 2009.
[21] Gallatin, G. "Resist blur and line edge roughness," *Proc. SPIE*, 2005, 5754.
[22] Liang, T. "Growth and printability of multilayer phase defects on extreme ultraviolet mask blanks," *J. Vac. Sci. & Technol.*, 2007, B25 (6), 2098.
[23] Goldberg, K; Naulleau, P; Mochi, I; Anderson, EH; et al., "Actinic extreme ultraviolet mask inspection beyond 0.25 numerical aperture," *J. Vac. Sci. & Technol. B*, 2008, 26 (6), 2220-4.
[24] ITRS litho roadmap: http://www.itrs.net/Links/2009ITRS/2009Chapters_2009 Tables/2009_Litho.pdf.
[25] Mori, I; Suga, O; Tanaka, H; Nishiyama, I; et al., "Selete's EUV program: progress and challenges," *Proc. SPIE*, 2008, 6921.
[26] Han, H; Cho, W; Goldberg, KA; Gullikson, EM; et al., "Determining the critical size of EUV mask substrate defects," *Proc*. SPIE, 2008, 6921.
[27] LaFontaine, B; Pawloski, A; et al., "Demonstration of phase-shift mask for extreme ultraviolet lithography," *Proc. SPIE*, 2006, vol 6151.
[28] Pawloski, AR; Acheta, A; Lalovic, I; LaFontaine, B; et al, "Characterization of line-edge roughness in photoresist using an image fading technique," *Proc. SPIE*, 2004, 5376, 414.
[29] Wallow, T; Higgins; C; Brainard; R; et al., "Evaluation of EUV resist materials for use at the 32 nm half-pitch node," *Proc. SPIE*, 2008, Vol. 6921.
[30] Zhang, G; Yang; PY; Liang; T; et al. "EUV mask process development integration," *Proc. SPIE*, 2006, vol. 6283.
[31] Wood, O; "Use of attenuated phase masks in exreme ulraviolet lithography," *JVST B,* 1997, Vol. 15, 2448.
[32] Levenson, MD; Viswanathan, NS; Simpson, RA. "Improving resolution in photolithography with a phase-shifting mask," *IEEE Transactions on Electron Devices,* 1982, Vols. ED-29., 1828-36.
[33] Goodman, D; Levenson, MD; Santini, H; et al., "Improved photolithographic resolution with a phase shifting mask," *Microcircuit Engineering*, 1982, 82, 93-98.
[34] Goldberg, KA; Mochi, I; Naulleau, P; et al., "EUV pattern defect detection sensitivity based on aerial image linewidth measurements," *JVST B*. 2009, vol 27, no 6.
[35] Yan, PY. "EUVL alternating phase-shift mask imaging evaluation," *Proc. SPIE*, 2002, Vol 4889.

[36] Jeong, CY; Kim, BH; et al., "The study of attenuated PSM structure for Extreme Ultra Violet Lithography with minimized mask shadowing effect," *Proc. SPIE*, 2008, vol 6921.

[37] Wurm, S; Kemp, K. "SEMATECH pushes extreme UV lithography forward," *Proc. SPIE*, 2006, http://spie.org/x8865.xml?ArticleID=x8865.

[38] Jonckheere, R; Den Heuvel, DV; Iwamoto, F; Stepanenko, N. "Investigation of EUV mask defectivity via full-field printing and inspection on wafer," *Proc. SPIE*, 2009, 7379.

[39] Hendrickx, E; Lorusso, GF; Jiang, J; et al., "Accurate models for EUV lithography," *Photomask Tech.* 2009, 14-17 September, 2009, Monterey, CA, USA.

[40] Murakami, K; Oshino, T; Kondo, H; et al., "Development progress of optics for extreme ultraviolet lithography at Nikon," J. *MicrolithographyMicrofabrication, and Microsystems*., 2009, vol 8, Issue 4.

[41] Lorusso, GF; Van Roey, F; Hendrickx, E; et al., "Flare in extreme ultraviolet lithography: metrology, out-of-band radiation, fractal point-spread function, and flare map calibration," *Jour. Microlith., Microfab., & Microsys*., 2009, Vol 8, Issue: 4.

[42] George, SA; Naulleau, PP; Rekawa, S; et al., "Estimating the out-of-band radiation flare levels for extreme ultraviolet lithography," *Jour. Microlithography, Microfabrication, and Microsystems*., 2009, vol 8, issue 4.

[43] Murakami, K; Oshino, T; Kondo, H; et al., "Development progress of optics for EUVL at Nikon," *Proc. SPIE*., 2009, vol 7271.

[44] Mizuno, H; McIntyre, G; Koay, CS; et al., "Flare evaluation of ASML α demo tool," *Proc. SPIE*, 2009, vol 7271.

[45] Wood, O; Chiew-Seng Koay, Petrillo, K; et al., "Integration of EUV lithography in the fabrication of 22-nm node devices," *Proc. SPIE*., 2009, vol 7271.

[46] Myers, AM; Lorusso, GF; Kim, I; et al., "Experimental validation of full-field extreme ultraviolet lithography flare and shadowing corrections," *Jour. Vac. Sc. & Tech. B*., 2008, vol 26, issue 6, pages: 2215-19.

[47] Goethals, AM; Jonckheere, R; Lorusso, GF; et al., "EUV lithography program at IMEC," *Proc. SPIE*., 2007, vol 6517.

[48] Uzawa, S; Kubo, H; Miwa, Y; et al., "Path to the HVM in EUVL through the development and evaluation of the SFET," *Proc. SPIE*., 2007, Vol 6517.

[49] Miura, T; Murakami, K; Kawai, H; et al., "Nikon EUVL development progress update," *Proc SPIE.,* 2009, Vol 7271.

[50] Goldberg, KA; Naulleau, PP; Barty, A; et al., "Performance of actinic EUVL mask imaging using a zoneplate microscope," *Proc SPIE.,* 2007, Vol 6730.

[51] Lorusso, GF; Goethals, AM; Jonckheere, R; et al., "Extreme ultraviolet lithography at IMEC: shadowing compensation and flare mitigation strategy," *Jour. Vac. Sc. & Tech. B*., 2007, vol 25(6), pages 2127-31.

[52] Goethals, AM; Gronheid, R; Van Roey, F; et al., "Progress in EUV resist performance," *Jour. Photopolymer Sc. & Tech*., 2006, vol 19(4), 501-6.

[53] Kim, I; Myers, A; Melvin, LS; III, B. Ward, et al., "Flare mitigation strategies in extreme ultraviolet lithography," *Microelectronics Engineering*, 2008, vol. 85, no. 5-6, 738-743.

[54] Krautschik, CG; Chandhok, M; Zhang, G; et al., "Implementing flare compensation for EUV masks through localized mask CD resizing," *Proc. SPIE*, 2003, 5037, 58-68.

[55] Krautschik, CG; Ito, M; Nishiyama, I; Okazaki, S. "Impact of EUV light scatter on CD control as a result of mask density changes," *Proc. SPIE*, 2002, vol. 4688.

[56] Lai, K; Wu, CH; Progler, C. "Scattered Light: The increasing problem for 193 nm exposure tools and beyond," *Proc. SPIE*, 2001, vol. 4246.

[57] Stover, JC; Bernt, ML; Church, EL. "Measurement and analysis of scatter from rough surface," *Proc. SPIE*, 1994, vol. 2260.

[58] Kinoshita, H; Yoshizumi, T; Osugi, M. "Study on critical dimension of printable phase defects using an EUV microscope," *Microelectronic Eng.,* 2009, 86(4-6), 505-8.

[59] Kinoshita, H; Wood, O. "EUV lithography -- an historical perspective," Chapter 1 in *EUV Lithography*, V. Bakshi, ed., 1-54, SPIE Press, Bellingham, WA 2008.

[60] Goldsmith, JEM; Berger, KW; Bozman, DR; et al., "Sub-100-nm imaging with an EUV 10x microstepper," *Proc. SPIE*, 1999, 3676.

[61] Tichenor, D; Kubiak, G; Replogle, W; et al., "EUV engineering test stand," *Proc. SPIE*, 2000, 3997.

[62] Goldberg, K; Naulleau; Denham, P; et al., "At-wavelength alignment and testing of the 0.3-NA MET optic," *J. Vac. Sci. Tech. B* 22(6), 2005, 2956-2961.

[63] Naulleau, P; Goldberg, K; Cain, J; et al., "EUV microexposures at the ALS using the 0.3-NA MET projection optics," *Proc. SPIE*, 2005, 5751.

[64] Naulleau, P; Anderson, C; Dean; K; et al., "Recent results from the Berkeley 0.3-NA EUV microfield exposures tool," *Proc. SPIE*, 2007, 6517.

[65] Cao, H; Wang, Y; Roberts, J. "EUV resist patterning performance from the Intel microexposure tool (MET)," *Proc. SPIE*, 2005, 5753.

[66] Tanaka, Y; Aoyama, H; Tawarayama; K. et al., "Effects of aberration and flare on lithographic performance of SFET," *Proc. SPIE*, 2008, 6921.

[67] http://www-cxro.lbl.gov/research/euvlithography/next.

[68] Meiling, H; Meijer, H; Banine, V; et al., "First performance results of the ASML α demo tool," *Proc. SPIE*, 2006, 6151.

[69] Lorusso, GF; Hermans, J; Goethals, AM; et al., "Imaging performance of the EUV α demo tool at IMEC," *Proc. SPIE*, 2008, 6921.

[70] Murakami, K; Oshino; T. Kondo, H; et al., "Development of EUV lithography tools at Nikon," *Proc. SPIE*, 2008, 7140.

[71] Wagner, C; Harned, N; Kuerz, P; et al., "EUV into production with ASML's NXE platform," *Proc. SPIE*, 2010, 7636.

[72] Shroff, YA; Goldstein, M; Rice, B; et al., "EUV pellicle development for mask defect control," *Proc. SPIE*, 2006, 6151.

[73] Choksi, N; Shroff, Y; Packard, D; Chen, Y; Oldham, WG; McCord, M; Pease, R; Markle, D. "Maskless extreme ultraviolet lithography," *J. Vac. Sci. Technol.*, Nov/Dec, 1999, B 17(6), 3047-3051.

[74] Chen, Y; Chu, C; Shroff, Y; Wang, JS; Oldham, WG., "Design and fabrication of tilting and piston micromirrors for maskless lithography," *Proc. SPIE*, vol. 5751, 2005, 1023-1037.

[75] Shroff, Y; Chen, Y; Oldham, WG. "Optical analysis of mirror based pattern generation," *Proc. SPIE*, vol. 5037, 2003, 550-559.

[76] Shroff, Y; Chen, Y; Oldham, WG. "Image optimization in maskless lithography," *Proc. SPIE*, vol. 5374, 2004, 637-647.

[77] Chen Y; Shroff Y. "The effects of wafer-scan induced image blur on CD control, image slope, and process window in maskless lithography," *Proc. SPIE*, vol. 6151, 2006, 61512D.
[78] Born, M; Wolf, E. *Principles of Optics*; Cambridge University Press: Cambridge, UK, 2002, 55-74.
[79] Wong, A. *Optical Imaging in Projection Microlithography*; SPIE Press: Bellingham, WA, 2005, 58-95.
[80] Goodman, JW. *Introduction to Fourier Optics*; McGraw-Hill, 1996, 32-125 & 258-264.
[81] Sheats, J. R; Smith, B. W. *Microlithography Science and Technology*; Marcel Dekker: New York, NY, 1998, 197-202.
[82] Levinson, HJ. *Principles of Lithography*; SPIE Press: Bellingham, WA, 2001, 365-366.
[83] Mack, C. *Fundamental Principles of Optical Lithography*; John Wiley & Sons, 2007, 58-62.
[84] Lin, BJ. *Optical Lithography: Here is Why*; SPIE Press: Bellingham, WA, 2010, 340-346.
[85] Spiller, E. *Soft X-Ray Optics*; SPIE Press: Bellingham, WA, 1994, 101-107.
[86] Attwood, D. *Soft X-Rays and Extreme Ultraviolet Radiation*; Cambridge University Press: Cambridge, UK, 1999, 44-53 & 98-106.
[87] Jackson, TN; Masnari, NA. "A novel submicron fabrication technique," *IEEE digital library;* 1979, 58-61.
[88] Hunter, WR; Holloway, TC; Chatterjee, PK; Tasch, AF. "A new edge-defined approach for submicrometer MOSFET fabrication," *IEEE Electron Device Letters*, vol. EDL-2, No. 1, January, 1981, 4-6.
[89] Sheats, J. R; Smith, B. W. *Microlithography Science and Technology*; Marcel Dekker: New York, NY, 1998,104-105.
[90] Andre, NT; Simon, LM. *MOSFETs: Properties, Preparations and Performance*; Nova Science Publishers: New York, NY, 2008, 253-273.
[91] Chen, Y.; Xu, X; Chen YM; Miao, L. "Self-aligned frequency tripling technology: cost-effective 20nm line/space (half pitch) patterning using 193nm dry lithography," *Applied Materials ET Conference*, June, 2009.
[92] Chen, Y.; Chen YM; Miao L.; Xu X.; Ngai C. S. "Self-aligned triple patterning to extend optical lithography for 1x patterning," *The International Symposium on Lithography Extensions*, Kobe, Japan, October 20-22, 2010.
[93] Carlson, A; Liu, TJ. "Low-variability negative and iterative spacer processes for sub-30-nm lines and holes," *J. Micro/Nanolith. MEMS MOEMS*, vol. 8(1), 011009, Jan-Mar 2009, 1-7.

In: CMOS Technology
Editor: Min-jun Kwon, pp.81-103
ISBN: 978-1-61761-325-8

Chapter 2

VARIABILITY AND RELIABILITY IN ULTRA-SCALED MOS DEVICES: EVALUATION AT THE NANOSCALE AND IMPACT ON DEVICE AND CIRCUIT FUNCTIONALITY

M. Nafría, R. Rodríguez, M. Porti, J. Martín-Martínez, M. Lanza and X. Aymerich [*]
Departament d'Enginyeria Electrònica, Universitat Autònoma de Barcelona (UAB), Bellaterra, Spain

ABSTRACT

In this work, the approaches adopted at (Universitat Autònoma de Barcelona) UAB to evaluate the nanoscale sources of device variability related to the MOSFETs gate oxide and to analyse the impact of device variability and aging on circuit performance and reliability are described. First, a Conductive Atomic Force Microscope is used to evaluate the effects of processing on the morphological and electrical characteristics of gate dielectrics at the nanoscale. As example, the dependence of the electrical properties of Al_2O_3/SiO_2 stacks on a thermal annealing at different temperatures is analyzed. Second, a reliability circuit simulation methodology to transfer the variability and aging effects in devices up to circuit level, which combines Montecarlo and SPICE simulations, is presented. The methodology is applied to evaluate the impact of threshold voltage time-dependent variability in differential amplifiers performance and reliability.

INTRODUCTION

Deep sub-micron technologies suffer from new variability and reliability problems, which are intrinsic to device scaling [1]. The MOSFET electrical characteristics show shifts from device to device, which derive from the particular fabrication conditions and whose

[*] Corresponding author: montse.nafria@uab.es

origin is related to the discrete nature of matter and charge. Several sources have been identified which lead to the observed variability of device parameters: (I) the random distribution of dopants in the substrate or Random Dopant Distribution (RDD); (II) the roughness in the gate of the transistor or Line Edge Roughness (LER, due to intrinsic imperfections of the photolithographic process); (III) the gate oxide roughness and (IV) the crystallization of the poly-gate and/or high-k dielectrics after high temperature annealings. It must be considered that, whatever the specific transistor architecture or the materials in the device, the continuous scaling warrants the susceptibility of these devices to the processing conditions [2]. Up to now, the study of the effects of the different variability sources in the performance of devices has been mostly centered in the effects of RDD [3] and LER [4] and is mainly based in the simulation of their I-V characteristics using atomistic and TCAD techniques, which take into account the discrete nature of matter and charge. These works have concluded that this variability sources will end in shifts in the threshold voltage of the MOSFETs. However, the variability sources related to the gate dielectrics have been less studied. The dielectric roughness and the crystallization of high-k materials, which strongly depend on the processing conditions [5-8], are properties of the dielectric at the nanoscale, so that their characterization requires techniques with a large lateral resolution. In this direction, the techniques based on the Atomic Force Microscope (AFM), such as the C-AFM (Conductive AFM), the SCM (Scanning Capacitance Microscope) and KPFM (Kelvin Probe Microscope) can reveal details of the electrical properties of the gate dielectrics at this scale. As examples, the C-AFM has been used to measure the oxide dielectric strength, local variations of its thickness, homogeneity of its conductivity… [8-10]. However, the use of these techniques with the aim of analyzing in detail the gate oxide process related variability sources has been scarcely explored [11].

Moreover, during the operation of the device within the circuit, aging mechanisms appear (associated to the high electric fields and temperatures in the device) that will progressively modify the electrical characteristics of devices. Among all the aging mechanisms in the IC, those related to the MOSFET gate dielectric have a special relevance, being some of the most important that must be faced to improve the device reliability. In the high-k based devices, the aging mechanisms observed in SiO_2 are even more harmful and additional mechanisms appear, which must be understood in order to correctly determine the device lifetime and minimize their effects [12, 13]. For example, the effects of Bias Temperature Instabilities (BTI) and Channel Hot Carrier (CHC) degradation have been also observed in high-k based MOSFETs [14]. When positive voltages are applied to the gate (Positive BTI or PBTI), specially at high temperatures, charge trapping occurs (which is thought to be related to the filling of pre-existing traps in the dielectric bulk), which causes variations in the threshold voltage and the degradation of the device current [15], being these effects larger than those in SiO_2 [15, 16]. The shifts in the threshold voltage in the case of negative voltages (Negative BTI, NBTI), which are thought to be related to an increase of trapped charge in the Si/SiO_2 interface or in the oxide bulk, have been also observed in high-k devices. The reduction of the reliability related to hot carriers, a much more serious problem in high-k dielectrics than in SiO_2 [12], also induces changes in the threshold voltage, a reduction of the mobility and the lifetime, among other effects [17, 18]. Finally, the dielectric breakdown takes place, which is detected as a large increment of the current through the gate and that has been associated to the generation of defects during the electrical stress [19]. One of the particularities of this mechanism is its extremely local nature (it happens in areas much smaller than the device area), so that the breakdown path localization along the channel considerably affects the electrical characteristics of the transistor [20]. All these aging mechanisms (dielectric breakdown, BTI, charge trapping...) are phenomena that take place at the nanometer scale, so that the analysis of these mechanisms (specially the dielectric breakdown) using AFM based microscopies, which provide resolutions in the same range of dimensions in which they take place, becomes natural [21-23].

Variability and device aging, which were second order effects in the past, are progressively gaining importance with each new technological node, independently of the materials and particular architecture of devices. Therefore, their effects on the electrical characteristics of devices, which will be time-dependent, will have to be determined and transferred up to the circuit level [24]. Since the time-dependent variability introduces randomness in the device electrical characteristics, a probabilistic design methodology is needed, which will have to take into account these variations and allow to optimize the circuit performance, yield and reliability, and that would have to consider from fabrication to applications. In this direction, the Reliability-Aware Design [25] Methodology has been proposed as an alternative, since its aim is to propagate the variability and degradation effects (that is, the time-dependent variability) through all the stages of IC manufacture. Several efforts have been already done in this direction [26, 27].

In this paper, the approaches adopted at the Universitat Autònoma de Barcelona, on the one hand, to characterize the nanoscale sources of device variability and, on the other, to include device variability and aging during circuit simulation are described. Our focus will be the variability sources and aging mechanisms related to the gate dielectric of MOSFETs. First, the Conductive Atomic Force Microscope is used to analyze at the nanoscale the effects of thermal annealings on the electrical properties (including their homogeneity) of high-k gate stacks. Second, a probabilistic simulation methodology is presented and used to determine the effects of device variability and aging on the performance and reliability of a differential amplifier.

NANOSCALE CHARACTERIZATION OF GATE OXIDE RELATED VARIABILITY

The nanoscale properties of gate dielectrics in MOSFETs, which will depend on the processing conditions [6-8], can be sources of device variability. Since, when using standard characterization techniques, the details of these oxide properties are hidden by the average behaviour of the device, in this section, Conductive Atomic Force Microscope (C-AFM) has been used to evaluate the impact of high-k crystallization (derived from annealing processes) on the electrical properties of $Al_2O_3/SiO_2/Si$ structures. Moreover, the effects of an electrical stress on the electrical conduction and charge trapping properties of the poly-crystalline high-k layers will be also studied.

Experimental Set-up and Sample Description

When working on bare dielectric surfaces on Si substrates, the conductive tip of the C-AFM (Figure 1) plays the role of the metal electrode of a MOS capacitor with an area mainly defined by the contact region between the tip and the sample. This area is of the order of few hundred nm^2 (depending on the tip used, [28]), so that the properties of nanometer oxide areas can be analysed. With the tips used in this work (silicon tips coated by a metallic layer of Pt/Ir), an electrical resolution of $\sim 300nm^2$ can be achieved.

The gate dielectrics have been electrically characterized by measuring current-voltage (I-V) characteristics, obtained when a ramped voltage test is applied at a fixed location of the oxide. In addition, current images can be collected, simultaneously to topography maps, when the tip-sample system is polarised. In C-AFM experiments (contact mode, working in air), positive voltages were always applied to the tip (electrons injected from the substrate) due to several advantages compared to negative polarity [29]. A Kelvin Probe Force Microscope (KPFM) has been used to measure the contact potential difference between the tip and the substrate (CPD). Since the CPD signal depends on the charge present in the oxide, KPFM measurements can provide information on the trapped charge in the dielectric.

Concerning the samples, Al_2O_3/SiO_2/p-Si structures for memory applications subjected to different annealing temperatures, T_A, have been analyzed [11]. Seven wafers which consist of a nominal 10 nm thick Al_2O_3 layer and a 1 nm SiO_2 interface layer on top of p-type Silicon substrate have been studied. After the Al_2O_3 deposition, they were annealed by a RTP (Rapid Thermal Process) in Nitrogen at different temperatures, T_A (750, 850, 900, 925, 950, 975 and 1000°C) during 20s. No gate electrode material was deposited on top.

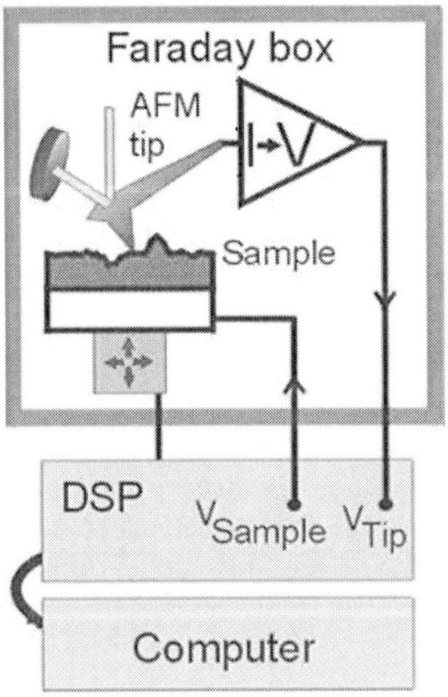

Figure 1. Experimental set-up of a C-AFM: a conductive tip is scanned over a bare dielectric surface. When a bias is applied between the tip and the sample a current flows through the structure. This current is measured with the C-AFM simultaneously to the surface topography. The high resolution of the technique allows to analyze areas of $\sim 300nm^2$

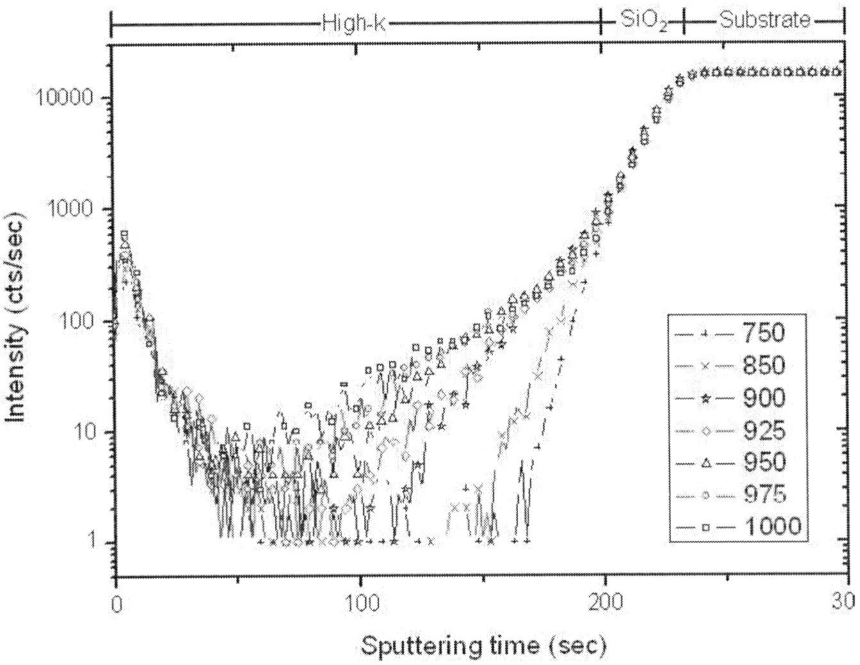

Figure 2. ^{28}Si ToF-SIMS depth profile for Al_2O_3 stacks, which allowed to determine the percent of diffused silicon in the sample depending on the depth, for all annealing temperatures

Physical Characterization

Before the electrical characterization of the stacks, a physical characterization was carried out, to analyze several morphological properties of the dielectric [11]. First of all, the content of diffused Silicon from the substrate into the Al_2O_3 layer due to the annealing process (an unintentional effect observed during the manufacturing process) was analyzed. To do so, the depth profile was measured using ^{28}Si signal ToF-SIMS (Time-of-Flight Secondary Ion Mass Spectrometry), whose magnitude is proportional to the density of Silicon in the material (Figure 2). After 225 seconds of sputtering, the ToF-SIMS process collected only ^{28}Si isotopes, which means that the substrate has been reached. As can be seen, two regions can be easily recognized in the gate dielectric stack. First of all, there is a region near the surface (from 0 to 50 seconds) where the density of Silicon decreases with the sputtering time. This was an expected effect because, as other works demonstrated [30], the Al_2O_3 insulator changes to conductor at the (1,1,0) surface. In the second region, the percent of Silicon increases when getting closer to the Substrate (for all annealing temperatures) until it reaches a maximum at 225 seconds. The most remarkable result is the dependence of the percent of Silicon on T_A. As the graphic shows, the higher the annealing temperature, the higher the content of Silicon diffused from the substrate into the Al_2O_3 stack.

The dependence of the crystallization of the high-k dielectric on the annealing temperature has been also investigated using Grazing Incidence X-Ray Diffraction (GIXRD). The results (not shown) demonstrate that the Al_2O_3 crystallizes at a temperature between 850°C and 900°C. These observations are further supported by TEM images. As an example, figure 3 shows a cross section TEM image of the sample annealed at (a) 850°C and (b) 950°C (c and d correspond to a zoom of a and b, respectively). Note that the different layers of the stack structure are clearly distinguished (SiO_2 interfacial layer and high-k dielectric). Moreover, from figures 3.b and 3.c it can be observed that the sample annealed at 950°C shows a poly-crystalline structure (the different intensities in grey observed in the high-k layer correspond to the different orientations of the nanocrystals). This behavior has been also observed for all samples annealed at temperatures of 900°C or higher (not shown), in agreement with the GIXRD analysis. In particular, the crystalline grains seem to have a diameter of 15-30 nm.

The surface of the samples has also been studied from AFM topography maps obtained in tapping mode. In particular, the samples annealed at 750 (figure 4a) and 1000 °C (figure 4b) were investigated. Although in this experiment the resolution of the set-up does not allow to distinguish the position of the crystals, table in figure 4 indicates an increase of the surface roughness after crystallization. This is compatible with the results in [31], where it was shown that ZrO_2 crystallized layers had a larger surface roughness than amorphous materials.

Finally, the actual physical thickness of the different layers was also determined from X-Ray Reflectometry (table I, columns 2 and 3) and TEM images (figures 3a-d). Paying attention to the thickness of both high-k and SiO_2 layers (Table I), it is possible to conclude that for T_A between 850°C and 900°C there is a reduction of the thickness of the high-k material, whereas for T_A between 900 and 925°C, there is a slight thickness increase in the case of the SiO_2 layer. The reduction of thickness of the high-k layer after crystallization was also observed in other works [32]. In [32], TEM images on n-channel MOSFET devices with HfO_2 showed that, after annealing, the gate oxide thickness appeared to be reduced about 2nm, which was interpreted as a "densification of HfO_2" or "consumption by interacting with Poly-Si". Since in our case no Poly-Si layer was deposited on top of the gate oxide, the decrease of the thickness could be related to the "densification" of the Al_2O_3 layer.

In summary, the physical characterization of the high-k stacks show that the annealing process leads to crystallization of the Al_2O_3 layer (at T_A between 850 and 900°C), a decrease of the Al_2O_3 layer thickness, an increase of the SiO_2 layer thickness, and Silicon diffusion from the substrate to the Al_2O_3/SiO_2 stack. These results will be taken into account when analyzing the electrical properties of the structures.

Table I. Experimental values of several magnitudes of the analyzed samples, as a function of the annealing temperature. EOT is the theoretical value calculated from experimental thicknesses

T_A (°C)	Physical thickness (nm)		Onset Voltage (V)		EOT (nm)	Deviation of the CPD (mV)
	Al_2O_3	SiO_2	Mean	Deviation		
750	14.6	1.0	13.5	0.12	7.31	6.12
850	14.6	1.0	12.7	0.20	7.29	6.04
900	12.6	0.9	12.4	0.44	6.32	6.77
925	12.4	1.1	13.7	0.28	6.50	7.05
950	12.4	1.2	13.4	0.32	6.58	7.73
975	12.0	-	12.9	0.47	-	7.02
1000	12.3	1.1	12.6	0.46	6.49	6.59

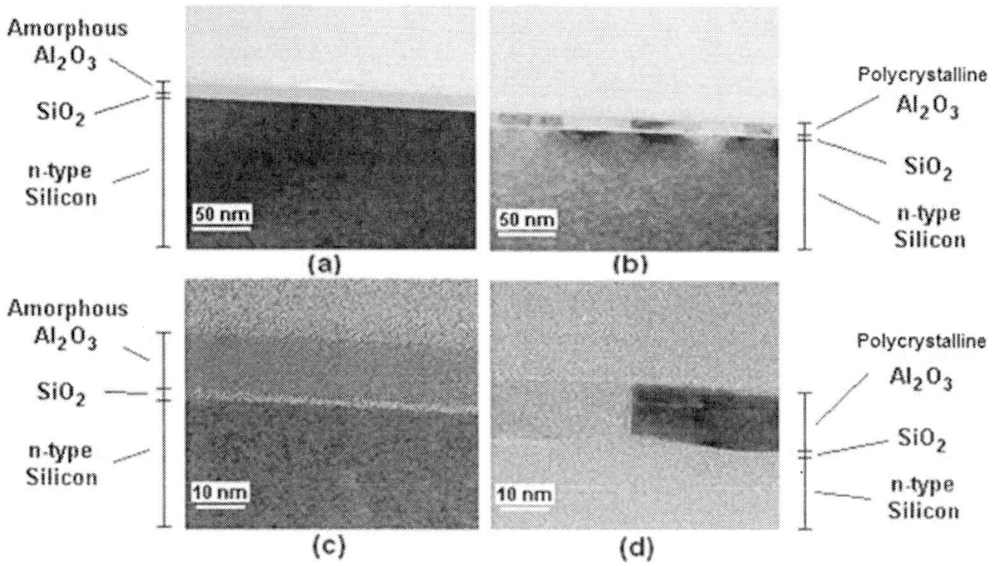

Figure 3. TEM images of structures annealed at (a) 850°C and (b) 950°C. (c) and (d) correspond to zooms of (a) and (b). For temperatures of 900°C and higher, the high-k layer shows poly-crystals

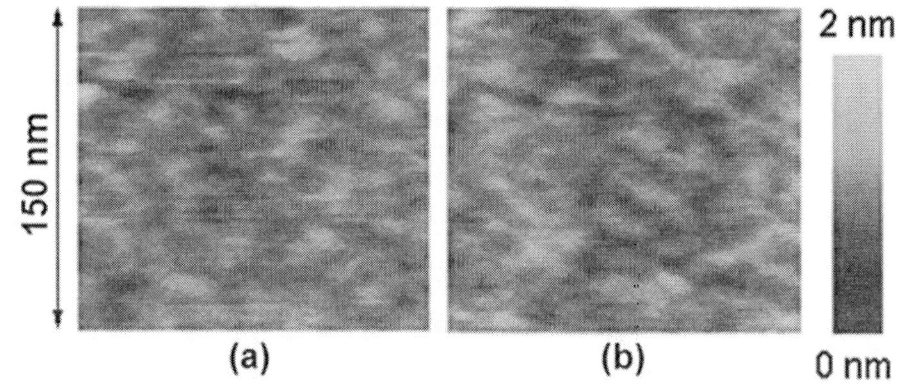

Topography roughness	
750 °C	*1000 °C*
0.14 nm	*0.22 nm*

Figure 4. Topographic maps of Al_2O_3 stacks annealed at 750 °C (a) and 1000 °C (b). The table shows their rms values. The size of the images is 150 x 150 nm^2

Electrical Characterization of the as-Grown High-K Stacks

In this section, the impact of the annealing, as a function of T_A, on the electrical conduction of the different samples has been analyzed at the nanoscale from I-V curves and current images obtained on fresh structures (before an electrical stress). Figure 5 shows the statistical distribution of the onset voltages (minimum voltage needed to measure current above the noise level), which can be considered indicative of the structure conductivity, i.e, the larger the onset voltage the smaller the conductivity, and Table I (columns 4 and 5) shows their statistical parameters (mean and rms values). From both, figure 5 and table I, it can be determined that the samples show three different electrical behaviors, depending on the annealing temperature. The first group corresponds to amorphous samples ($T_A \leq 850$°C), the second one to crystalline stacks ($T_A \geq 925$°C) and the third one corresponds to the sample annealed at 900°C, which exhibits unique properties. Amorphous oxides ($T_A \leq 850$°C) show smaller onset voltage dispersion than crystalline structures, indicating a more homogeneous conductivity over the sample area. Moreover, except for the sample annealed at 900°C, the electrical conductivity is lower for crystalline structures annealed at the lowest T_A. Finally, it can be observed that, for both groups, crystalline and amorphous oxides, a higher annealing temperature results in a smaller mean value (larger conductivity) and a higher deviation of the onset voltage (larger inhomogeneity) [11]. The explanation for the particular electrical behavior of the sample annealed at 900°C could be related to changes in the physical thickness of the layers of the stack during annealing. As columns 2 and 3 in Table I show, a decrease of the physical Al_2O_3 thickness is observed between 850°C and 900°C and a minimum in the physical SiO_2 thickness is reached at 900°C. This means that the 900°C annealed sample, although crystalline (see section 1), has the smallest equivalent oxide thickness (Table I, column 4) and, therefore, the highest conductivity. Consequently, the sudden changes in the physical thickness of the layers around the crystallization temperature (which lead to the minimum EOT) could explain the unique properties observed in the sample annealed at 900°C. However, this thickness reduction cannot justify the difference in the average conductivity observed between amorphous samples at T_A=850°C and crystalline samples at the lowest T_A (T_A=925°C). Note that the EOT of the stacks (Table I, column 4) that have crystallized is smaller. However, the onset voltage is larger (that is, the conductivity is smaller). Therefore, the reduction in the electrical conductivity after crystallization (respect to the amorphous) cannot be related to changes in the thickness of the layers, but to the crystallinity of the stack. On the other hand, in addition to the reduction of the average electrical conduction, the rms value and, therefore, the inhomogeneity, is higher in crystalline stacks. This increase of the inhomogeneity could be related to the existence of grain boundaries between the polycrystals, which raises the leakage current due to trap assisted tunneling (TAT) through the defects present on them [32].

Finally, when comparing amorphous ($T_A \leq 850$°C) or crystalline samples ($T_A \geq 900$°C) separately, a higher annealing temperature results in a smaller onset voltage mean value (larger conductivity) and a higher deviation (larger inhomogeneity). Generally, the annealing process is used to eliminate defects in the dielectric and improve its performance but, in this case, the increase of the conductivity for larger T_A could be explained from the increase of the amount of Silicon. The Si atoms could act as tunneling centers that lead to an increase of the tunneling current.

The structures have been also analyzed with KPFM, which can provide information about the presence of charge and trapping centers in the stack. Table I, last column, shows the deviation obtained from the CPD images of each of the investigated structures. The results do not show a progressive increase of the rms value of CPD images with the annealing temperature (and, therefore, the diffused Si content), but a sudden increase around the crystallization temperature. Therefore, the charge observed from KPFM images could be related to some kind of defects or trapping sites generated after the crystallization (instead of those related to the diffusion of Si), maybe around the grain boundaries, which could also enhance the non-inhomogeneity observed in current images.

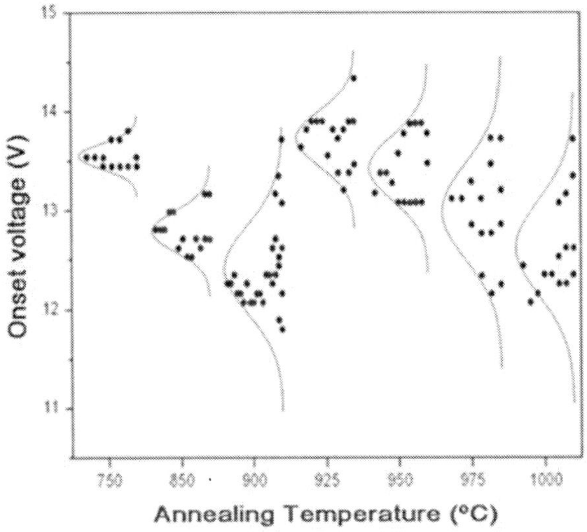

Figure 5. Statistical distribution of the onset voltage obtained from the I-V curves, as a function of T_A.

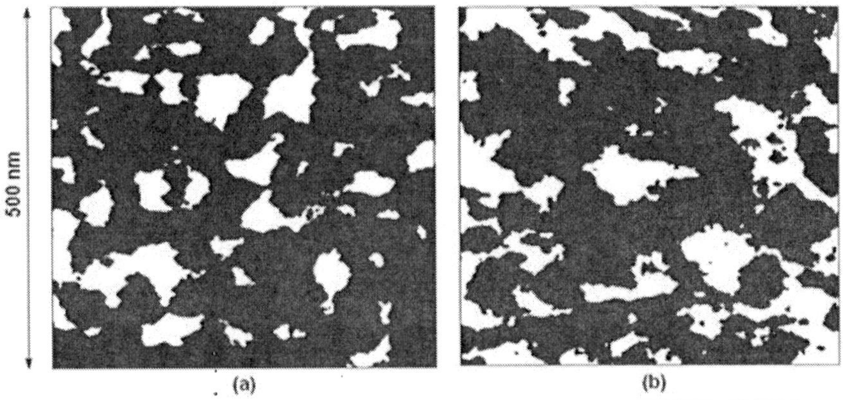

Annealing Temperature	# Grains	Density (#/μm²)	Diameter (nm)	
			Mean	Deviation
900 °C	78	312	19.5	19.0
950 °C	90	360	13.0	16.2

Figure 6. Grain analysis of current images obtained for the samples annealed at 900 °C and 950 °C

The results presented until now demonstrate that the crystallization of the Al_2O_3 leads to a larger inhomogeneity of the sample conduction, which could be attributed to the different electrical properties of nanocrystals and grain boundaries. Taking advantage of the large lateral resolution of the CAFM, a more detailed analysis has been performed to explore this point. To do so, current images of samples which have crystallized at different temperatures have been measured and the areas with smaller conductivity have been evaluated. Figure 6 corresponds to current images of the crystalline samples annealed at 950°C (a) and 1000°C (b). The black areas in both images correspond to the regions with a current above 0.2pA, while the white areas show a current lower than the noise level (all images are 500nm x 500nm). Figure 6 also includes the results of the statistical analysis of both images, indicating the number and density of grains (white regions) and their size. Note that the average size of the grains ranges from 15 to 20nm, which perfectly agree with the results obtained from TEM images (figure 3) for the sizes of the Al_2O_3 nanocrystals. Moreover, we have also observed that the average size seems to be independent of T_A. Therefore, these results suggest the idea that the regions with a smaller conductivity could be related to the grains of the crystalline structure: the nanocrystals are more insulating whereas the grain boundaries show a larger conductivity.

A qualitative parallelism between this phenomenon and electromigration could be established. In a metal (like an aluminum wire), the lack of a homogeneous crystalline structure leads to the appearance of grain boundaries where the metal ions are bonded more weakly than in a regular crystal. Therefore, due to the electron current density, these atoms become separated from the grain boundaries and are transported in the direction of the current. This direction is also influenced by the grain boundaries themselves, because atoms tend to move along them, leading to electromigration [33]. In our case, the crystals are related to the high-k material, whereas the electrons are the elemental charges that tend to flow principally through the paths open along the grain boundaries of the stack. It is important to emphasize that this is a qualitative result, since the resolution of the technique is not high enough to resolve the width of the grain boundaries because CAFM tips are typically coated with a layer of Co/Cr or Pt/Ir which drastically reduces its lateral resolution to ~20 nm.

Charge Trapping after Electrical Stress

In addition to the study of the electrical conductivity of as-grown high-k materials, the change of the electrical properties of the high-k stacks during operation must be also characterized. In this section, the impact of an electrical stress on the electrical conduction and charge trapping of the Al_2O_3 layers will be analyzed at the nanoscale, before breakdown. In particular, differences between amorphous and crystalline structures will be evaluated. To do so, sequences of current images have been collected, on amorphous and crystalline Al_2O_3 samples. First, a 500nm x 500nm area was scanned by applying a large enough constant voltage to induce degradation. Afterwards two zooms out were done, and larger areas were scanned (1000x1000nm^2 and 1500x1500nm^2). Figure 7 shows a sequence of three scans measured on the samples annealed at 850°C (amorphous, a, b and c) and 975°C (crystalline, d, e and f). The first scan corresponds to images (a) and (d) and, the last scan, to images (c) and (f). The size of the images is (a and d) 500 nm x 500 nm, (b and e) 1μm x 1μm and (c and f) 1.5 μm x 1.5 μm. The applied voltage was 11.5 V in all cases. This procedure allows to compare areas that have been subjected to different stresses or, in other words, areas that have experienced different degradation levels: fresh sample (the area in the third image that was scanned only once, peripheral area), low degradation (area scanned twice) and large degradation (three times, center of the image). Note that the comparison of the different regions in the 3rd image is strongly reliable, since the measurement was done under the same experimental conditions, that is, the data is coming from the same image. Therefore, this experiment allows to minimize most of the experimental parameters that can adversely affect CAFM current measurements: variations of the tip conductivity and the contact area when using different tips, their wear out, fluctuations of the thickness when different samples are used, etc.

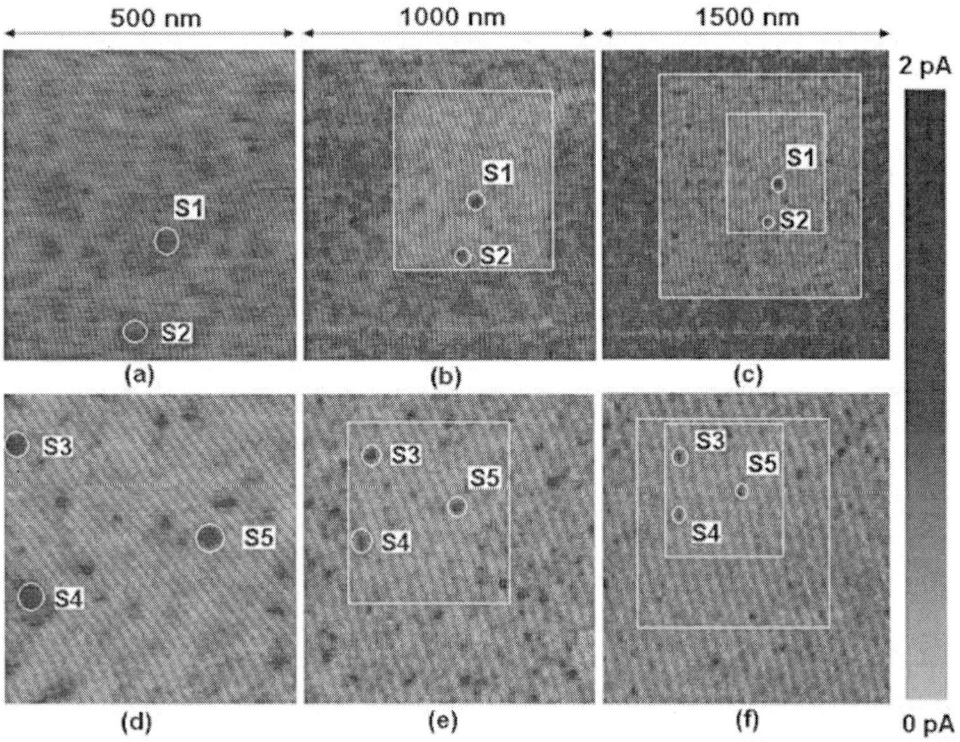

	Current on an amorphous sample (pA)			Current on a crystalline sample (pA)			
	Background	S1	S2	Background	S3	S4	S5
First Scan	0.42	1.4	0.99	0.10	5.8	2.5	4.0
Second scan	0.13	1.2	0.51	0.07	1.2	0.87	1.1
Third scan	0.05	0.65	0.48	0.05	0.67	0.32	0.62

Figure 7. First scans (a and b) and two consecutive zooms out (b/e and c/f) on samples annealed at 850°C (a, b and c) and 975°C (d, e and f). Their sizes are (a and d) 500 nm x 500 nm, (b and e) 1μm x 1μm and (c and f) 1,5 μm x 1,5 μm. The applied voltage was 11.5 V for all samples. The table shows the evolution of the maximum current driven by the spots S1-S5 and on background areas for the sample annealed at 850°C and 975°C

Comparing images 7.a and 7.d, which correspond to the first image (fresh area) measured on amorphous and crystalline structures, respectively, similar results to those shown in the previous section are obtained. On crystalline samples, background conduction is smaller (table of Figure 7, first scan). However, the leaky sites of crystalline structures (spot S3, S4 and S5) have a larger conductivity compared to those of amorphous samples (S1 and S2). As discussed in the previous section, the larger inhomogeneity in the structure annealed at 975°C (compared to the amorphous oxide) could be attributed to the differences in the conductivity between the crystals (background) and grain boundaries (leaky sites).

The effect of the stress has been analyzed from the images measured during the zooms out. On the amorphous sample (Figure 7.b), the central area (which was previously pre-stressed, Figure 7.a) shows smaller currents than the rest of the scanned regions. A similar behavior can be observed in image 7.c, where three concentric areas can be distinguished: a first central area with the smallest current value (three scans), another second area with a larger current (two scans) and the peripheral and most conductive area (only one scan, that is, a fresh area). The quantitative values of the background current on the three concentric areas are shown in the table of Figure 7. In the amorphous sample, the background current decreases significantly as the stress proceeds, making the structure less conductive. This behavior, as already pointed out for SiO_2 layers [34] or high-k dielectrics, can be related to negative charge trapping in the native defects or in traps generated during the stress. The trapped charge increases the injection barrier, reducing the conductivity or, equivalently, increasing the electrical thickness of the oxide, and making it less conductive. In the case of crystalline structures (Figure 7.d-f and table), the decrease in the background conductivity is less important when compared to amorphous samples. This result suggests a smaller impact of the stress at the positions where crystals are present.

Although, in crystalline samples, regions with background currents (which can be probably related to positions with a crystal under the CAFM tip), seem to be more resistive and robust to an electrical stress than in amorphous oxides, this behavior cannot be extended to the weak spots (leaky sites). As an example, the table in Figure 7 shows the maximum current in different spots and its evolution with the stress on amorphous (spot S1 and S2) and crystalline samples (S3, S4 and S5). Note that the weak spots in the crystalline structure show, before the stress ("first scan"), larger leakage currents compared to the amorphous gate oxide. However, after the stress ("second scan" and "third scan"), the reduction of current is larger in crystalline structures. The higher conduction level (before the stress) and the larger current reduction (after the stress) in the crystalline layers compared to amorphous oxides can be interpreted in terms of "dielectric strength". The leaky sites of crystalline samples are, from an electrical point of view, weaker than those in amorphous oxides (because the dielectric is thinner or due to the presence of defects that enhance tunneling). That is the reason why, initially, their conductivity is higher. As the stress proceeds, a larger amount of charge is trapped in the as-grown or generated defects on the leaky positions, leading to a higher reduction of the current compared to amorphous oxides. These leaky regions could be related to the grain boundaries between nanocrystals: since, initially, the current specially flows through the grain boundaries [35, 36], the defect generation and, therefore, the charge trapping mainly occurs in those locations, leading to a higher reduction of the conductivity compared to the background areas. In amorphous samples, no distinction between crystals and grain boundaries can be observed, so that the fresh conduction and stress effects are randomly distributed in the gate area. It must be remarked that if these leaky regions were related to grain boundaries, they should be observed as a line or a curve in the current image instead of a spot. However, the large radius of the tips used and the water meniscus present between the tip and the sample when measuring in air conditions result in a low resolution, so that the grain boundaries cannot be resolved. Therefore, since a direct correlation between grain boundaries and leaky sites can not be established, other hypothesis on their origin could not be discarded at this moment and further experiments should be performed to verify this point. Note, however, that other works have reported the relation between leaky sites and grain boundaries [35, 36]. The presence of grain boundaries in high-k stacks increases the inhomogeneity of the stack conduction so that it can be a source of variability, which at device level can lead to a statistically distributed electrical behavior of devices. Further work is needed to translate the CAFM/KPFM results into the electrical parameters of the device.

TIME-DEPENDENT VARIABILITY OF DEVICES AND CIRCUITS

Device mismatch will have to be accounted for during the design phase of the integrated circuit, since it can strongly affect the circuit performance. Morevover, the device parameters and their mismatch can be modified (progressively or suddenly) during the circuit operation, due to different aging mechanisms. It has been shown that the aging mechanisms related to the gate oxide (such as BTI and CHC degradation) lead to shifts of the threshold voltage during circuit operation and a decrease of the saturation current [15]. Actually, device aging can be considered as time-dependent variability (or in other words, reliability) and, as such, its effects will have to be transferred from device to circuit levels. Some efforts have been done to evaluate the effects of device variability related to the fabrication process on circuit performance [37], but the effects of device time-dependent variability at circuit level have been less studied [38, 39]. In this section, variability and BTI aging of devices have been studied and transferred up to circuit level. To do so, MOSFET gate oxide wear-out has been induced by electrically stressing isolated nMOS and pMOS transistors and also CMOS inverters to evaluate the degradation effects devices and circuits. The experimental results on the stressed devices and circuits have been reproduced in a circuit simulator by means of the BSIM4 SPICE model using a modified model parameter set. On the other hand, considering the results obtained at device level, a three-stage simulation methodology has been developed, based on Montecarlo and SPICE simulation [24], which allows to consider the time-dependent variability of devices during the circuit design phase. To illustrate the procedure, the effect of BTI on the reliability of a CMOS differential amplifier has been evaluated, considering simultaneously the variability (of V_{th}) due to the fabrication process and the aging mechanisms. This will allow analyzing how this variability affects the amplifier reliability.

Variability and Aging Experimental Characterization

Firstly, the effect of variability and aging in isolated transistors has been experimentally studied. nMOS and pMOS devices with SiON as gate dielectric (EOT=2.2nm), channel length of 0.13μm and channel width of 0.3μm for nMOS and 0.6μm for pMOS have been considered. Their operation voltage was 1.2V. These transistors were degraded at room temperature with a high DC voltage at the gate (3.9V for nMOS and -3.9V for pMOS) and the other terminals grounded (Figure 8 left). The stress was periodically interrupted to register the I_D-V_{GS} and I_D-V_{DS} characteristics. From these curves, V_{th} and the saturation current at operation voltage (I_{SAT}) have been extracted. To determine the average and standard deviation of the device electrical magnitudes, 8 devices of each transistor type have been considered.

A CMOS inverter, as the building block of most digital circuits, has been chosen as test structure to experimentally analyze the effect of gate oxide degradation at circuit level. The studied CMOS inverters had W_N/L_N=3μm/0.13μm for nMOS and W_P/L_P=6μm/0.13μm for pMOS and were fabricated with the same technology than the isolated transistors described in the previous paragraph. During the stress, the inverters were oversupplied at V_{DD} = 3.2V and a unipolar square voltage waveform with an amplitude of V_{DD} and a frequency of 500KHz was set at the input (Figure 8 right). Note that when using this stress voltage configuration several transistor aging mechanisms are present (BTI and CHC) [40]. The stresses were periodically stopped to measure the inverter voltage transfer characteristic (VTC) to study the impact of the dielectric degradation on the inverter functionality.

Figure 9 shows typical I_D-V_G (left) and I_D-V_D (right) characteristics measured on the fresh (non-degraded) nMOS and pMOS devices and after 10.000 seconds of stress. The gate oxide wear-out induced by the electrical stress leads to a decrease in the saturation current and an increase in V_{th}, whose values depend on the substrate type. This phenomenology has been attributed to PBTI/NBTI effects in nMOS/pMOS transistors induced by the stress [41].

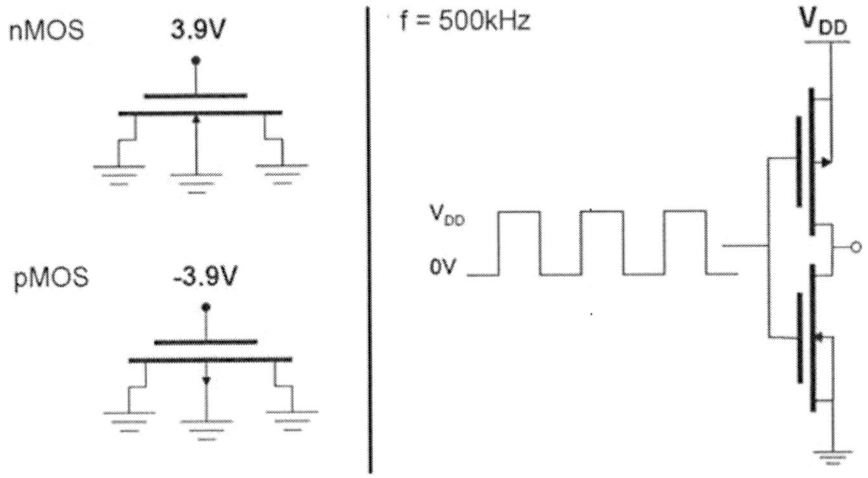

Figure 8. Stress scheme used to accelerate the gate oxide degradation in isolated transistors (left) and CMOS inverters (right).

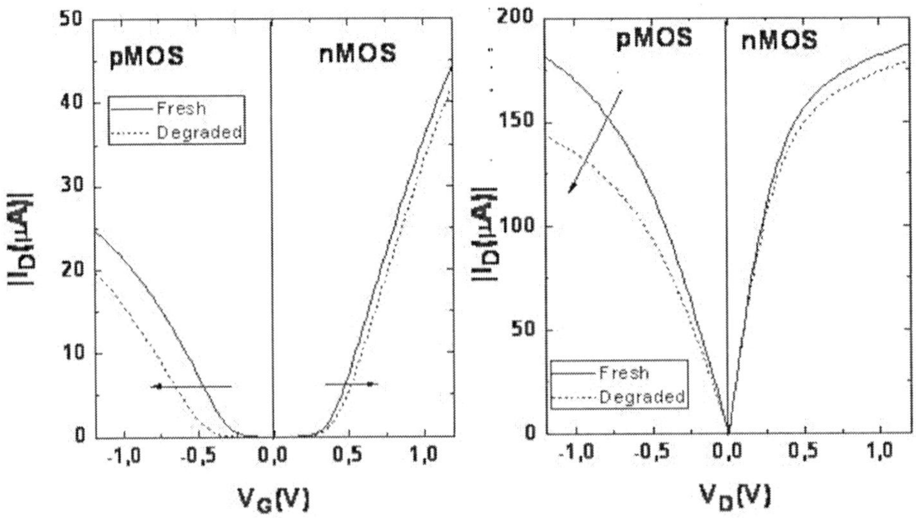

Figure 9. Typical I_D-V_G (left) and I_D-V_D (right) curves obtained in fresh and stressed nMOS and pMOS transistors. The gate oxide wear-out induced by the stress provokes a decrease of the transistor saturation current and an increase of the threshold voltage

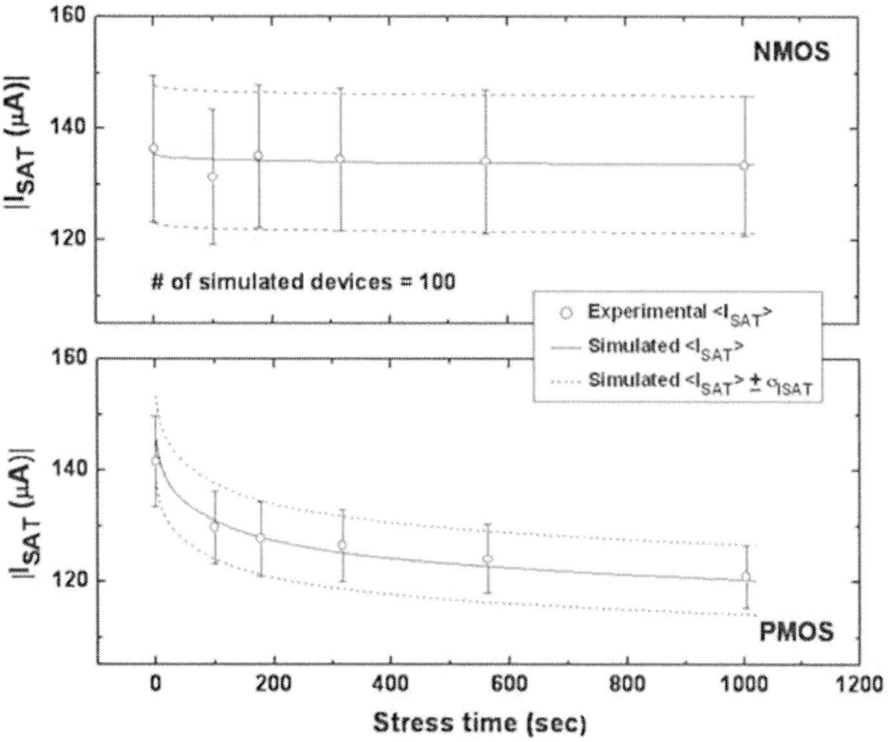

Figure 10. Time evolution of I_{SAT} for NMOS (top) and PMOS devices (bottom). Dots are the experimental average values (in 8 devices) and bars are the σ of the distribution. Continuous lines correspond to the mean value (of 100 devices) and dashed lines to the dispersion of I_{SAT} obtained in MC simulations.

The observed average I_{SAT} evolution with the stress time is shown in Figure 10 (circles). A much larger degradation is observed in pMOS transistors (17% I_{SAT} decay after 1000s of stress), compared to nMOS devices. Concerning to the I_{SAT} variability, the errorbars in Figure 10 indicate the standard deviation of the I_{SAT} distribution (σ_{ISAT}) for nMOS and pMOS transistors (a normal distribution has been assumed). Note that σ at t=0 is related to the fabrication process. Although the number of devices used to evaluate σ can be considered statistically reduced, the assumption of a normal distribution is supported by the results obtained in a much larger number of devices [42], which suggest that current and V_{th} distributions after the stress are still normal despite the V_{th} shift (ΔV_{th}) caused by BTI is not [42,43]. In our measurements any or negligible variation of σ_{ISAT} with the stress time, t_S, is observed in both transistor types.

The impact of the stress on the inverter response has been characterized from the VTC curves measured for the stress waveform indicated in Figure 8 (right). As an example Figure 11a shows the experimental VTC curves (symbols) measured on the fresh inverters (solid circles), and after 10.000s of 500 kHz pulsed stress (open circles). The stress causes a shift in the VTC curve. However, the shift direction depends on the different degradation induced in the nMOS/pMOS devices, depending on the stress waveform and time [40]. Although no changes were observed in the inverter output values, the shift in the VTC will have an effect in the parameters that characterize the performance of the circuit, from digital and analog points of view. The noise margins (for digital applications) of the circuit can be importantly affected. For analog applications, the VTC derivative is indicative of the circuit gain when it is used as an amplifier. This derivative is shown in Figure 11b, where a shift of the curve and a decrease of the peak (in absolute value) is observed after stress, which are indicative of a shift and reduction of the maximum gain of the circuit.

The previous results show that aging mechanisms related to the transistor gate oxide affect the MOSFET characteristics and have also an impact on circuits functionality. Therefore, to evaluate how these mechanisms affect the reliability of the circuit during the initial design phase, models for the effects of device aging, which can be included in circuit simulators, are needed. On the other hand, in the next technological nodes, the variability of the transistor electrical parameters must be accounted for, and in particular, the temporal evolution of the variability because it can affect the circuit performance. To try to clarify these points, in the next section, electrical models will be presented which reproduce the electrical characteristics of transistors affected by failure mechanisms associated to the gate dielectric. After that, the variability and its temporal evolution will also be included in the models. In this way, using these models in a circuits simulator, the response of circuits with degraded transistors will be obtained.

SPICE Modelling of the Aging Mechanisms and Montecarlo Simulation of Device Variability

To include the gate oxide degradation effects in a circuit simulator, fresh and degraded transistor characteristics have been reproduced using the BSIM4 SPICE model [44] which in previous works has been demonstrated to be a valid model to describe the wear-out effects in nMOS [20]. Aurora software has been used to extract the parameters of the fresh and worn-out devices. The BSIM4 parameter set obtained from the stressed devices can be introduced in a circuit simulator to simulate the performance of circuits that include degraded transistors. This procedure has been used to reproduce the performance of the stressed inverters through simulation (lines in Figure 11). The good fitting of the VTC curve and its derivative allows to conclude that the BSIM4 model is also able to describe the degradation of both pMOS and nMOS devices in situations where several aging mechanisms coexist (when the proper set of parameters is used) and that can accurately characterize the effects of the stress at circuit level. To fit the I_D-V_{GS} and I_D-V_{DS} curves of the damaged transistors only two parameters of the complete BSIM4 parameter set that describes the electrical performance of the fresh transistor need to be modified: U0 and VTH0, which are related to the transistor mobility and threshold voltage, respectively. U0 and VTH0 have been evaluated for all the damaged isolated transistors each time the stress was interrupted and their average, spread and time dependence calculated. The evolution with the stress time of these parameters follows a power law (not shown) [24], which allows to estimate the degraded parameters of the stressed device from the fresh ones for any time. As observed for σ_{ISAT}, the standard deviations of U0 and VTH0 BSIM4 parameters, σ_{U0} and σ_{VTH0}, do not show remarkable changes with time, in the studied stress time interval.

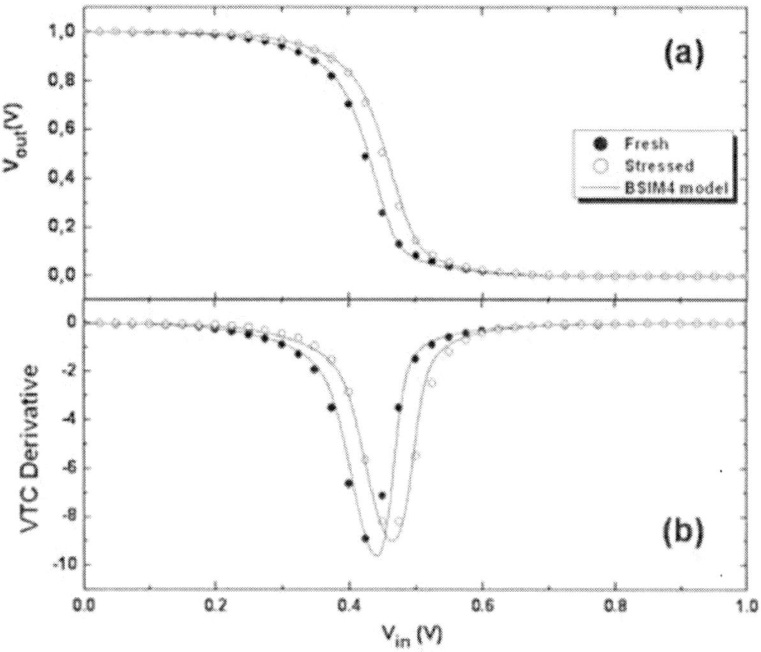

Figure 11. (a) Fresh and degraded (after 10000s of 500kHz pulsed stress) inverter VTC. (b) Derivatives of the VTCs (which are related to the gain of the amplifier built with the CMOS inverter) before and after the stress. Symbols correspond to the experimental data. Continuous lines correspond to SPICE simulations using the BSIM4 model to describe the electrical characteristics of the fresh and degraded NMOS and PMOS

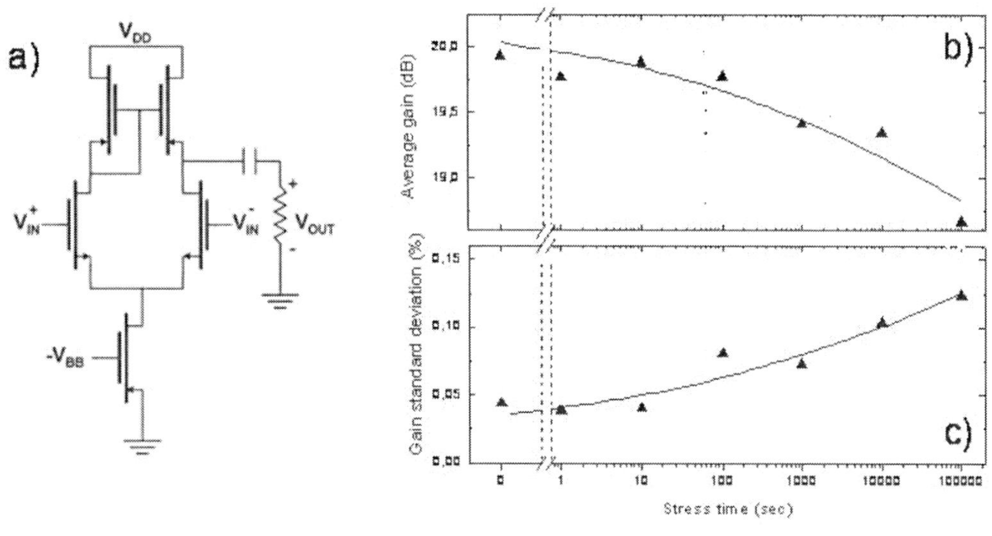

Figure 12. The wear-out and variability effects have been studied in a differential amplifier (a). Combined SPICE and Monte Carlo simulations allow to calculate the average gain (b) and its standard deviation (c) as a function of the stress time

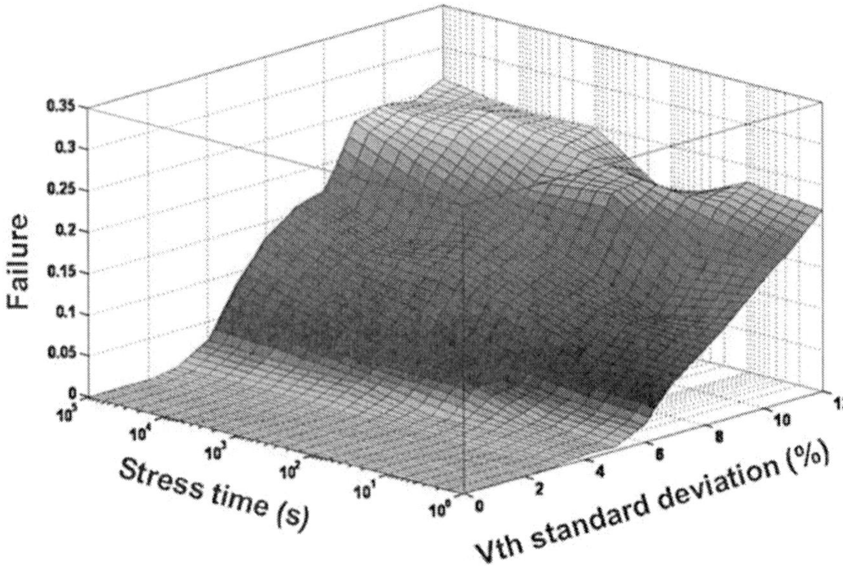

Figure 13. Cumulative failure distribution of differential amplifiers as a function of stress time and initial V_{th} standard deviation

From the average, standard deviation and time dependence of U0 and VTH0, SPICE and Montecarlo (MC) simulations have been combined to reproduce the I_{SAT} variations experimentally observed (Figure 10). To do so, the BSIM4 parameters of the fresh devices have been considered and U0 and VTH0 have been modified to describe the performance of the stressed device. First, taking into account the fresh U0 and VTH0 average values and their standard deviations (σ_{U0} and σ_{VTH0}), 100 values of U0 and VTH0 have been normal-randomly generated. The new BSIM4 parameter set has been introduced in a circuit simulator to reproduce the fresh (t=0) transistor I_D-V_{DS} characteristics and extract I_{SAT}. I_{SAT} has been averaged on the 100 devices (a number large enough to get MC simulation convergence) and the standard deviation calculated. Second, to reproduce the I_{SAT} evolution, the same procedure has been repeated for different stress times in an iterative process, considering the time dependence of U0 and VTH0, to get the values of the degraded transistor BSIM4 parameters. Continuous lines in Figure 10 show the mean I_{SAT} value obtained from the simulations and the dashed lines represent the margin $<I_{SAT}> \pm \sigma_{ISAT}$. Figure 10 demonstrates that it is only necessary to modify two physically meaningful SPICE parameters (actually their average and standard deviation) to correctly reproduce the experimentally observed I_{SAT} evolution and its variability.

SPICE Simulations: from Device Time-Dependent Variability to Circuit Reliability

To study the impact of the MOSFET time-dependent variability related to the gate oxide wear-out on circuits functionality, SPICE circuit simulations have been done, considering the device level results. nMOS differential pairs biased with a pMOS current mirror and a pMOS as a current limiting mechanism (Figure 12a) have been simulated at different stress times considering the device data (Montecarlo generated) obtained in the previous section. The amplifier gain has been taken as figure of merit of their performance. 500 Bode diagrams for each stress time have been generated taking into account the device variability. The degraded NMOS and PMOS electrical characteristics have been described by the previously obtained BSIM4 model parameter set. However, in this case, the experimental average of U0 and VTHO have been considered but σ_{U0} and σ_{VTH0} have been chosen so as to obtain a 5% (a standard value) V_{th} variability (σ_{Vth}), for the nMOS and pMOS transistors. Figure 12b shows the amplifier average gain and Figure 12c shows its standard deviation. A large gain reduction is obtained if V_{th} variability is considered. As an example, after 10^5 seconds of stress the average gain decreases around 1dB. However, if the same simulation is repeated without taking into account the transistors characteristics variability ($\sigma_{VTH0}=\sigma_{U0}=0$), no remarkable changes in the amplifier gain are observed. These results suggest that the V_{th} variability plays an important role in the circuit performance, not only at $t_s=0$, but also after long operation times. Moreover, note that the gain standard deviation continuously increases with the stress time.

To analyze the time dependent variability impact in circuit performance and reliability in more detail, a study of the wear out effects in the amplifier gain has been done as a function of stress time and σ_{Vth}. 7 different stress times (ranging from 0, i.e., fresh device, up to 10^5s), and 12 V_{th} standard deviations (from 0 to 12%) have been considered. For each stress time-σ_{Vth} pair, 500 amplifiers have been simulated (giving a total of $4.2{\cdot}10^4$ circuits). To determine the circuit reliability, the next failure criterion has been chosen: the circuit fails if the gain decays 3dB from the obtained value in the fresh amplifier, without device V_{th} variability. Figure 13 shows the 3D plot of the cumulative failure distribution, F, plotted versus σ_{Vth} and t_S. As can be observed, for low σ_{Vth} (below 4-5%), F is low during the whole simulated stress time interval (neither yield nor reliability problem). For large enough σ_{Vth} (>10%), F is so high at $t_s=0$ (without stress) that it is only slightly influenced by the electrical stress (yield problem). However, for intermediate σ_{Vth} values, F is small at the beginning of the stress, but increases quickly with time (reliability problem). This result indicates that the V_{th} variability resulting from the fabrication process has a strong influence not only on the circuit yield, but also on its reliability.

CONCLUSIONS

In this work, the approaches adopted at the Universitat Autònoma de Barcelona to characterize the effects of processing on the MOSFETs gate oxide electrical properties at the nanoscale and to transfer the effects of process related variability and aging from device to circuit level have been described.

Since the sources of process related variability have to be sought at the nanoscale, advanced characterization methods with a large lateral resolution have been used to analyze the gate oxide electrical properties in this range of dimensions. Conductive Atomic Force Microscope (CAFM) has been used to determine the impact of a thermal annealing on the nanoscale electrical conduction of Al_2O_3/SiO_2/p-Si structures, as a function of temperature. In particular, the conductivity and charge trapping of the Al_2O_3 stacks have been studied before and after an electrical stress. Previously to the electrical characterization, a physical analysis has been performed to investigate the morphology of the high-k layer. The results show that the samples crystallize at an annealing temperature (T_A) between 850 and 900°C and that Si diffuses from the substrate (increasing with T_A). Regarding the electrical properties of the fresh stacks, I-V curves and current images show that the conductivity of amorphous samples ($T_A < 900$°C) is more homogeneous than that of crystalline ones. In addition, for each group of samples (amorphous and crystalline) separately, it has been observed that the higher the annealing temperature (or Silicon content), the smaller the homogeneity and the higher the electrical conductivity, which have been associated to the larger Si content. A statistical analysis of the current images registered on crystalline samples has been compared to the measurements obtained with TEM. The statistical analysis shows that the mean size of the less conductive areas (in current images) is similar to the dimensions of the crystals measured with TEM. Therefore, these results suggest that the regions with a smaller conductivity could be related to the grains of the crystalline structure: the polycrystals are more insulating whereas the grain boundaries show a larger conductivity. The charge trapping of amorphous and crystalline samples was also investigated after an electrical stress. The results suggest that, although the crystals are more resistive and harder (from an electrical point of view) than amorphous oxides, the grain boundaries of the crystalline samples seem to behave as leaky sites with a higher conductivity and a smaller "dielectric strength" than non-crystallized structures: grain boundaries could act as conductive paths which also favor charge trapping after the stress, reducing the reliability of the gate stack. Therefore, crystallization and Si content (which depend on the annealing temperature) strongly contribute to the inhomogeneity increase of the conduction and trapping properties of the stacks and reduce the reliability of MOS devices due to the weaker dielectric strength of the grain boundaries. How these nanoscale oxide properties are reflected at device level has to be the aim of future works.

Isolated transistors and CMOS inverters have been subjected to electrical stress to study the effects of gate oxide degradation on MOSFET and circuit performance. At device level, due to the oxide wear-out, the threshold voltage increases and I_{SAT} decreases (both in absolute value). BSIM4 SPICE model has been used to fit the I-V characteristics of the stressed devices. We have shown that the methodology is able to reproduce the experimental observations (in devices and circuits) can be reproduced by only changing the averages of two of the physically meaningful parameters of the BSIM4 model parameter set obtained in fresh devices. To translate the time-dependent threshold voltage variability, related to the fabrication process and aging mechanisms, at device level into effects in circuit performance and reliability, a three-stage simulation methodology has been developed, based on Montecarlo and SPICE simulation. The simulation technique has been used to evaluate the gain of differential amplifiers based on NMOS transistors biased with a PMOS current mirror, taking into account the stress effects and the variability observed at device level. As the stress proceeds, the gain average value decreases and a larger spread is obtained. In addition, the initial threshold voltage variability strongly affects the circuit lifetime: intermediate values of σ will reduce the reliability of circuits that showed an acceptable yield. Therefore, these results show that the device mismatch related to the variability of the fabrication process has a strong influence not only on the circuit yield but also on its reliability.

Acknowledgments

This work has been partially supported by the Spanish MICINN (TEC2007-61294/MIC and HA2007-0029) and the Generalitat de Catalunya (2009SGR-783). Numonyx (Agrate, Italy) and Qimonda (Dresden, Germany) are acknowledged for sample provision and the University of Applied Sciences in Deggendorf (Germany) for allowing the use of their facilities to make the CAFM measurements.

References

[1] International Technology Roadmap for Semiconductors, Semiconductor Industry Association, http://public.itrs.net.

[2] Alam, M. (2008). *Microelectronics Reliability, vol. 48*, 114-1122.

[3] Bukhori, M. F., Roy, S. & Asenov, A. (2008). *Microelectronics Reliability, vol. 48*, 1549-1552.

[4] Xingsheng, W., Roy, S. & Asenov, A. (2008). IEEE Int. Solid-state Dev. *Research Conf.*, 190-193.

[5] Hegde, R. I., Chonko, M. A. & Tobin, P. J. (1996). *Journal Vacuum Science and Technology B, vol. 14*, 3299.

[6] Park, H., Jo, M., Choi, H., Hasan, M., Choi, R., Kirsch, P. D., Kang, C. Y., Lee, B. H., Kim, T. W., Lee, T. & Hwang, H. (2008). *IEEE Elect. Dev. Lett, vol. 29*, 54.

[7] Yanev, V., Rommel, M., Lemberger, M., Petersen, S., Amon, B., Erlbacher, T., Bauer, A. J., Ryssel, H., Paskaleva, A., Weinreich, W., Fachmann, C., Heitmann, J. & Schroeder, U. (2008). *Applied Physics Letters, vol. 92*, 252910.

[8] Lanza, M., Porti, M., Nafría, M., Benstetter, G., Frammelsberger, W., Lodermeier, E., Ranzinger, H. & Jaschke, G. (2007). *Microelectronics Reliability, vol. 47*, 1424.

[9] Olbrich, A., Ebersberger, B. & Boit, C. (1998). *Appl. Phys. Lett., vol. 73*, 3114.

[10] Fiorenza, P., Lo Nigro, R., Raineri, V., Lombardo, S., Toro, R. G., Malandrino, G. & Fragala, I. L. (2005). *Appl. Phys. Lett., vol. 87*, 231913.

[11] Lanza, M., Porti, M., Nafría, M., Aymerich, X., Benstetter, G., Lodermeier, E., Ranzinger, H., Jaschke, G., Wilde, L. & Michalowski, P. P. (2009). *Microelectronic Engineering, vol. 86(7-9)*, 1921-1924.

[12] Zafar, S., Kumar, A., Gusev, E. & Cartier, E. (2005). *IEEE Trans. Dev. Mater. Reliab., vol. 5*, 45.

[13] Ribes, G., Mitard, J., Denais, M., Bruyere, S., Monsieur, F., Parthasarathy, C., Vincent, E. & Ghibaudo, G. (2005). *IEEE Trans. Dev. Mater. Reliab., vol. 5*, 5.

[14] Amat, E., Kauerauf, T., Degraeve, R., Rodríguez, R., Nafría, M., Aymerich, X. & Groeseneken, G. (2009). IEEE Trans. Dev. *Mater. Reliab., vol. 9*, 454-458.

[15] Chang, K., Chang, F. M., Ruzyllo, J. (2006). *Solid State Electronics., vol. 50*, 1670-1672.

[16] Degraeve, R., Aoulaiche, M., Kaczer, B., Roussel, P., Kauerauf, T., Sahhaf, S. & Groeseneken, G. (2008). *Proc. IPFA*, 1-6.

[17] Groeseneken, G. (2001). *IEEE Trans. Dev. Mat. Reliab., vol. 5*, 23-32.

[18] Cimino, S., Pantisano, L., Aoulaiche, M., Degraeve, R., Kwak, H., Groeseneken, G., Paccagnella, A. (2005). *Proc. IRPS*, 275-279.
[19] Sahhaf, S., Degraeve, R., Roussel, P. J., Kauerauf, T., Kaczer, B. & Groeseneken, G. (2007). *Proc. IEDM*, 501-504.
[20] Fernández, R., Martín-Martínez, J., Rodríguez, R., Nafría, M. & Aymerich, X. (2008). IEEE Trans. on *Elec. Dev., vol 55*, 997-1004.
[21] Fiorenza, P., Polspoel, W. & Vandervorst, W. (2006). *Appl. Phys. Let., vol. 88*, 222104.
[22] Polspoel, W. & Vandervorst, W. (2007). *Microelec. Eng., vol. 84*, 495-500.
[23] Aguilera, L., Porti, M., Nafria, M. & Aymerich, X. (2006). *IEEE Elec. Dev. Lett., vol. 27(3)*, 157-159.
[24] Martín-Martínez, J., Rodríguez, R., Nafría, M. & Aymerich, X. (2009). *IEEE Trans. Dev. Mat. Reliab., 9*, 305.
[25] Groeseneken, G., Degraeve, R., Kaczer, B. & Roussel, P. (2007). *Proc. Int. Symp. Physical and Failure Analysis of Integrated Circuits*, (IPFA), 1.
[26] Dierickx, B., Miranda, M., Dobrovolny, P., Kutscherauer, F., Papanikolaou, A. & Marchal, P. (2007). *Proc. Int. Workshop on Physics of Semiconductor Devices* (IWPSD).
[27] Paul, B. C., Kang, K. & Kufluoglu, H. (2005). *IEEE Elect. Dev. Lett., 26*, 560.
[28] Porti, M., Nafría, M. & Aymerich, X. (2003). *IEEE Trans. Elect. Dev., 50*, 933.
[29] Day, H. C. & Allee, D. R. (1993). *Appl. Phys. Lett., 62*, 2691.
[30] Pinto, H., Nieminen, R. M. & Elliott, S. D. (2004). *Phys Rev B, 70*, 125402-1-125402-11.
[31] Weinreich, W., Wilde, L., Kucher, P., Lemberger, M., Yanev, V., Rommel, M., Bauer, A. J., Erben, E., Heitmann, J., Schroder, U. & Oberbeck , L. (2009). *J. Vac Sci Tech B, 27(1)*, 364-368.
[32] Kim, Y., Gebara, G., Freiler, M., Barnett, J., Riley, D., Chen, J., Torres, K., JaeEun, L., Foran, B., Shaapur, F., Agarwal, A., Lysaght, P., Brown, G. A., Young, C., Borthakur, S., Hong-Jyh, L., Nguyen, B., Zeitzoff, P., Bersuker, G., Derro, D., Bergmann, R., Murto, R. W., Hou, A., Huff, H. R., Shero, E., Pomarede, C., Givens, M., Mazanėz, M. & Werkhoven, C. (2001). *IEEE IEDM 20.2.1*, 455.
[33] Cao, C. X., Fan, L. Y. & Zhang, W. (2008). *Analyst, 133(9)*, 1139-1157.
[34] Porti, M., Nafría, M., Aymerich, X., Olbrich, A. & Ebersberger, B. (2002). *J Appl. Phys, 91(4)*, 2071-2079.
[35] Yu, X., Huang, J., Mingbin, Y. & Chunxiang, Z. (2007). *IEEE Elect. Dev. Lett, 28(5)*, 373-374.
[36] Paskaleva, A., Yanev, V., Rommel, M., Lemberger, M. & Bauer, A. J. (2008). *J Appl. Phys., 104*, 024108.
[37] Cheng, B., Roy, S., Brown, A., Millar, C. & Asenov, A. (2008). *Proc. Int. Conf. Solid-State and Integrated-Circuit Technology (ICSICT), 389.*
[38] Hua, W., Miranda, M., Catthor, F. & Dehaene, W. (2007). *IEEE Trans. Dev. Mat. Reliab., 7*, 581-591.
[39] Fischer, T., Olbrich, A., Georgakos, G., Lemaitreand, B. & Schmitt-Landsiedel, D. (2007). *Adv. Radio Sci., 5*, 321.
[40] Martin-Martinez, J., Gerardin, S., Amat, E., Rodriguez, R., Nafria, M., Aymerich, X., Paccagnella, A. & Ghidini, G. (2009). *IEEE Trans. Elec. Dev., 55*, 2155.

[41] Denais, M., Huard, V., Parthasarathy, C., Ribes, C., Perrier, F., Revil, N. & Bravaix, A. (2004). *IEEE Trans. Dev. Mater. Rel.*, *4*, 715.

[42] Fischer, T., Amirante, E., Hofmann, K., Ostermayr, M., Huber, P. & Schmitt-Landsiedel, D. (2008). *Eur. Solid-State Dev. Conf.*, *51*.

[43] Rauch, S. (2007). *IEEE Trans. Dev. Mater. Rel.*, *7*, 524.

[44] http://www-device.eecs.berkeley.edu/~bsim3/bsim4_intro.html.

In: CMOS Technology
Editor: Min-jun Kwon, pp.105-134
ISBN: 978-1-61761-325-8

Chapter 3

LINEAR AND NON-LINEAR APPLICATIONS OF CMOS DVCC

Sudhanshu Maheshwari, Mohd. Samar Ansari and Syed Atiqur Rahman
Department of Electronics Engineering,
Aligarh Muslim University, Aligarh, India

ABSTRACT

The differential voltage current conveyor (DVCC) has recently emerged as a versatile building block for the realization of a variety of analog signal processing functions including, but not limited to, analog filters, amplifiers, integrators, etc. Most of these applications can be classified as *linear* although *non-linear* applications like the voltage-mode comparator, function generators and oscillators are also reported in the technical literature.

This chapter presents several new and traditional applications of DVCC including both linear and non-linear ones. Amongst the linear applications, voltage-mode and current-mode analog filters, amplifiers and integrators are reviewed. Similarly, amongst the available non-linear ones, oscillators and negative resistance converter are reviewed whereas DVCC-based digital logic gates, linear equation solver and circuits for solving linear & quadratic programming problems are proposed. Real device simulations on some of the newly presented circuits are also given. The material presented is intended to explore full potential of DVCC covering a broad range of practical circuit applications.

I. INTRODUCTION

Over the past few decades, a variety of active building blocks (including, but not limited to, Current Conveyors, Current-Controlled Conveyors, Operational Transconductance Amplifiers, Current Differencing Transconductance Amplifier, Dual-X Current Conveyor, etc.) and their applications have been added to the technical literature [1-5]. Amongst these, the Second Generation Current Conveyor (CCII) has received the most attention of circuit designers. This is because of the fact that the CCII proves to be a versatile building block that can be used to implement a variety of high-performance circuits which are simple to construct. The CCII has a disadvantage that only one of the input terminals exhibits a high input impedance (Y terminal). This disadvantage becomes evident when the CCII is required to handle differential signals, as in the case of an instrumentation amplifier. Differential signal processing also has the advantage of better rejection of common-mode signals. The first step towards removing the said disadvantage came in 1989 when Pal [6] proposed a new form of the CCII and called it the modified CCII. Pal's modified CCII and the resultant DVCC II are presented in Figure 1 below.

Although enjoying differential inputs, the DVCC II of Figure 1 was not taken up very enthusiastically by the research fraternity. The DVCC remained a neglected active building block till 1997 when Elwan and Soliman [7] came up with a novel realization. This DVCC implementation, shown in Figure 2, utilized the dominant monolithic integration technology viz. CMOS.

Elwan and Soliman [7] were also the first to give a formal matrix equation (1) describing the functionality of the five port DVCC.

$$\begin{bmatrix} V_X \\ I_{Y1} \\ I_{Y2} \\ I_{Z1} \\ I_{Z2} \end{bmatrix} = \begin{bmatrix} 0 & 1 & -1 & 0 & 0 \\ 0 & 0 & 0 & 0 & 0 \\ 0 & 0 & 0 & 0 & 0 \\ 1 & 0 & 0 & 0 & 0 \\ -1 & 0 & 0 & 0 & 0 \end{bmatrix} \begin{bmatrix} I_X \\ V_{Y1} \\ V_{Y2} \\ V_{Z1} \\ V_{Z2} \end{bmatrix} \quad (1)$$

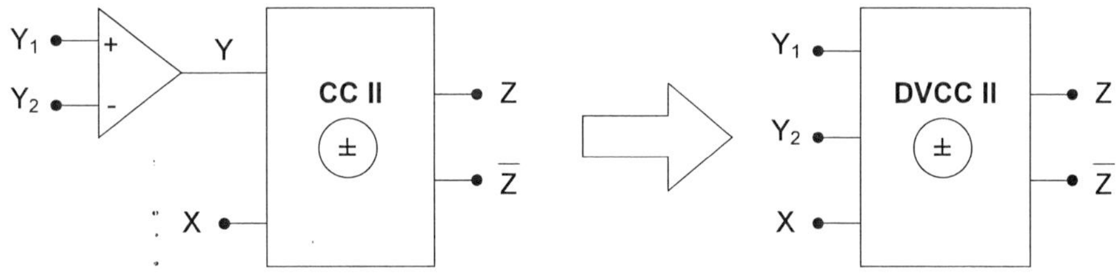

Figure 1. Modified Current Conveyor and its symbolic equivalent [6]

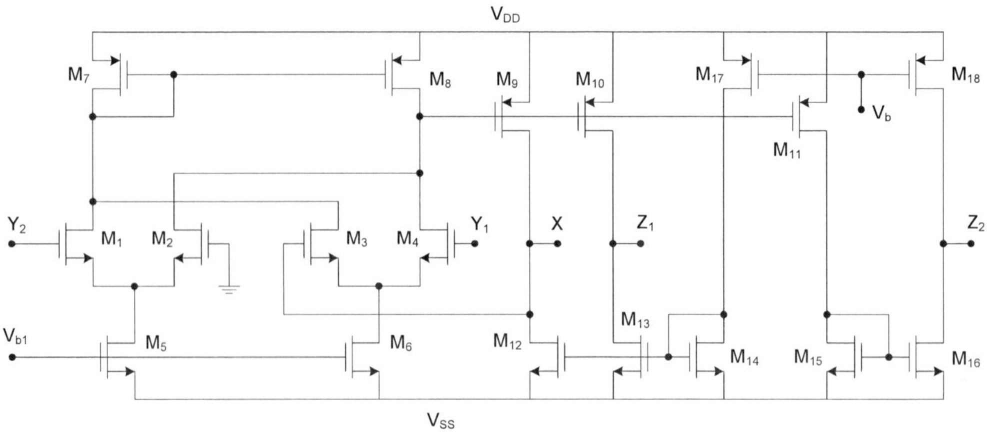

Figure 2. CMOS implementation of the DVCC [7]

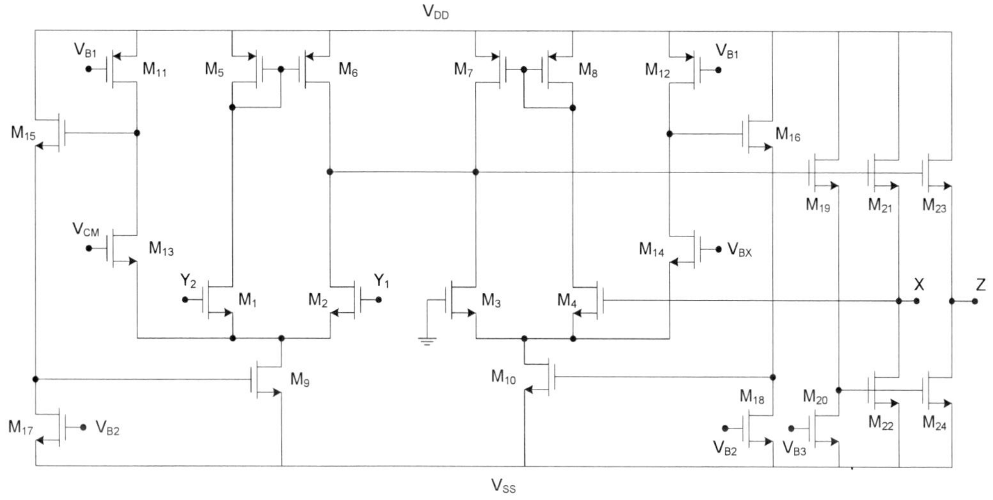

Figure 3. Low-Voltage Low-Power CMOS DVCC [9]

Several new and improved CMOS implementations were reported in the technical literature after the circuit of Figure 2. These included a DVCC realization based on Quasi-Floating Gate (QFG) transistors proposed by Moradzadeh and Azhari [8]. Mahmoud presented a low-voltage low-power implementation of the DVCC which is reproduced in Figure 3 for reference [9].

Another popular version of the DVCC, which is a slightly modified form of the circuit of Figure 2, has been used in [10] and is reproduced in Figure 4. Other noteworthy DVCC realizations can be attributed to Hu *et al.* [11], Siripruchyanan and Jaikla [12] (a bipolar implementation) and, more recently, Hassan and Mahmoud [13].

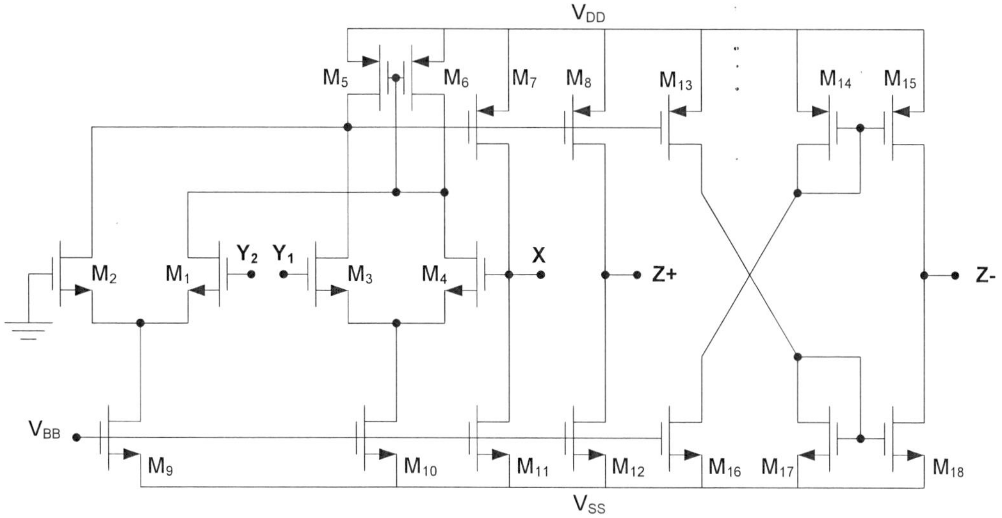

Figure 4. Improved CMOS implementation of DVCC [10]

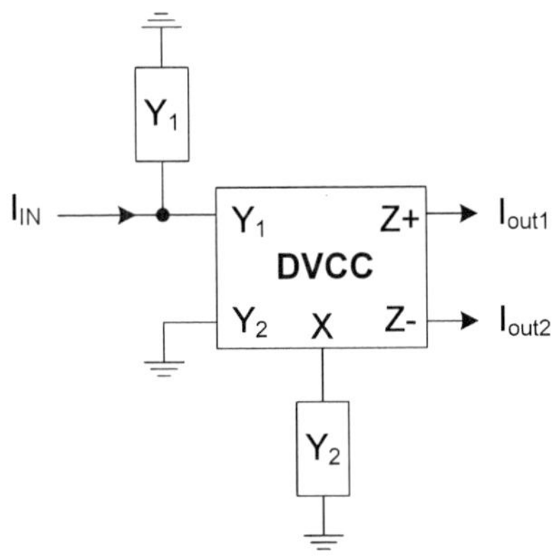

Figure 5. DVCC based basic current processing block [15]

II. Survey of Some Existing DVCC Based Circuits

The literature is replete with both linear and non-linear applications of DVCC [6-27]. Amongst the linear DVCC-based circuits that are reported, some are discussed here.

A. Amplifiers

A voltage-mode amplifier using DVCC was presented in [14]. Another amplifier circuit that needs a mention is the versatile building block shown in Figure 5 [15]. It can be made to work as a current-mode summing amplifier by proper selection of the admittances Y_1 and Y_2. For instance, if the admittances are $Y_1 = 1/R_1$ and $Y_2 = 1/R_2$, the current-mode amplifier can be realized with a gain equal to (R_1/R_2).

A DVCC-based instrumentation amplifier was presented in [13] and is reproduced in Figure 6 below. The circuit takes a differential input voltage, multiplies it with a gain, and produces a single-ended output voltage. The relation between the output voltage and the differential input voltage was shown to be

$$V_{OUT} = \frac{R_2}{R_1} V_{IN-DIFF} \tag{2}$$

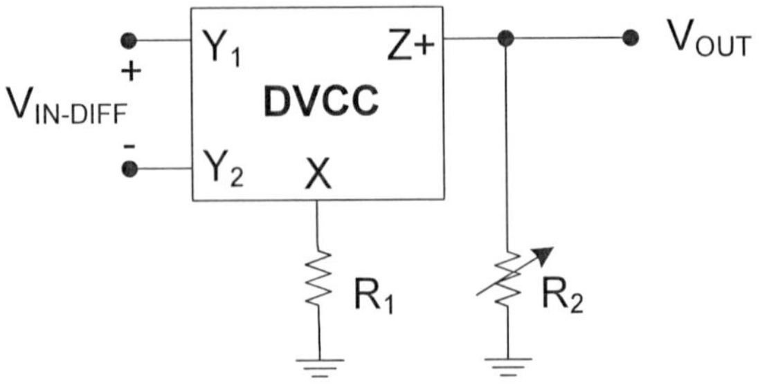

Figure 6. DVCC based Instrumentation Amplifier [13]

B. Integrators

A voltage-mode integrator using DVCC was presented in [14]. A current-mode integrator circuit was obtained from the basic current processing block, shown in Figure 6, in [15] by a judicious selection of the admittances Y_1 and Y_2. By choosing $Y_1 = sC_1$ and $Y_2 = 1/R_2$ and the transfer function was shown to be

$$\frac{I_{out1}}{I_{in}} = \frac{-I_{out2}}{I_{in}} = \frac{1}{sC_1R_2} \tag{3}$$

C. Filters with Bi-Linear Transfer Functions

Another application area where the DVCC has been put to good use is the design of voltage-mode and current-mode analog filters. First, a current-mode first-order all-pass filter section is discussed. The circuit of Figure 7, attributed to Minaei [16], presents the DVCC based first order AP (all-pass) filter. The advantage of using a DVCC can be understood by the fact that due to the differential input at Y-terminals, a very simple all-pass filter realization is obtained. Another noteworthy point is that the circuit is cascadable and employs only grounded passive components. This makes the circuit attractive for monolithic integration. Selecting R_2=$2R_1$, the current transfer function for the AP filter of Figure 7 [16] is given in (4) below. By using (4), the gain and phase response of the filter were shown to be those given in (5) and (6).

A current-mode all-pass section using a new 'modified' DVCC was reported in [17] and is reproduced in Figure 8. The DVCC utilized in this case had two Z+ outputs and no Z- output. Moreover, one of the Z+ stages had a gain of 2. This was achieved by tweaking the aspect ratios of the transistors comprising the second Z+ stage.

$$\frac{I_{out}}{I_{IN}} = \frac{R_2}{2R_3}\frac{1-scR_2}{1+scR_2} \tag{4}$$

$$Gain, K = \frac{R_2}{2R_3} \tag{5}$$

$$Phase, \varphi(\omega) = 180^0 - 2\tan^{-1}(\omega CR_2) \tag{6}$$

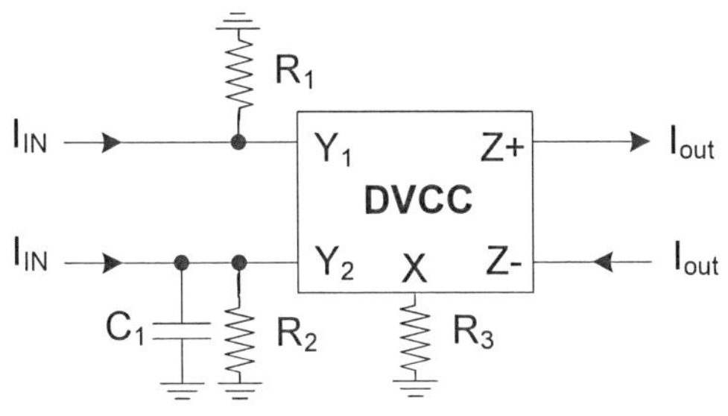

Figure 7. First-Order Dual-Input All-Pass Filter using a Single DVCC [16]

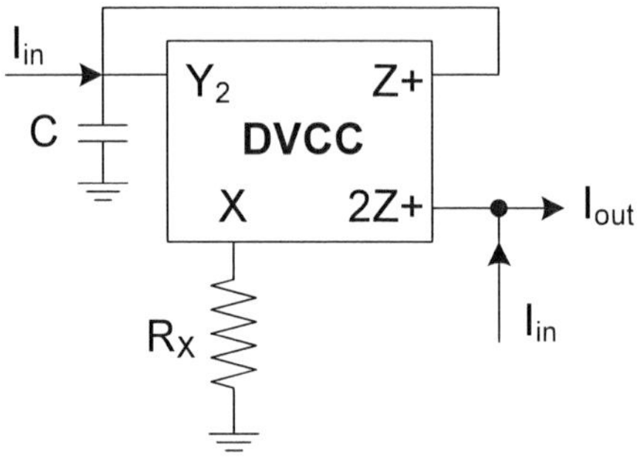

Figure 8. Current-mode All-Pass Filter using a Modified DVCC [17]

Assuming ideal DVCC operation, the current transfer gain for the circuit of Figure 8 [17] is reproduced below.

$$\frac{I_{out}}{I_{in}} = \frac{s-(1/CR_X)}{s+(1/CR_X)} \tag{7}$$

Equation (7) is the standard transfer function of an all-pass filter. The circuit of Figure 8 therefore provides a constant unity gain at all frequencies and frequency dependent phase with a value

$$\varphi(\omega) = 180^0 - 2\tan^{-1}(\omega C R_X) \tag{8}$$

The circuit [17] with only two passive components, both grounded, is quite compact and ideal for monolithic implementation. Besides, the circuit also exhibits high output impedance making it suitable for current-mode cascading.

Another class of first order single voltage input filter circuits based on DVCC that provide six independent outputs simultaneously; three of which are voltage outputs and three as current outputs were proposed by Maheshwari [18]. Shown in Figure 9, the DVCC based circuit takes a single voltage-mode input and is able to provide low-pass, high-pass and all-pass filter functions both in the voltage- as well as current-mode.

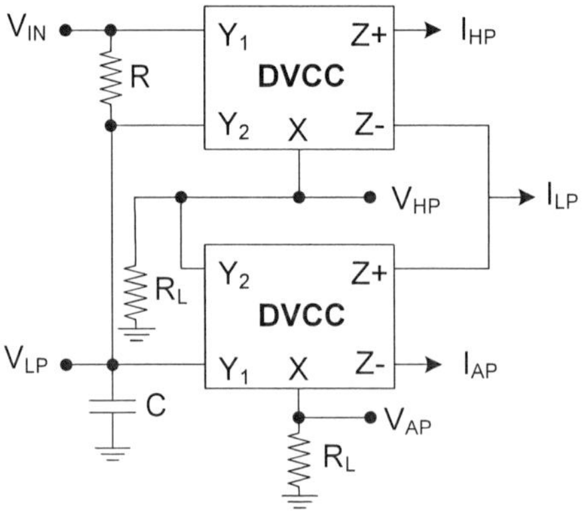

Figure 9. DVCC based first order filter with single input and six outputs [18]

The various voltage-mode and current-mode filter transfer functions for the circuit [18] are reproduced below.

$$\frac{V_{LP}}{V_{IN}} = \frac{1/RC}{s+1/RC} \tag{9}$$

$$\frac{V_{HP}}{V_{IN}} = \frac{s}{s+1/RC} \tag{10}$$

$$\frac{V_{AP}}{V_{IN}} = \frac{s-1/RC}{s+1/RC} \tag{11}$$

$$\frac{I_{LP}}{V_{IN}} = \frac{1}{R_L}\left[\frac{1/RC}{s+1/RC}\right] \tag{12}$$

$$\frac{I_{HP}}{V_{IN}} = \frac{1}{R_L}\left[\frac{s}{s+1/RC}\right] \tag{13}$$

$$\frac{I_{AP}}{V_{IN}} = \frac{1}{R_L}\left[\frac{s-1/RC}{s+1/RC}\right] \tag{14}$$

It is to be noted that six first order responses (9 – 14) are simultaneously obtained from the circuit with a total active and passive component count of five, which makes the circuit unique till date.

A voltage-mode all-pass section was also reported in [19] and is reproduced in Figure 10. The canonical first-order all-pass section employed a single grounded capacitor and two DVCCs, whose R_X was tuned through bias voltage, thereby ensuring a resistor-less realization. The idea of controlling the R_X for DVCC was in fact first introduced in this work [19]. This development proved to be a significant one for realizing active-C tunable networks using DVCC.

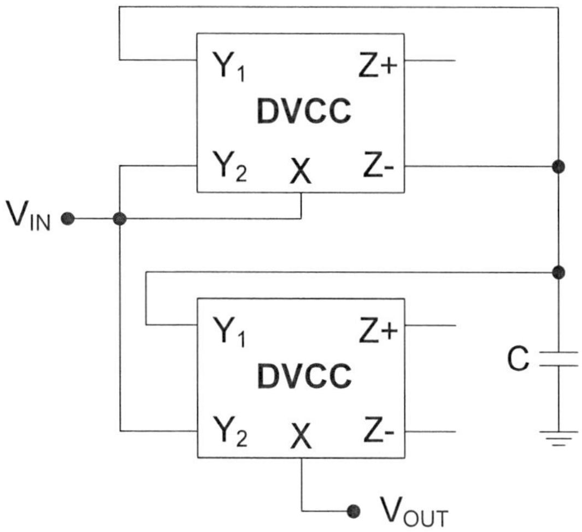

Figure 10. The first voltage-controlled voltage-mode first-order all-pass section using DVCC[19]

The transfer function of the circuit [19] was shown to be

$$\frac{V_{OUT}}{V_{IN}} = \frac{s-1/CR_X}{s+1/CR_X} \tag{15}$$

As can be seen from (15), the circuit of Figure 10 indeed acts as a voltage-mode all-pass filter and exhibits a gain of unity for all frequencies (ideally) and has a frequency dependent phase characteristics given by

$$\varphi(\omega) = -2\tan^{-1}(\omega CR_X) \tag{16}$$

D. Biquadratic Filters

Elwan and Soliman [7] proposed a CMOS continuous time filter for analog signal processing in the current-mode domain. The electronically tunable band-pass filter, shown in Figure 11 [7], was shown to have the following transfer function

$$\frac{I_{BP}}{I_{IN}} = \frac{s\frac{G_1}{C_1}}{s^2+s\frac{G_1}{C_1}+\frac{G_1G_2}{C_1C_2}} \tag{17}$$

Where

$$G_1 = 2K_1(V_{C1} - V_T) \tag{18}$$

$$G_2 = 2K_2(V_{C2} - V_T) \tag{19}$$

The center frequency and quality factor of the current-mode band-pass filter of Figure 11 were given as

$$\omega_o = \sqrt{\frac{G_1G_2}{C_1C_2}} \tag{20}$$

$$Q = \sqrt{\frac{C_1G_2}{G_1C_2}} \tag{21}$$

Equations (20) and (21) suggest that the circuit does not exhibit non-interactive control over ω_o and Q, which is a serious drawback.

Next, the differential all-pass and notch filter circuit is discussed [20]. From Figure 12, it can be seen that the circuit accepts differential voltage-mode input (V_{id}) and provides a differential output (V_{od}). By selecting $R_{z1} = R_{z2} = R_z$ and $C_{z1} = C_{z2} = C_z$, the differential-mode gain of the circuit of Figure 12 [20] can be written as

$$A_{dm} = \frac{s^2C_ZR_ZC_XR_X - s(2C_XR_Z - C_ZR_Z - C_XR_X)+1}{s^2C_ZR_ZC_XR_X + s(C_ZR_Z+C_XR_X)+1} \tag{22}$$

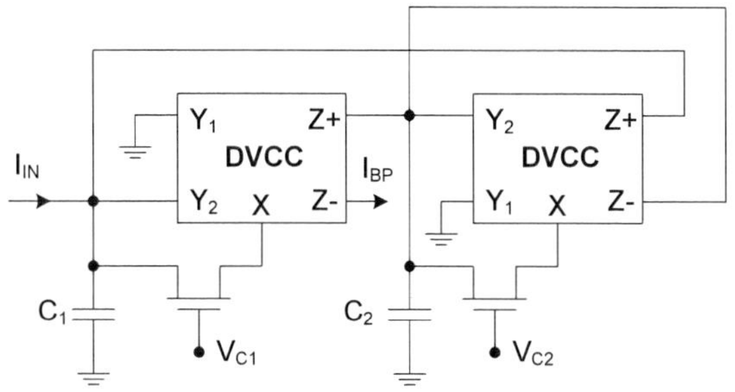

Figure 11. DVCC based Current-mode MOSFET-C Bandpass Filter [7]

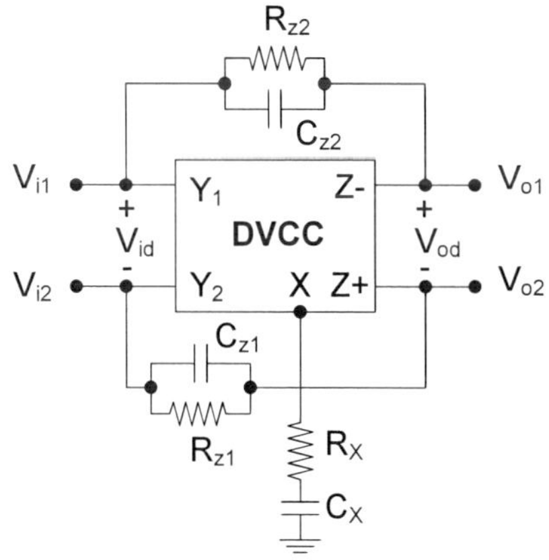

Figure 12. Differential Voltage-mode All-Pass and Notch Filter Circuit [20]

From (14), it is evident that voltage-mode all-pass filter function is realized by choosing $C_X = 2C_Z$ and $R_Z = 2R_X$. Similarly, the notch can be realized by selecting $C_X = C_Z$ and $R_Z = R_X$.

Chen and Shen proposed a three-input, five-output, universal capacitor-grounded voltage-mode filter using DVCCs [21]. The circuit, shown in Figure 13, employs two differential voltage current conveyors as active elements together with two grounded capacitors and four resistors as passive elements. The proposed configuration can be used as either in the single-input five-output or three-input two-output configuration. The circuit was shown to simultaneously realize five different generic filtering signals: lowpass, bandpass, highpass, bandreject, and allpass while enjoying the advantage of all grounded capacitors. Depending on the status of the three biquad input voltages, V_{i1}, V_{i2}, and V_{i3}, different filter functions are reproduced in the table below.

Table I. Obtaining different filter functions from the circuit of figure 13

Node Voltages Filter Function	V_{i1}	V_{i2}	V_{i3}	V_{out}
Low-Pass	V_{in}	0	0	V_{o2}
Band-Pass	0	V_{in}	0	V_{o3}
High-Pass	0	0	V_{in}	V_{o3}
Band-Stop	V_{in}	0	0	V_{o3}
All-Pass	V_{in}	V_{in}	0	V_{o3}

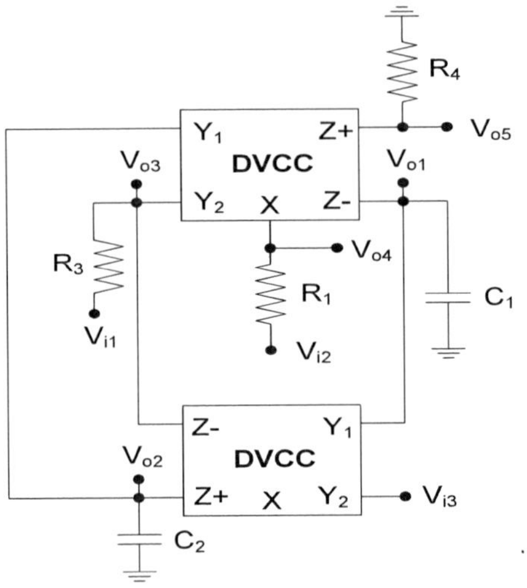

Figure 13. DVCC based voltage-mode biquad [21]

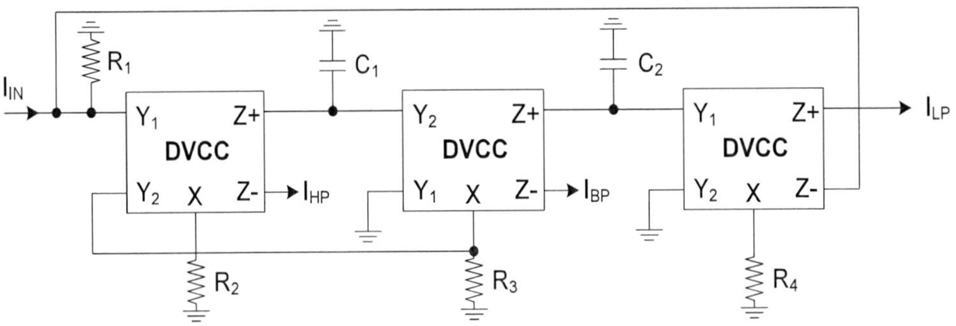

Figure 14. Current-mode KHN biquad based on DVCC [15].

A multi-functional filter based on the Kerwin-Huelsman-Newcomb (KHN) topology and utilizing DVCCs was also proposed by Ibrahim, Minaei and Kuntman [15]. The circuit, shown in Figure 14, operates in the current-mode domain and employs three DVCCs as active elements together with two capacitors and four resistors as passive elements, all of which are grounded. The circuit simultaneously provides the three basic filter functions, viz. band-pass, high-pass and low-pass functions. The notch and all-pass functions can be obtained by connecting appropriate output currents directly without using additional active elements.

The circuit of Figure 14 was shown to yield the following current-mode transfer functions

$$\frac{I_{HP}}{I_{IN}} = \frac{(R_1/R_2)s^2}{s^2+(\frac{1}{C_1R_2})s+\frac{R_1}{R_2R_3R_4C_1C_2}} \tag{23}$$

$$\frac{I_{BP}}{I_{IN}} = \frac{(R_1/R_2R_3C_1)s}{s^2+(\frac{1}{C_1R_2})s+\frac{R_1}{R_2R_3R_4C_1C_2}} \tag{24}$$

$$\frac{I_{LP}}{I_{IN}} = \frac{(R_1/R_2R_3R_4C_1C_2)}{s^2+(\frac{1}{C_1R_2})s+\frac{R_1}{R_2R_3R_4C_1C_2}} \tag{25}$$

The natural angular frequency ω_o and quality factor of the filters can be expressed as

$$\omega_0 = \sqrt{\frac{R_1}{R_2 R_3 R_4 C_1 C_2}} \tag{26}$$

$$Q = \sqrt{\frac{R_1 R_2 C_1}{R_3 R_4 C_2}} \tag{27}$$

It can be seen that the three basic filter functions (HP, BP and LP) are obtained simultaneously, all at high impedance outputs. Also, for $R_1=R_2$, a notch function can be realized by connecting high-pass and low-pass outputs, i.e. by adding the corresponding currents without using additional active elements. Moreover, for $R_1=R_2=R_3$, an all-pass function can be obtained by connecting high-pass, low-pass and band-pass outputs. Equations (26) and (27) further show that non-interactive control of ω_o and Q is only possible by varying resistance ratio. Other realizations of the KHN biquad using three or four DVCCs appear in [22].

Another universal biquadratic filter that needs to be mentioned is the one proposed by Horng *et al.* [23]. The circuit, which is reproduced in Figure 15, employs three DVCCs and six passive components to yield five voltage-mode filter functions viz. low-pass, high-pass, band-pass, band-stop and all-pass simultaneously. An attractive feature of the circuit from monolithic implementation point of view is that all the capacitors are grounded.

The transfer functions, resonance angular frequency and quality factor for the circuit of Figure 15 are reproduced below.

$$\frac{V_{O1}}{V_{IN}} = \frac{G_1 G_2 G_3}{s^2 C_1 C_2 G_3 + s C_1 G_1 G_2 + G_1 G_2 G_3} \tag{28}$$

$$\frac{V_{O2}}{V_{IN}} = \frac{-s C_1 G_1 G_3}{s^2 C_1 C_2 G_3 + s C_1 G_1 G_2 + G_1 G_2 G_3} \tag{29}$$

$$\frac{V_{O3}}{V_{IN}} = \frac{-s^2 C_1 C_2 G_3}{s^2 C_1 C_2 G_3 + s C_1 G_1 G_2 + G_1 G_2 G_3} \tag{30}$$

$$\frac{V_{O4}}{V_{IN}} = \frac{s^2 C_1 C_2 G_3 + G_1 G_2 G_3}{s^2 C_1 C_2 G_3 + s C_1 G_1 G_2 + G_1 G_2 G_3} \tag{31}$$

$$\frac{V_{O5}}{V_{IN}} = \frac{s^2 C_1 C_2 G_3 - s C_1 G_1 G_2 \frac{G_3}{G_4} + G_1 G_2 G_3}{s^2 C_1 C_2 G_3 + s C_1 G_1 G_2 + G_1 G_2 G_3} \tag{32}$$

$$\omega_0 = \sqrt{\frac{1}{R_1 R_2 C_1 C_2}} \tag{33}$$

$$Q = \frac{1}{R_3}\sqrt{\frac{R_1 R_2 C_2}{C_1}} \tag{34}$$

Equations (33) and (34) show that ω_o and Q can be tuned independent of each other, thereby making the circuit useful.

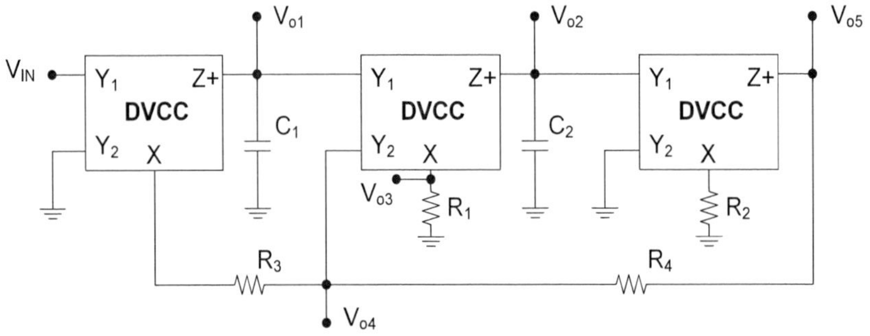

Figure 15. Voltage-mode Universal Biquadratic Filter employing DVCCs [23]

E. Sinusoidal Oscillators and Function Generators

Another class of DVCC-based analog circuit applications that will be considered includes sinusoidal oscillators and function generators. The first such circuit is the four phase clock generator [19]. As can be seen in Figure 16, an all-pass section is used to provide a 90-degree phase shift to an input sinusoidal signal and then two DVCC based comparators are used to obtain four distinct square waveforms in phase quadrature. The use of DVCC as comparator was shown for the first time in [19], which led to further new applications, as would also be shown later in this chapter.

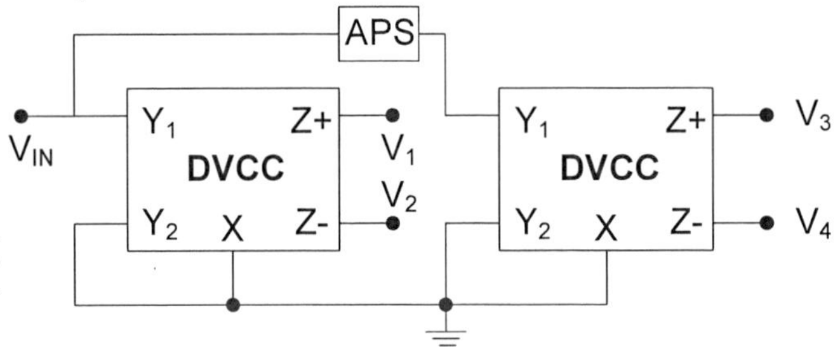

Figure 16. Four phase clock generator using an all-pass section [19]

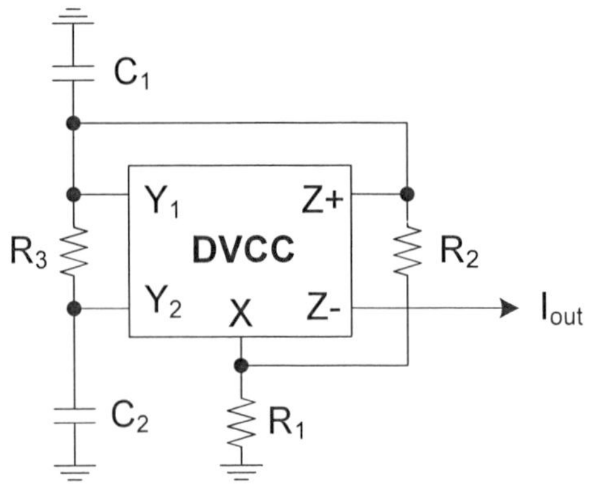

Figure 17. Current-mode Single Resistance Controlled Oscillator using DVCC [24]

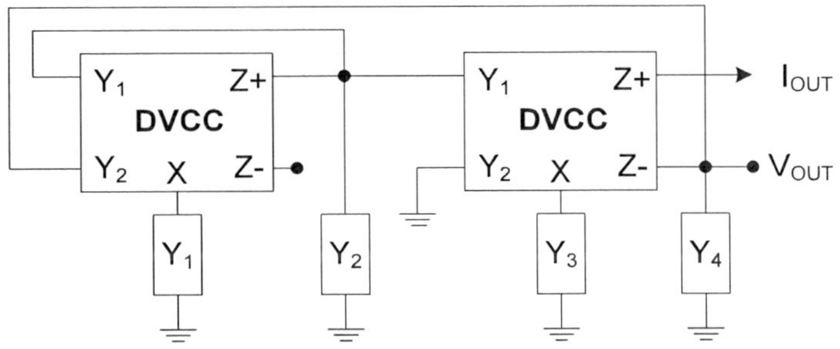

Figure 18. Single element controlled mixed-mode sinusoidal oscillator based on DVCC [25]

A DVCC-based circuit which belongs to the category of single element controlled oscillators is discussed next. The circuit, attributed to Gupta and Senani [24] and shown in Figure 17, employs only a single DVCC to provide a current-mode sinusoidal output which can be controlled by varying a single resistance. It needs to be mentioned that the particular circuit was the first single-resistance controlled current-mode oscillator using DVCC which employed all grounded capacitors.

The frequency and condition of oscillation for the circuit are reproduced below.

$$\text{FO: } f_0 = \frac{1}{2\pi}\sqrt{\frac{2}{R_2 R_3 C_1 C_2}} \tag{35}$$

$$\text{CO: } R_1 < R_3\left[\frac{C_2}{C_1+C_2}\right] \tag{36}$$

Since R_1 does not feature in FO above, the CO may be set by varying R_1 and then FO can independently be selected by choosing a suitable value of R_2.

Another second order sinusoidal oscillator based on the DVCC is discussed next. By proper selection of the admittances Y_1 through Y_4, eight different oscillator structures were obtained from the circuit shown in Figure 18 [25]. For instance, by setting

$$Y_1 = G_1;\ Y_2 = G_2 + sC_2;\ Y_3 = G_3;\ Y_4 = sC_4$$

a mixed-mode sinusoidal oscillator is obtained where the frequency and condition of oscillation are given as

$$\text{FO: } f_0 = \frac{1}{2\pi}\sqrt{\frac{G_1 G_3}{C_2 C_4}} \tag{37}$$

$$\text{CO: } G_1 = G_2 \tag{38}$$

An evident advantage of the above circuit is that the FO (frequency of oscillation) can be adjusted by tuning a single element (capacitor or resistor) without disturbing the CO (condition of oscillation) while the CO can be adjusted by tuning another element (resistor or capacitor) without disturbing the FO thereby ensuring independent amplitude and frequency control of oscillations.

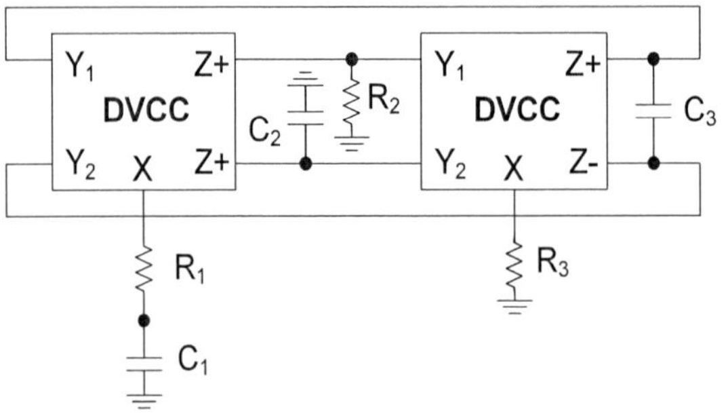

Figure 19. Voltage-mode fully-differential sinusoidal oscillator [26]

Next, a fully differential sinusoidal oscillator is discussed. Khan and Beg [26] proposed a DVCC based oscillator capable of producing a fully differential sinusoidal waveform. A voltage-mode fully-differential all-pass filter section is used to realize a voltage-mode fully-differential sinusoidal quadrature oscillator by cascading it with a differential integrator as shown in Figure 19. The use of one floating capacitor poses integration problems. Moreover, the R_1-C_1 series combination connected at the X-terminal limits the frequency generated by the circuit.

For the circuit of Figure 19, the frequency and condition of oscillation were found out to be [26]

$$\text{FO: } f_0 = \frac{1}{2\pi}\sqrt{\frac{1}{R_2 R_3 C_2 C_3}} \tag{39}$$

$$\text{CO: } R_1 C_1 = R_3 C_3 \tag{40}$$

Unlike the class of second order oscillators, a third order quadrature oscillator was recently proposed as an application of a new first order topology by Maheshwari [13]. The DVCC paper [13] was first of its kind, encompassing analog circuit applications ranging from first-order to third-order. It is to be noted that Maheshwari [13, 27] was first to propose third order quadrature oscillators based on DVCC. The circuit in [13] provided four quadrature voltage outputs, whereas the circuit in [27] was shown to provide seven distinct voltage and current outputs. The idea of voltage controlled DVCC [19] was further utilized in [27] to realize a voltage controlled quadrature oscillator of third order.

III. Proposed Circuits

This section presents some new applications of the DVCC. Some neural circuits based on the DVCC are proposed which find applications in solving simultaneous linear equations, linear programming problems and quadratic programming problems. Further, new applications of the DVCC in realizing digital logic functions like AND/NAND and OR/NOR are also explored.

A. Linear Equation Solver

The quest for methods to solve systems of linear equations has been going on for almost two millennia [28-30]. The state-of-art linear equation solvers use software algorithms. However, since most computer algorithms are sequential, the time to arrive at the solution is governed by the problem size and the hardware on which the software is run. For real-time systems, such a solution time of the order of milliseconds may not be appropriate. For improved solution time, the inherent parallel processing of neural networks may be utilized [30]. This section presents a neural circuit employing non-linear feedback to solve systems of linear equations.

Let the set of simultaneous linear equations to be solved be

$$a_{11}V_1 + a_{12}V_2 - b_1 = 0 \qquad (41) \qquad (41)$$

$$a_{21}V_1 + a_{22}V_2 - b_2 = 0 \qquad (42) \qquad (42)$$

Where V_1 and V_2 are the two variables and a_{11}, a_{12}, a_{21}, a_{22}, b_1 and b_2 are constants. Figure 20 shows the proposed hardware for solving a set of 2 simultaneous linear equations in 2 variables.

As can be seen from Figure 20, individual equations from the set of equations to be solved are passed through non-linear synapses. CMOS Differential Voltage Current Conveyors (DVCCs) are used as voltage-mode comparators and opamps are utilized to emulate the operation of neurons. The DVCC is able to provide multiple current outputs simultaneously, at the Z+ ports, which are scaled suitably to generate weighted currents at the input of the neurons. For this purpose, use is made of the W/L ratios of the MOSFETs employed in the output stage of the DVCC. R_{p1} & R_{p2} are the parasitic resistances and C_{p1} & C_{p2} are the parasitic capacitance that present themselves at the inputs of the Opamps used in the realization of neurons. These parasitic components are included to model the dynamic nature of the Opamps. For the DVCC, we know that

$$V_X = V_{Y1} - V_{Y2} \qquad (43) \qquad (43)$$

$$I_X = \frac{V_{Y1} - V_{Y2}}{R_X} \qquad (44) \qquad (44)$$

For the DVCC, comparator action can be achieved by directly grounding the X-terminal. The use of DVCC as a voltage-mode comparator was first given in [19]. For such a case, the current in the X-port will saturate and can be written as

$$I_X = I_m \tanh\beta\,(V_{Y1} - V_{Y2}) \qquad (45) \qquad (45)$$

By virtue of DVCC action, this current will be transferred to the Z+ ports. For DVCC-1, the output currents at the Z+ ports can be written as

$$I_{z1}{}^{+} = s_{11}I_{X1} = s_{11}I_m \tanh\beta\,(a_{11}V_1 + a_{12}V_2 - b_1) \qquad (46) \qquad (46)$$

$$I_{Z2}{}^{+} = s_{21} I_{X1} = s_{21} I_m \tanh \beta \,(a_{21} V_1 + a_{22} V_2 - b_2) \quad (47) \qquad (47)$$

Where s_{11} and s_{21} are the scaling factors introduced during the current copying process from the X-port to the Z-port. Ideally, these factors are unity. Similarly, for the second comparator, we may write

$$I_{Z1}{}^{+} = s_{12} I_{X2} = s_{12} I_m \tanh \beta \,(a_{11} V_1 + a_{12} V_2 - b_1) \quad (48) \qquad (48)$$

$$I_{Z2}{}^{+} = s_{22} I_{X2} = s_{22} I_m \tanh \beta \,(a_{21} V_1 + a_{22} V_2 - b_2) \quad (49) \qquad (49)$$

Where β is the open-loop gain of the comparator (practically very high) and I_m is the maximum (saturated) current at the output of the comparator. Writing node equation for the nodes 'A' and 'B' in Figure 20, we get

$$C_{p1} \frac{du_1}{dt} = s_{11} I_m \tanh \beta \left(a_{11} V_1 + a_{12} V_2 - b_1\right) + s_{12} I_m \tanh \beta \left(a_{11} V_1 + a_{12} V_2 - b_1\right) - \frac{u_1}{R_{p1}} \quad (50)$$

$$C_{p2} \frac{du_2}{dt} = s_{21} I_m \tanh \beta \,(a_{11} V_1 + a_{12} V_2 - b_1) + s_{22} I_m \tanh \beta \,(a_{11} V_1 + a_{12} V_2 - b_1) - \frac{u_2}{R_{p2}} \quad (51)$$

Moreover, it is known that in the case of symmetric connections, the equations of motion for this network of analog processing units corresponds to a stable system, in which the output voltages of all amplifiers become constant [31]. Also, the stable states of the network are the local minima of a quantity known as the computational energy 'E' of the system (more commonly referred to as 'Lyapunov function' or 'Energy Function') [32].

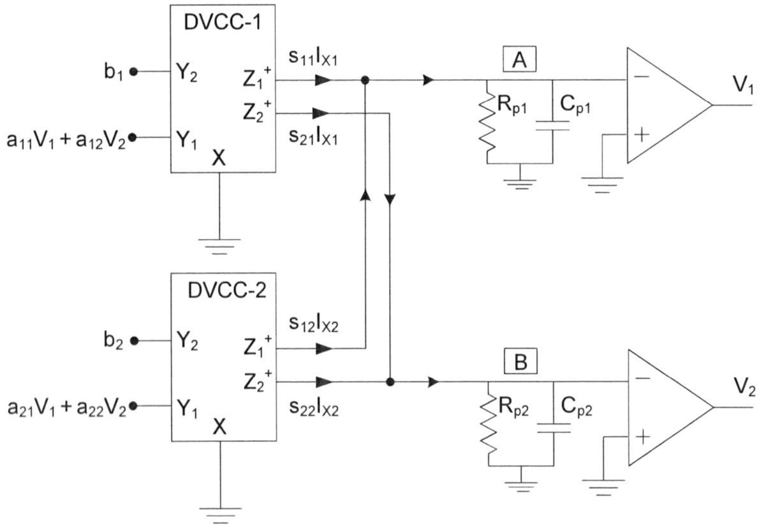

Figure 20. Proposed DVCC-based CMOS compatible circuit to solve a system of simultaneous linear equations in 2 variables

For the network of Figure 20, it can be shown that the Energy Function 'E' of is related to the voltages V_1 and V_2 by

$$\frac{\partial E}{\partial V_1} = I_m\, a_{11}\tanh\beta(a_{11}V_1 + a_{12}V_2 - b_1) + I_m\, a_{12}\tanh\beta(a_{21}V_1 + a_{22}V_2 - b_2) - \left[\frac{u_1}{R_{p1}}\right] \quad (52)$$

$$\frac{\partial E}{\partial V_2} = I_m\, a_{21}\tanh\beta(a_{11}V_1 + a_{12}V_2 - b_1) + I_m\, a_{22}\tanh\beta(a_{21}V_1 + a_{22}V_2 - b_2) - \left[\frac{u_2}{R_{p2}}\right] \quad (53)$$

Also, if 'E' is the Energy function, it must satisfy the following conditions for stability [33].

$$\frac{\partial E}{\partial V_1} = C_{p1}\frac{du_1}{dt} \quad (54)$$

And

$$\frac{\partial E}{\partial V_2} = C_{p2}\frac{du_2}{dt} \quad (55)$$

Comparing equations (50) & (52), and (51) & (53) according to (54) and (55), we get

$$s_{11} = a_{11};\ s_{12} = a_{12};\ s_{21} = a_{21};\ s_{22} = a_{22}$$

Although Figure 20 shows the proposed circuit as applicable to a system of simultaneous linear equations in two variables, the technique can be readily extended to solve *n*-simultaneous linear equations in *n*- variables. Moreover, for obtaining weighted (scaled) current outputs use was made of the (W/L) ratios of the transistors comprising the Z+ stages of the DVCCs used in Figure 20. This results in a particular circuit implementation 'hard-wired' to solve a given set of equations. Methods to scale the currents by a mechanism external to the circuit need to be investigated to impart flexibility to the circuit.

B. Linear Programming Circuit

In a linear programming problem (LPP), a linear function is optimized subject to certain linear constraints [34-36]. Linear programming can be applied to various fields of study. It is used most extensively in business and economics, but can also be utilized for some engineering problems. Industries that use linear programming models include transportation, energy, telecommunications, and manufacturing. It has proved useful in modeling diverse types of problems in planning, routing, scheduling, assignment, and design.

Let the function to be minimized be

$$F = P_1V_1 + P_2V_2 \quad (56)$$

Subject to

$$a_{11}V_1 + a_{12}V_2 \leq b_1 \tag{57}$$

$$a_{21}V_1 + a_{22}V_2 \leq b_2 \tag{58}$$

Where V_1, and V_2 are the variables and a_{ij}, b_i and c_{ij} are constants. The proposed neural-network based circuit to minimize (56) subject to (57, 58) is presented in Figure 21. As can be seen from Figure 21, individual equations from the set of constraints are passed through non-linear synapses which are realized using DVCC-based unipolar comparators. The outputs of the comparators are fed to neurons having weighted inputs. These weighted neurons are realized by using opamps where the scaled currents coming from various comparators act as weights. The parasitic components that need to be included to model the dynamic nature of the opamps are also considered (as was done for the previous circuit) and are shown in Figure 21.

Node equation for nodes 'A' and 'B' give the equations of motions of the first and second neurons as

$$C_{p1}\frac{du_1}{dt} = s_{11}I'_{X1} + s_{21}I'_{X2} + \left[\frac{P_1}{R_1}\right] - u_1\left[\frac{1}{R_{eq1}}\right] \tag{59}$$

Where, $s_{11}I'_{X1}$ and $s_{21}I'_{X2}$ represent the currents $s_{11}I_{X1}$ and $s_{21}I_{X2}$ after passing through diodes as shown in Figure 21 and

$$\frac{1}{R_{eq1}} = \frac{1}{R_{p1}} + \frac{1}{R_1} \tag{60}$$

And,

$$C_{p2}\frac{du_2}{dt} = s_{12}I'_{X1} + s_{22}I'_{X2} + \left[\frac{P_2}{R_2}\right] - u_2\left[\frac{1}{R_{eq2}}\right] \tag{61}$$

Where, $s_{12}I'_{X1}$ and $s_{22}I'_{X2}$ represent the currents $s_{12}I_{X1}$ and $s_{22}I_{X2}$ after passing through diodes as shown in Figure 21 and

$$\frac{1}{R_{eq2}} = \frac{1}{R_{p2}} + \frac{1}{R_2} \tag{62}$$

In equations (59) and (61), u_i is the internal state of the i-th neuron and s_{ij} is the current scaling factor at the j-th output of i-th DVCC. As is shown later in this section, these weights are governed by the constraint inequalities (57, 58). Using (46, 47) in (59) and (61) results in (63) and (64) given below.

$$C_{p1}\frac{du_1}{dt} = \frac{1}{2}s_{11}gV_m[\tanh\beta\,(a_{11}V_1 + a_{12}V_2 - b_1) + 1] + \frac{1}{2}s_{21}\,gV_m[\tanh\beta\,(a_{21}V_1 +$$
$$a22V2 - b2 + 1 + P1R1 - u11Req1 \tag{63}$$

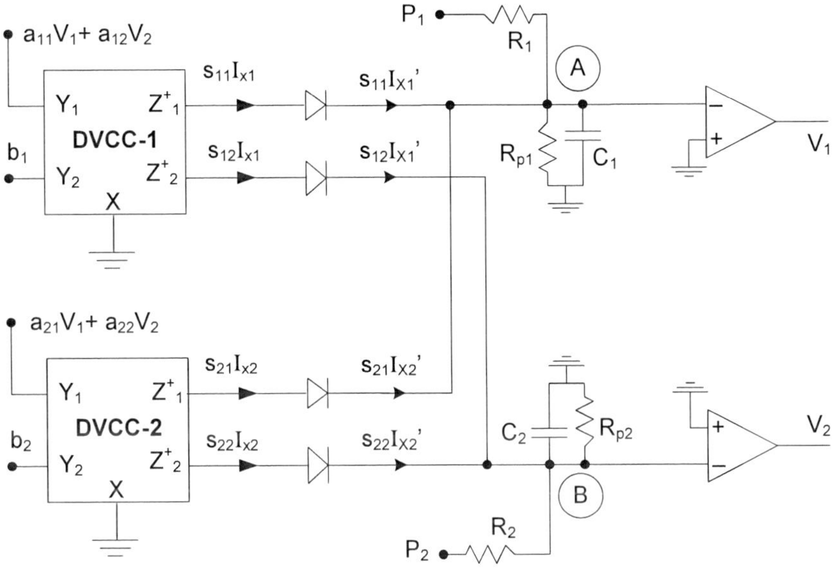

Figure 21. Proposed neural network for solving a linear programming problem in 2 variables subject to 2 linear constraints

$$C_{p2}\frac{du_2}{dt} = \frac{1}{2}s_{12}gV_m\left[\tanh\beta\left(a_{11}V_1 + a_{12}V_2 - b_1\right) + 1\right] + \frac{1}{2}s_{22}\,gV_m\left[\tanh\beta\left(a_{21}V_1 +\right.\right.$$
$a22V2 - b2+1+ P2R2-u21Req2$ (64)

Moreover, it can be shown that the network of Figure 21 may be associated with an 'energy' function of the form

$$E = \frac{V_m}{\beta}\ln\cosh\,\beta\left(a_{11}V_1 + a_{12}V_2 - b_1\right) + \frac{V_m}{\beta}\ln\cosh\beta\left(a_{21}V_1 + a_{22}V_2 - b_2\right) + P_1V_1 +$$
$P2V2+Vm2a11V1+a12V2+a21V1+a22V2-1Req10V1u1dV1- 1Req20V2u2dV2$ (65)

From (68), it follows that

$$\frac{\partial E}{\partial V_1} = P_1 + V_m\,a_{11}\tanh\beta(a_{11}V_1 + a_{12}V_2 - b_1) + V_m a_{21}\tanh\beta(a_{21}V_1 + a_{22}V_2 - b_2) - u_1\left[\frac{1}{R_{eq1}}\right] \quad (66)$$

Also, if 'E' is the Energy Function, it must satisfy the following condition for stability [33].

$$\frac{\partial E}{\partial V_i} = K\,C_i\frac{du_i}{dt} \quad (67)$$

Where 'K' is a constant of proportionality and has the dimensions of resistance. Comparing (63) and (66) according to (67) yields

$$s_{11} = {}^{a_{11}}/_{gK},\ s_{21} = {}^{a_{21}}/_{gK} \tag{68}$$

$$R_1 = K \tag{69}$$

Analysis on similar lines can be performed to obtain the values of the synaptic weights for the second neuron and are presented in (70) and (71) below.

$$s_{12} = {}^{a_{12}}/_{gK},\ g_{22} = {}^{a_{22}}/_{gK} \tag{70}$$

$$R_2 = K \tag{71}$$

The values of the current scaling factors (s_{ij}) of Figure 21 can be calculated by choosing a suitable value of '*K*' and then using (68, 70).

C. Quadratic Programming Circuit

Quadratic Programming problems are very important in the field of optimization [37-39]. They arise in many applications such as constrained least mean square estimation and the classical newsvendor problem. Besides its wide applications, quadratic programming is also forms a basis for solving some general nonlinear programming problems as most higher order problems can usually be approximated by a second order function. Traditional methods for solving quadratic programming problems typically involve an iterative process, but long computational time limits their usage. There is an alternative approach to solution of this problem. It is to exploit the artificial neural networks which can be considered as an analog computer relying on a highly simplified model of neurons [31].

Let the quadratic function to be minimized be

$$F = \frac{1}{2}C_{11}{V_1}^2 + C_{12}V_1V_2 + \frac{1}{2}C_{22}{V_2}^2 \tag{72}$$

Subject to

$$a_{11}V_1 + a_{12}V_2 \leq b_1 \tag{73}$$

$$a_{21}V_1 + a_{22}V_2 \leq b_2 \tag{74}$$

Where V_1, and V_2 are the variables and a_{ij}, b_i and c_{ij} are constants. The proposed neural-network based circuit to minimize (72) subject to (73, 74) is presented in Figure 22.

As was done in the case of linear programming circuit, individual equations from the set of constraints are passed through non-linear synapses which are realized using DVCC-based unipolar comparators. The outputs of the comparators are fed to neurons having weighted inputs. These weighted neurons are realized by using opamps where the scaled currents coming from various comparators act as weights. Weighing is also provided by the selffeedback resistances (R_{11} & R_{22} for neurons 1 & 2 respectively) and the cross-feedback resistances (R_{21} and R_{12}).

Node equation for nodes 'A' and 'B' give the equations of motions of the first and second neurons as

$$C_{p1}\frac{du_1}{dt} = s_{11}I'_{X1} + s_{21}I'_{X2} + \left[\frac{V_1}{R_{11}} + \frac{V_2}{R_{21}}\right] - u_1\left[\frac{1}{R_{eq1}}\right] \tag{75}$$

Where, $s_{11}I'_{X1}$ and $s_{21}I'_{X2}$ represent the currents $s_{11}I_{X1}$ and $s_{21}I_{X2}$after passing through diodes as shown in Figure 22 and

$$\frac{1}{R_{eq1}} = \frac{1}{R_{p1}} + \frac{1}{R_{11}} + \frac{1}{R_{21}} \tag{76}$$

And,

$$C_{p2}\frac{du_2}{dt} = s_{12}I'_{X1} + s_{22}I'_{X2} + \left[\frac{V_1}{R_{12}} + \frac{V_2}{R_{22}}\right] - u_2\left[\frac{1}{R_{eq2}}\right] \tag{77}$$

Where, $s_{12}I'_{X1}$ and $s_{22}I'_{X2}$ represent the currents $s_{12}I_{X1}$ and $s_{22}I_{X2}$ after passing through diodes as shown in Figure 22 and

$$\frac{1}{R_{eq2}} = \frac{1}{R_{p2}} + \frac{1}{R_{12}} + \frac{1}{R_{22}} \tag{78}$$

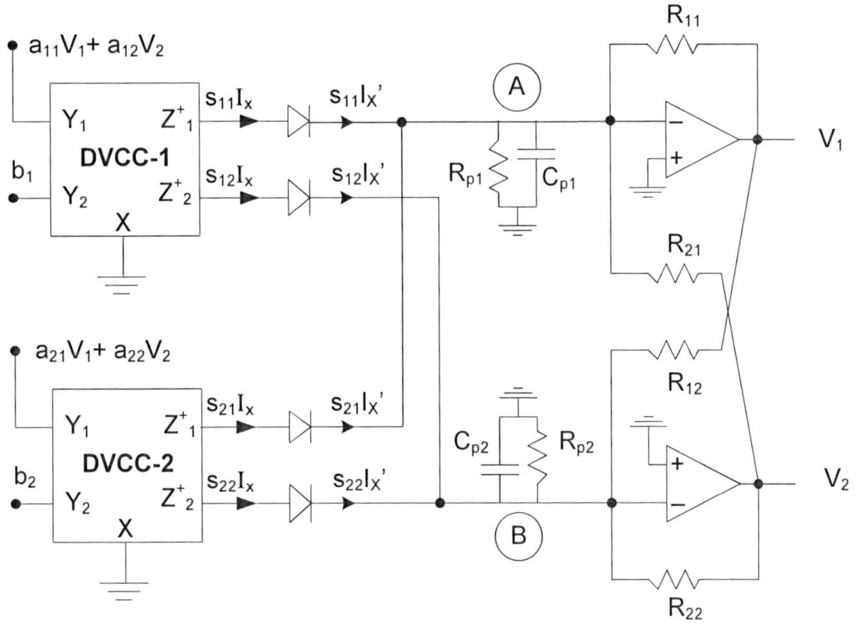

Figure 22. Proposed circuit to solve quadratic programming problem in 2-variables with 2 linear constraints

In equations (78) and (80) u_i is the internal state of the i-th neuron and s_{ij} is the current scaling factor at the j-th output of i-th DVCC. As is shown later in this section, these weights are governed by the constraint inequalities (73, 74). Using (46, 47) in (75) and (77) results in (79) and (80) given below.

$$C_{p1}\frac{du_1}{dt} = \frac{1}{2}s_{11}gV_m\left[\tanh\beta\left(a_{11}V_1 + a_{12}V_2 - b_1\right) + 1\right] + \frac{1}{2}s_{21}\,gV_m\left[\tanh\beta\left(a_{21}V_1 + \right.\right.$$
$$a22V2 - b2+1+ V1R11+ V2R21-u11Req1 \qquad (79)$$

$$C_{p2}\frac{du_2}{dt} = \frac{1}{2}s_{12}gV_m\left[\tanh\beta\left(a_{11}V_1 + a_{12}V_2 - b_1\right) + 1\right] + \frac{1}{2}s_{22}\,gV_m\left[\tanh\beta\left(a_{21}V_1 + \right.\right.$$
$$a22V2 - b2+1+ V1R12+ V2R22-u21Req2 \qquad (80)$$

Moreover, it can be shown that the network in Figure 22 can be associated with an Energy Function 'E' of the form

$$E = \frac{1}{2}C_{11}{V_1}^2 + C_{21}V_1V_2 + \frac{1}{2}C_{22}{V_2}^2 + \frac{V_m}{2\beta}\ln\cosh\,\beta\left(a_{11}V_1 + a_{12}V_2 - b_1\right) + \frac{V_m}{2\beta}\ln\cosh\beta\left(a_{21}V_1 + a_{22}V_2 - b_2\right) + \frac{V_m}{2}\left(a_{11}V_1 + a_{12}V_2 + a_{21}V_1 + a_{22}V_2\right) - \frac{1}{R_{eq1}}\int_0^{V_1} u_1\,dV_1 - \frac{1}{R_{eq2}}\int_0^{V_2} u_2\,dV_2 \qquad (81)$$

From (81), it follows that

$$\frac{\partial E}{\partial V_1} = c_{11}V_1 + c_{21}V_2 + \frac{1}{2}V_m\,a_{11}\tanh\beta(a_{11}V_1 + a_{12}V_2 - b_1) + \frac{1}{2}V_m a_{21}\tanh\beta(a_{21}V_1 +$$
$$a22V2-b2)- u11Req1 \qquad (82)$$

Also, if 'E' is the Energy Function, it must satisfy the following condition for stability [33].

$$\frac{\partial E}{\partial V_i} = K\,C_i\frac{du_i}{dt} \qquad (83)$$

Where 'K' is a constant of proportionality and has the dimensions of resistance. Comparing (79) and (82) according to (83) yields

$$a_{11} = gKs_{11};\; a_{21} = gKs_{21} \qquad (84)$$

$$c_{11} = \frac{K}{R_{11}};\; c_{21} = \frac{K}{R_{21}} \qquad (85)$$

The values of the current scaling factors (s_{ij}) of Figure 22 can be easily calculated by setting the value of 'K' to be equal to $1/g$ and then using (84). The values of the weight resistances connected to the first neuron, R_{11} and R_{21}, can be obtained by using (85). Analysis on similar lines can be performed to obtain the values of the synaptic weights for the second neuron.

D. Implementation of Logic Gates Using DVCC

Another application area where the DVCC can be put to use is Digital Logic Gates. Although implementations of various digital logic gates are abundant in various technologies including CMOS, the use of analog building blocks like the DVCC has never been considered before. The circuit for 2-input AND/NAND gate, presented in Figure 23, is a first attempt to showcase the full potential of the DVCC and is by no means a comparison with the existing implementations, which employ fewer transistors.

The working of the AND gate of Figure 23 can be understood as follows. As long as V_2 is low, transistor M_1 remains off leading to $I_X = 0$ (irrespective of the voltage applied at Y_1 terminal) which causes I_Z and consequently V_{AND} to be zero. When V_2 becomes high, transistor M_1 conducts and the current I_X will be governed by the presence or absence of V_1. Under such a condition, for the case of V_1 being low (zero), I_X will be zero and consequently, V_{AND} is low. For the case when V_1 is high, the output voltage V_{AND} will be high. The input and output for the AND gate are related as

$$V_{AND} = \begin{cases} \frac{1}{2}\mu_n C_{ox}\frac{W_1}{L_1}R_1(V_2 - V_t)^2, \; V_1, V_2 \rightarrow HIGH \\ 0, \; Otherwise \end{cases}$$

Similarly, the circuit for OR/NOR gates can be obtained if the Y_3 terminal of the DVCC is also utilized [13]. An analysis similar to that done for the AND/NAND circuit shows that the circuit of Figure 24 does indeed works as an OR/NOR gate. It may be mentioned that V_{OR}/V_{NOR} may be needed to be passed through an additional comparator if a binary digital output is required.

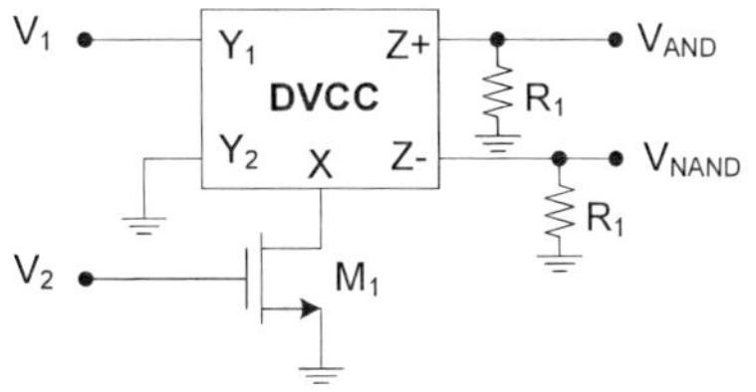

Figure 23. AND/NAND gate realization using DVCC

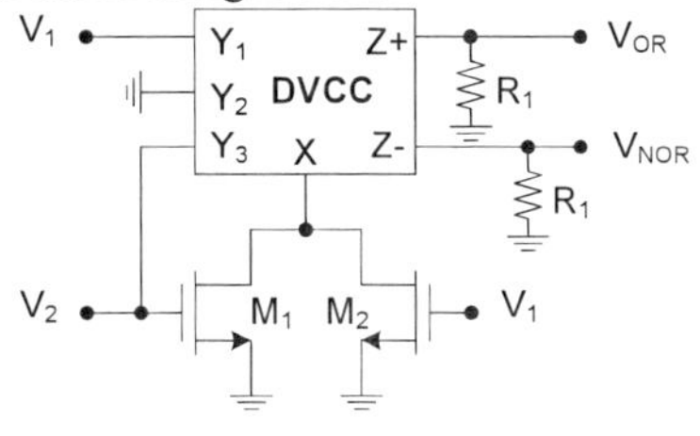

Figure 24. OR/NOR gate realization using DVCC

IV. SIMULATION RESULTS

This section deals with the simulation results for the proposed circuits. Simulations were carried out using PSPICE program. The circuit for DVCC shown in Figure 4 was utilized and standard 0.5 micron CMOS parameters were used for simulation purposes. For the opamp, use was made of the LMC7101A CMOS opamp from National Semiconductor. The sub-circuit file for this opamp is available in Orcad Model Library.

The proposed circuit for solving linear equations was tested for solving the following set of 2 simultaneous linear equations in 2 variables.

$$\begin{bmatrix} 2 & 1 \\ 1 & 1 \end{bmatrix} \begin{bmatrix} V_1 \\ V_2 \end{bmatrix} = \begin{bmatrix} 2 \\ 1.5 \end{bmatrix} \tag{86}$$

For the above set of equations, the required weighted currents that need to be generated are given as: $s_{11} = 2$, $s_{12} = 2$, $s_{21} = 1$ and $s_{22} = 1$. Scaling of the Z-port currents was done by varying the aspect ratios of the transistors in the output stage of the DVCC. The circuit used for PSPICE simulation is presented in Figure 25. As can be seen, some additional circuitry is required to generate the proper input voltages of the comparators [30]. The values of the different resistances shown in Figure 23 are: $R_{e11} = 2K$; $R_{e12} = 4K$; $R_{e13} = 4K$; $R_{e21} = 3K$; $R_{e22} = 3K$; $R_{e23} = 3K$.

PSPICE simulations were performed for the chosen set of equations and the results are given in Figure 26. As can be verified from the figure, the simulation results are found to agree exactly with the algebraic solution.

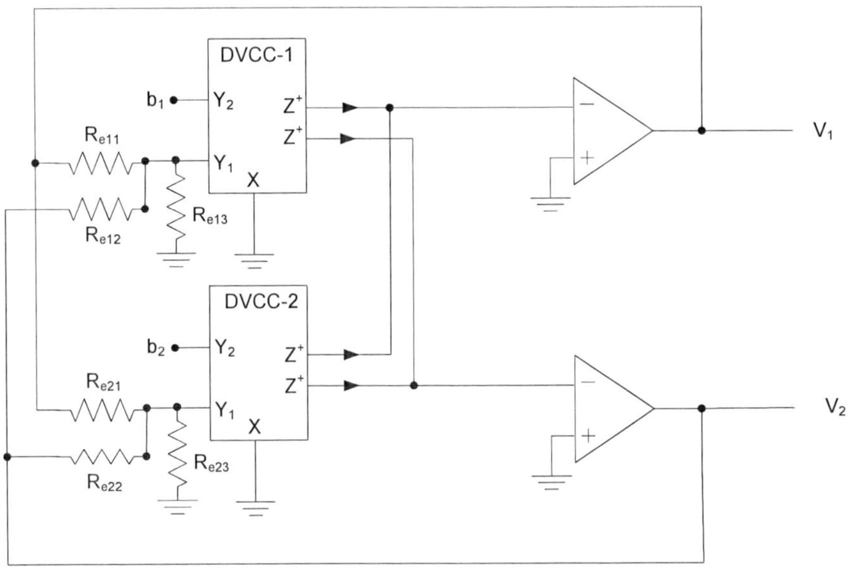

Figure 25. The proposed circuit applied to solve 2 linear equations in 2 variables

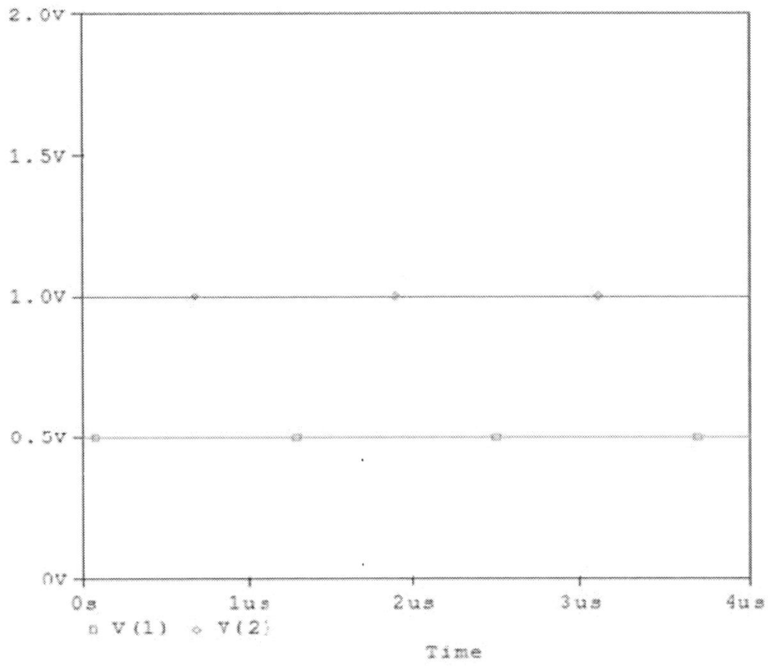

Figure 26. PSPICE simulation results of the circuit to solve (86)

Next, the proposed network for solving linear programming problems was tested using PSPICE simulation. The circuit was used to minimize the objective function

$$5V_1 + 2V_2 \tag{87}$$

subject to

$$\left.\begin{aligned} V_1 + V_2 &\geq 1 \\ V_2 &\leq 1 \end{aligned}\right\} \tag{88}$$

The values of resistances acting as the weights on the neurons are obtained from (69, 71). Towards that end, first the value of g was computed for the DVCC which was found to be 1.091 milli-mhos. Choosing $K = {}^{1}/_{g}$ allows us to fix the current scaling factors as given below:

$$s_{11} = a_{11};\ s_{12} = a_{12};\ s_{21} = a_{21};\ s_{22} = a_{22}$$

Routine mathematical analysis of (87, 88) yields: $V_1 = 0$, $V_2 = 1$. The results of PSPICE simulation are presented in Figure 27. From the plots of the neuron output voltages, it can be seen that V(1) = 0 V and V(2) = 1 V which correspond exactly to the algebraic solution thereby confirming the validity of the approach.

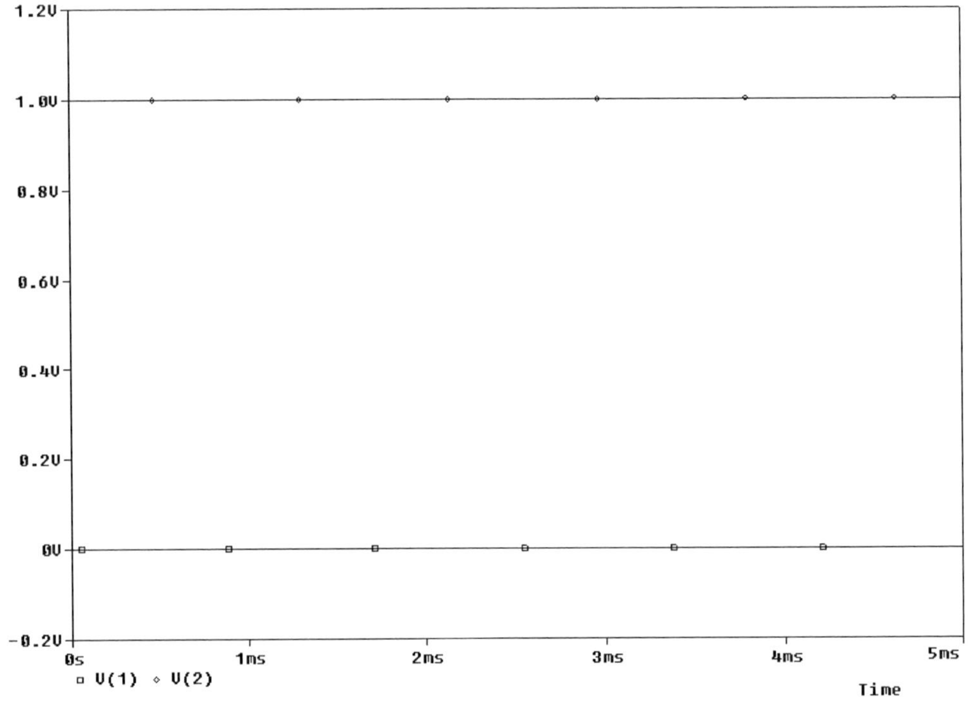

Figure 27. Simulation results for the proposed circuit for solving linear programming problems (87, 88)

The proposed neural circuit for solving quadratic programming problems was tested by minimizing the objective function

$$3V_1^{\,2} + 4V_1V_2 + 5V_2^{\,2} \tag{89}$$

subject to

$$\left.\begin{aligned} V_1 - V_2 &\leq -1 \\ V_1 + V_2 &\leq 1 \end{aligned}\right\} \tag{90}$$

The values of resistances acting as the weights on the neurons are obtained from (85). The corresponding values of the weight resistances are found to be: R_{11} =305.53Ω; R_{21} =229.14Ω; R_{12} =229.14Ω; R_{22} =183.318Ω. The values of the current scaling factors were found out to be: $s_{11} = 1$; $s_{12} = -1$; $s_{21} = 1$; $s_{22} = 1$.

Routine mathematical analysis of (89, 90) yields: V_1= -0.584, V_2= 0.416. The results of PSPICE simulation are presented in Figure 28. From the plots of the neuron output voltages, it can be seen that V(1) = -0.58 V and V(2) = 0.41 V which are very near to the algebraic solution thereby confirming the validity of the approach.

Next, the AND gate of Figure 23 was verified. The results of simulation are presented in Figure 29 where the voltages V_1 and V_2 are the inputs to the AND gate and voltage V_4 is the AND-ed output.

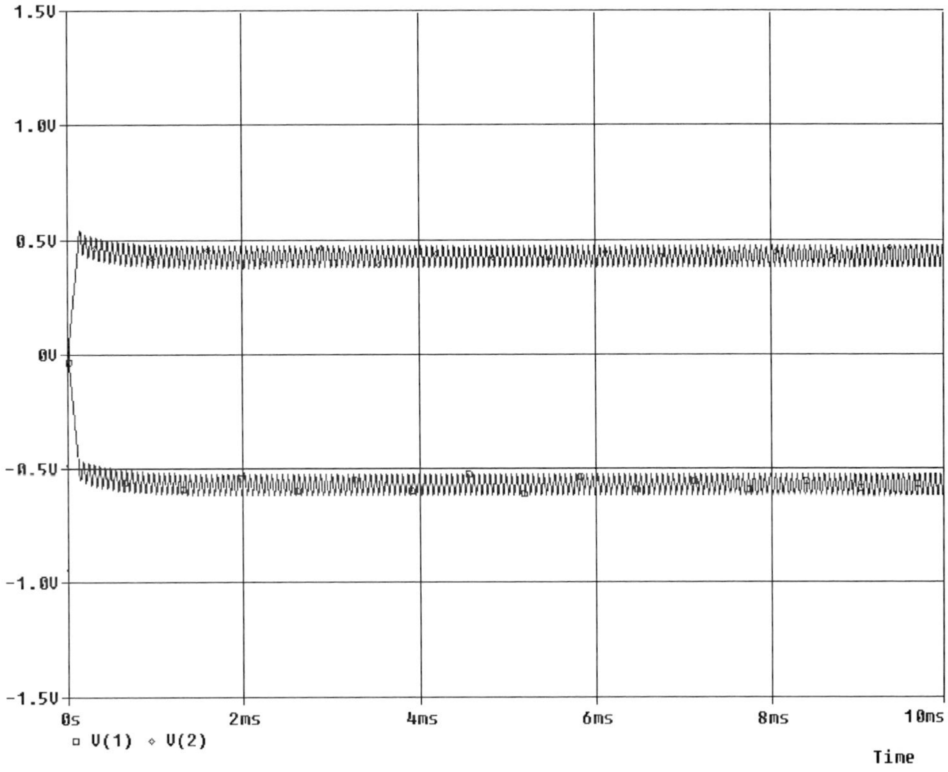

Figure 28. Simulation results for the chosen 2 – variable problem (89, 90)

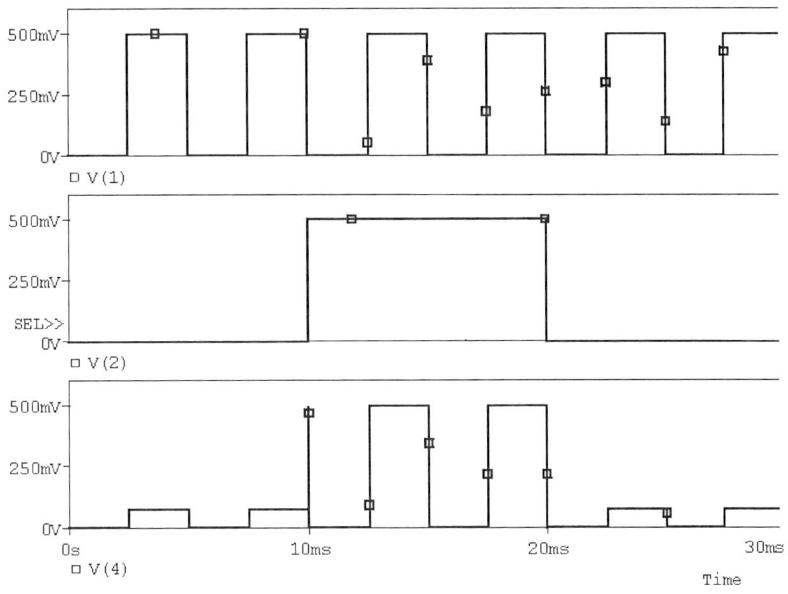

Figure 29. Simulation results for the proposed AND gate using DVCC

An important point that needs to be discussed is the VLSI implementability issue of the existing and proposed CMOS circuits. It is well known that in standard CMOS technology, processing variations prohibit the fabrication of individual resistances with precise values. It is, however, possible to fabricate a pair of resistors having the same value, although the exact value may not necessarily be predicted. Considering this fact, one way of implementing resistances in CMOS analog circuits is to tweak the design such that the critical (performance deciding) factors are governed by resistance-ratios [40]. Wherever this is not possible, use may be made of MOS-based resistors. The function of a resistor may be obtained using MOSFETs to realize voltage-controlled transconductances or by linearizing the transistor's region of operation in the triode region [41].

CONCLUSION

This chapter was based on circuit applications of differential voltage current conveyor in CMOS technology. A survey of existing linear and non-linear circuits was presented. Some new applications of DVCC were also proposed. The chapter explores a wide variety of analog and a few digital applications and is expected to both provide a deep insight into the subject and also further enhance the existing knowledge.

REFERENCES

[1] Smith, K. C. & Sedra, A. S. (1968). "The current-conveyor-a new circuit building block," *IEEE Proc.*, *Vol. 56*, Pages 1368-69.

[2] Maheshwari, S. & Khan, I. A. (2005). "Current controlled third order quadrature oscillator," *IET Circuits Devices Syst.*, *Vol. 152*, No. 6, Pages 605-607.

[3] Khan, I. A. & Ahmed, M. T. (1986). "Electronically tunable first-order OTA-capacitor filter sections," *Int. J. of Electronics*, *Vol. 61*, No. 2, Pages 233-237, August.

[4] Biolek, D. (2003). "CDTA–Building block for current-mode analog signal processing", *Proc. ECCTD Poland*, Pages 397-400.

[5] Minaei, S. & Yuce, E. (2008). "A new full-wave rectifier circuit employing single dual-X current conveyor," *Int. J. of Electronics*, *Vol. 95*, No. 8, Pages 777-784, August.

[6] Pal, K. (1989). "Modified current conveyors and their applications," *Microelectronics Journal*, *Vol. 20*, No. 4, Pages 37-40.

[7] Elwan, H. O. & Soliman, A. M. (1997). "Novel CMOS differential voltage current conveyor and its applications," *IEE Proc. Circuits Devices Syst.*, *Vol. 144*, No. 3, Pages 195-200, Jun.

[8] Moradzadeh, H. & Azhari, S. J. (2008). "High performance low-voltage QFG-based DVCC and a novel fully differential SC integrator based on it," *IEICE Electronics Express*, *Vol 5*, No. 23, Pages 1017-1023, Dec.

[9] Mahmoud, S. A. (2008). "Low Voltage Wide Range CMOS Differential Voltage Current Conveyor and its Applications," *Contemporary Engineering Sciences*, *Vol. 1*, No. 3, Pages 105-126.

[10] Minaei, S. & Ibrahim, M. A. (2005). "General configuration for realizing current-mode first-order all-pass filter using DVCC," *Int. J. of Electronics*, *Vol. 92*, No. 6, Pages 347-356.

[11] Hu, J., Xu, T., Zhang, W. & Xia, Y. (2005). "A CMOS Rail-to-Rail Differential Voltage Current Conveyor and its Applications," IEEE Proc. ICCCAS, *Vol. 2*, Pages 1079-1083.

[12] Siripruchyanan, M. & Jaikla, W. (2007). *"Floating Capacitance Multiplier Using DVCC and CCCIIs,"* Proc. Int. Symp. on Comm. and Info. Tech. (ISCIT '07).

[13] Hassan, T. M. & Mahmoud, S. A. (2010). "New CMOS DVCC realization and applications to instrumentation amplifier and active-RC filters," *Int. J. Electronics and Communications*, *Vol. 64*, Pages 47-55.

[14] Maheshwari, S. (2009). "Analogue signal processing applications using a new circuit topology," *IET Circuits Devices Syst.*, *Vol.3*, No.3, Pages 106-115.

[15] Ibrahim, M. A., Minaei, S. & Kuntman, H. (2005). "A 22.5 MHz current-mode KHN-biquad using differential voltage current conveyor and grounded passive elements," *Int. J. Electronics and Communications*, *Vol. 59*, No. 5, Pages 311-318, Jul.

[16] Minaei, S. & Temizyurek, C. (2003). "Dual Input All-Pass Filter using DVCC," *Proc. Int. Symposium on Signals, Circuits and Systems (SCS)*, *Vol. 2*, Pages 477-480.

[17] Maheshwari, S. (2007). "Novel cascadable current-mode first order all-pass sections," *Int. J. Electronics*, *Vol. 94*, No. 11, Pages 995-1003, Nov.

[18] Maheshwari, S. (2008). "CMOS Compatible Single Input Five/Six Outputs First Order Filter Circuits," *J. Active Passive Elec. Dev.*, *Vol. 3*, Pages 125-133.

[19] Maheshwari, S. (2008). "A Canonical Voltage-Controlled VM-APS with a Grounded Capacitor," *Circuits, Systems, and Signal Processing*, *Vol. 27*, No. 1, Pages 123-132, Feb.

[20] Ibrahim, M. A., Minaei, S. & Kuntman, H. (2006). "DVCC based differential-mode all-pass and notch filters with high CMRR," *Int. J. Electronics*, *Vol. 9*, No. 4, Pages 231-240, Apr.

[21] Chen, H. P. & Shen, S. S. (2007). "A Versatile Universal Capacitor-Grounded Voltage-Mode Filter Using DVCCs," *ETRI Journal*, *Vol. 29*, No. 4, Pages 470-476, Aug.

[22] Soliman, A. M. (2008). "Generation and classification of Kerwin–Huelsman–Newcomb circuits using the DVCC," *Int. J. Circ. Theor. and Appl.*, *Vol. 37*, Pages 835-855, Jun.

[23] Jiun-Wei Horng et al, (2006). "High Input Impedance Voltage-Mode Universal Biquadratic Filter with One Input and Five Outputs Using Current Conveyors," *Circuits, Systems, Signal Processing*, *Vol. 25*, No. 6, Pages 767-777.

[24] Gupta, S. S. & Senani, R. (2000). "Grounded-Capacitor Current-mode SCRO: Novel application of DVCC," *Electronics Letters*, *Vol. 36*, No. 3, Pages 195-196.

[25] Kumar, P., Keskin, A. U. & Pal, K. (2007). "DVCC-Based Single Element Controlled Oscillators Using All-Grounded Components and Simultaneous Current-Voltage Mode Outputs," *Frequenz*, *Vol. 61*, Pages 141-144.

[26] Khan, I. A. & Beg, P. (2009). "Fully Differential Sinusoidal Quadrature Oscillator Using CMOS DVCC," *Proc. Int. Conf. On Communication, Computer and Power (ICCCP'09)*, Muscat, Pages 196-198, Feb.

[27] Maheshwari, S. (2009). "Quadrature oscillator using grounded components with current and voltage outputs," *IET Circuits Devices Syst.*, *Vol.3*, No.4, Pages 153-160.

[28] Gustafson, J. L. (2006). "The Quest for Linear Equation Solvers and the Invention of Electronic Digital Computing," *Proc. JVA'06, IEEE John Vincent Atanasoff 2006 International Symposium on Modern Computing.*

[29] Xia, Y., Wang, J. & Hung, D. L. (1999). "Recurrent Neural Networks for Solving Linear Inequalities and Equations," *IEEE Trans. Circuits and Systems–1: Fundamental Theory and Applications*, *Vol. 46*, No. 4, Pages 452-462, April.

[30] Ansari, M. S. & Rahman, S. A. (2009). "A Novel Current-mode Non-Linear Feedback Neural Circuit for Solving Linear Equations," *Proc. IMPACT – 2009*, Aligarh, India, Pages 284 – 287, Available on IEEE *Xplore*, DOI: 10.1109/MSPCT.2009.5164189

[31] Hopfield, J. J. & Tank, D. W. (1985). "'Neural' computation of decisions optimization problems," *Biological Cybernetics*, *Vol. 52*, Pages 141-152.

[32] Tank, D. W. & Hopfield, J. J. (1986). "Simple neural optimization networks: an A/D converter, signal decision network, and linear programming circuit," *IEEE Trans. Circ. Syst*, *Vol. 33*, Pages 533-541.

[33] Rahman, S. A. Jayadeva, & Dutta Roy, S. C. (1999). "Neural network approach to graph colouring," *Electronics Letters*, *Vol. 35*, Pages 1173-1175.

[34] Wang, J. (1993). "Analysis and Design of a Recurrent Neural Network for Linear Programming", *IEEE Trans. Circuits and Systems–1: Fundamental Theory and Applications*, *Vol. 40*, No. 9, Pages 613-618, Sept.

[35] Xia, Y. (1995). "A New Neural Network for Solving Linear Programming Problems and its Application," *IEEE Trans. Neural Networks*, *Vol.7*, No.2, Pages 94-103, March.

[36] Malek, A. & Yari, A. (2005). "Primal–dual solution for the linear programming problems using neural networks," *Applied Mathematics and Computation*, *Vol. 167*, Pages 198-211.

[37] Kennedy, M. P. & Chua, L. O. (1988). "Neural networks for nonlinear programming," *IEEE Trans. Circ. Syst.*, *Vol. 35*, Pages 554-562.

[38] Wu, X. Y., Xia, Y. S., Li, J. & Chen, W. K. (1996). "A high performance neural network for solving linear and quadratic programming problems," *IEEE Trans. Neural Networks*, *Vol. 7*, No. 3, Pages 643-651.

[39] Malek, A. & Alipour, M. (2007). "Numerical solution for linear and quadratic programming problems using a recurrent neural network," *Appl. Math. Computation*, *Vol. 192*, Pages 27-39.

[40] Uyemura, J. P. (2006). "Chip Design for Submicron VLSI: CMOS Layout and Simulation," *Cengage Learning*, First Ed., Pages 381-382.

[41] Mahmoud, S. A., Elwan, H. O. & Soliman, A. M. (1997). "Generation of CMOS Voltage-Controlled Floating Resistor," *Microelectronics Journal*, *Vol. 28*, Pages 627-640.

In: CMOS Technology
Editor: Min-jun Kwon, pp.135-158
ISBN: 978-1-61761-325-8

Chapter 4

COMPACT MODELING OF MULTI-GATE MOSFET INCLUDING HOT-CARRIER EFFECTS

T. Bentrcia[1] and F. Djeffal[2,*]
[1]Department of Physics, University of Batna, 05000, Algeria
[2]LEA, Department of Electronics, University of Batna, 05000, Algeria
[2]LEPCM, University of Batna, 05000, Algeria

1. INTRODUCTION

A global view of today's world permits to notify that many hardware and software based systems, are characterized by complex behavior and often have special features and structures requiring high complexity approaches for modeling [1, 2]. In such situation, this may result in many unwanted phenomena disturbing the correct function particularly when taking into account the VLSI constraints. In order to avoid the heavy costs associated with maintenance and diagnostic operations [3], a deep understanding of the degradation process should be established so that the reliability of such systems can be enhanced significantly. Despite that the recent progress in different fields has made resulted products more powerful than ever, the increasing level of complexity implied that reliability problems not only will continue to exist, but also are likely to require ever more innovative solutions [4].

The demand for high-performance devices has been experiencing a steady rise in the sophistication degree of semiconductor manufacturing techniques, which in turn have driven the associated Metal Oxide Semiconductor Field Effect Transistor (MOSFET) size close to their physical limits. Theoretical analyses make it clear that we should be able to rearrange atoms and molecules one by one to get more efficient tools at a tiny scale [5]. Therefore, the initial stage consists in the seek of new design approaches for devices miniaturization scheme, the adopted approach based on keeping the internal electric fields constant had to be abandoned due to several practical reasons including [6]:

Loss of compatibility with previous generation circuits,

* Corresponding author: E-mail: faycaldzdz@hotmail.com Tel/Fax: 0021333805494.

Decrease in noise margins because of the non-scaling of the threshold voltage and the subthreshold slope,

Decrease in operating speeds in sub-micron devices due to the non-scaling of parasitic capacitances.

As a result, the design of alternate scaling schemes becomes an inevitable choice to remedy the previous cited disadvantages. In these schemes, despite the obvious correlation between the consequent increase in internal electric fields in aggressively scaled MOSFETs and the benefit of increased carrier velocities leading to a considerable amelioration in switching speed, it also presents a major reliability concern for the long term operation. As device feature size moves into deep submicron region, the benefits of higher electric fields saturate while the corresponding reliability problems get worse [7]. Such tradeoff between both criteria has initiated a tremendous race towards developing strategies providing guidelines for devices downscaling including degradation effect, which is reflected by a shift in the main measured parameters describing the device operating under different regimes circumstances.

We will start this chapter by reviewing the various types of defects existing in CMOS-based devices. We will then present a surface-potential-based approach for the analytical modeling of main device parameters in subthreshold regime and later we will develop accurate models to handle the device operation in saturation regime. The obtained results provide comprehensive description of multigate MOSFETs behavior including hot-carriers effects, and offer new insights for future CMOS-based circuits design.

2. Defects in CMOS-Based Devices

In spite of the effort paid up to now by the scientific community concerning Silicon and silicon based devices widely used in many fields of physics, there still remain a lot of aspects not clarified, related to the behavior of impurities, and a global understanding of their local structure and properties became increasingly important due to the reduction in chip sizes and to the increase of the operation speed [8].

Since the reliability concepts are the most popular framework allowing the good description of the system life cycle based on failures occurrence; which is mainly caused by the original existence or generation of defects in the device, it would be of great utility to analyze different evolution stages. The reliability of semiconductor devices is represented by the failure rate curve known as the "bathtub curve", depicted in figure 1. The curve can be divided into three regions [9]:

Initial failures are considered to occur when a latent defect is formed, then becomes manifest under the stress of operation. Generally, the period of initial failures is defined as the first six months to one year of a product cycle, the failure rate tends to decrease with time because only devices having latent defects will fail, and these devices are gradually removed.

Random failures in which the failure rate is constant, such mode covers soft errors, electrical noise, electrostatic discharge, and other problems. All of these problems will occur at random under the stress of external factors.

Wear-out failures occur due to the ageing of devices from wear and fatigue. This period indicates the life of different failure modes of semiconductors such as Hot carriers, electromigration and time dependant dielectric breakdown.

In the screening process, if we assume a certain length of time (t_0) before device shipment, the lifetime can be forecasted by the probability density function. With the Weibull distribution density function [10], the rate of accumulated failures (t: t_0) available after screening becomes:

$$F(t_0 : t) = 1 - \exp\left(-\frac{(t_0 + t)^m - t_0^m}{\eta^m}\right) \quad (1)$$

where m and η are scale and shape parameters of the Weibull distribution.

Therefore, it's possible to estimate the device lifetime basing on probabilistic exploitation of such distribution density function. The reliability based analysis highlights the necessity to improve the device lifetime through the enlargement of the random failures interval, where the failure rate is approximately constant. Hence, it's a mandatory requirement to assimilate existing defects in both extreme regions, so that abusive effects could be eliminated or at least reduced.

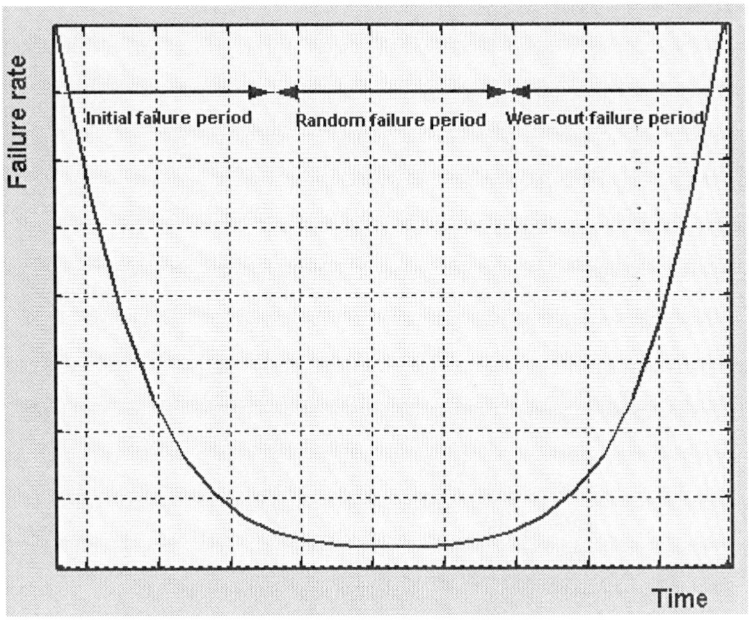

Figure 1. Failure rate curve (Bathtub Curve) of semiconductors devices

2.1. Bulk Defects

This type of defects is closely related to the bulk structure of a semiconductor, in the sense that the trap defects are distributed over the whole volume. The resulted effect on the behavior of future produced devices has been studied intensively in literature. However, with the advance of MOSFETs technology at nanoscale level, the gap becomes more apparent and additional efforts then before should be focused on the subject.

2.1.1. Origin of the problem

The availability of semiconductor materials and their use as a basis of nanodevices have bring to the front many disadvantages related from a part to the inherit structure and from another part to production process constraints. In single crystal semiconductors, deep level used impurities have usually a metallic nature, but they can be crystal imperfections (dislocations, precipitates, vacancies, and interstitials) [11, 12]. Such defects are in general avoidable, but occasionally they are intentionally introduced to affect a device characteristic (switching time, resistivity). For amorphous semiconductors, defects are mainly due to structural imperfections. Manufacturing environment conditions are also a key medium altering the concentration of existing impurities in the semiconductor; methods used for crystal growth such as epitaxy can give rise to deep level traps. Another recognized cause concerns several properties of employed resources; the higher static charge levels in a semiconductor clean room than in a conventional room. In fact, it is not uncommon for a clean room with no safeguards against static charge to achieve levels of tens of kilovolts on a variety of critical surfaces [13].

The existence of foreign atoms or crystal defects perturbs the periodicity of the single crystal, hence resulting in discrete energy levels into the band gap, as illustrated in figure 2. They can be considered as recombination centers when there are excess carriers, and as generation centers when the carrier density is below its equilibrium value.

The thermodynamics of defects plays an important role in the deduction of some specific aspects, the thermodynamic parameters including band gaps, energies of defect formation and migration, are not the Eigen values of a Schrödinger equation describing the crystal. Such parameters are defined statistically in terms of reactions occurring among ensembles of all possible configurations of the system.

The governing law of the equilibrium concentration of point defects introduced into a perfect semiconductor crystal is given for a neutral point defect X^0 by the expression [14]:

$$\frac{[X^0]}{[S]} = \theta_{X^0} \exp\left[\frac{-G^f_{X^0}}{kT}\right] \tag{2}$$

where $[S]$ is the concentration of available lattice sites in the crystal, θ_{X^0} is the number of degrees of internal freedom of the defect on a lattice site, and $G^f_{X^0}$ is the standard Gibbs free energy. The parameters k and T respectively represent Boltzmann's constant and temperature.

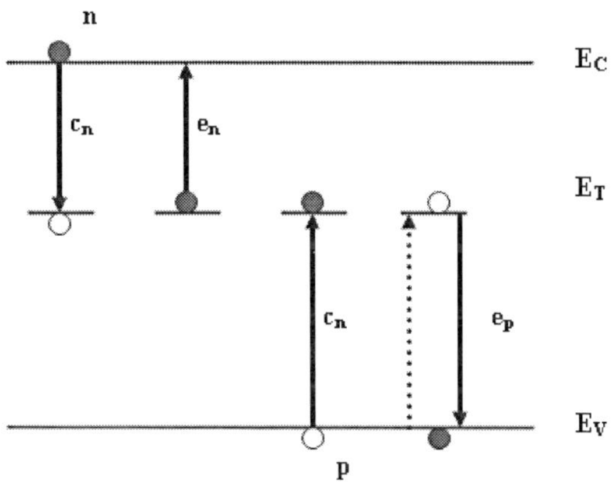

Figure 2. Electron energy band diagram for a semiconductor with deep-level impurities in addition to capture and emission process

In order to get an accurate model of the real defects equilibrium inside the bulk semiconductor, the enthalpy of formation need not refer only to elementary point defects such as interstitial, but also supports more complicated defect cluster formation encompassing a wide variety of defects including the divacancy, diinterstitial and vacancy-dopant pair, where we find in literature numerous methods for calculating the formation enthalpy of a defect pair and more progress still take place [15, 16].

2.1.2. Characterization methods

In order to study defect properties in CMOS-based structures, several experimental techniques have been used. Some of these approaches can be extended to deal with new materials and organic semiconductors [17]. There is no universal approach suited to all situations, so continual improvement consists a major concern. However, two basic methods have been used abundantly in experimental applications [18], the first on is deep level transient spectroscopy (DLTS) developed by Lang in 1974, it exhibits its highest performances under ideal circumstances where it can reach defect concentrations as low as 6-10 times the carrier concentration of the semiconductor. By mean of this technique, it's also possible to identify radiation induced defects.

The key feature of DLTS consists in the limitation of emission rate so that only values of the capacitance transient within this range are extracted by the measurement apparatus. With the temperature variation, a peak in the instrument response is observed at the temperature associated with the trap emission rate. The sensitivity of DLTS when used in the capacitive mode decreases linearly with the distance towards the junction. Small uncertainties in the determination of the position of a DLTS peak can cause significant errors in the evaluation of the parameters of electronic states deep within the band gap; it is difficult to distinguish defects that are very close in energy.

Although DLTS is spectroscopic in nature giving trap energies, it has some undesirable properties: namely the large effort required in sample preparation and the limitation to samples of certain doping concentrations [19].

The second more versatile technique known as positron annihilation spectroscopy (PAS), is the spectroscopy of gamma (γ) rays emerging from the annihilation of positrons and electrons. The method is based on using the energy and momentum conservation during the annihilation process to study solids taking the advantage of parameters sensitivity to lattice imperfections. When a positron is trapped in an open-volume defect, such as vacancies and dislocations, the annihilation parameters are modified in a characteristic way. Its lifetime increases due to the lower electron density. Momentum conservation leads to a small angular spread of the collinear γ –quanta or a Doppler shift of the annihilation energy [20]. Most positron lifetimes for the important semiconductors and lifetimes for various vacancy-type defects have been experimentally determined. Neutral and negative vacancy-type defects, as well as negative ions, are the dominant positron traps in semiconductors. Temperature-dependent lifetime measurements may distinguish between both defect types.

The intricate challenge facing the practical implementation of the method is the high costs of equipment not readily available to most researchers.

2.2. Hot Carrier

In the last years, numerous studies have been revised and enhanced to get more adequate models regarding the hot carrier induced device degradation. In fact, the phenomenon has been observed or implemented under special forms in some types of Random Access Memory (RAM) cells [21, 22]. With the recent trends in nanodevices manufacturing, condensed efforts are invested to gain higher immunity performances without altering the optimal function of the device.

2.2.1. Origin of the problem

Under the bias conditions of a MOSFET structure, large electric fields are created along the device channel, as a result, carriers gain sufficiently high energies that allow their injection into the surrounding dielectric films such as the gate and sidewall oxides. The accumulation of mobile carriers at interfaces over long periods of operation can dramatically trigger many harmful physical processes causing the circuit in the worst case to fail. The hot carrier degradation process in a DG MOSFET device generates a second region having distinct properties influencing the behavior of the device as shown in figure 3.

A more vigilant analysis indicates that the process takes initially place near the drain side where a drain substrate junction is created due to the formation of a pinch-off region. The lateral electric field exhibits a sharp peak in this region increasing with the shrink of channel length, as stated by figure 4. The tunneling effect caused by the application of high drain biases, results in carriers penetration through the considered region and exhibition of non-equilibrium energy distributions. As a consequence, the concentration of energetic carriers available for injection into the oxide requires the accurate calculation of electric field distribution in silicon because of the strict dependency between both quantities.

In both n and p-channel MOSFETs, vulnerabilities towards the injection of hot carriers have been detected, electrically active defects are generated in the oxide and at the Si-SiO_2 interface [23, 24, 25]. At advanced stages of the process, the energy gained by the carriers in the high field region of the silicon substrate induces the break of bonds associated with extrinsic or intrinsic defects in the oxide, and such rearrangement in its atomic structure is the generator factor of the device instabilities observed during the hot-carrier injection.

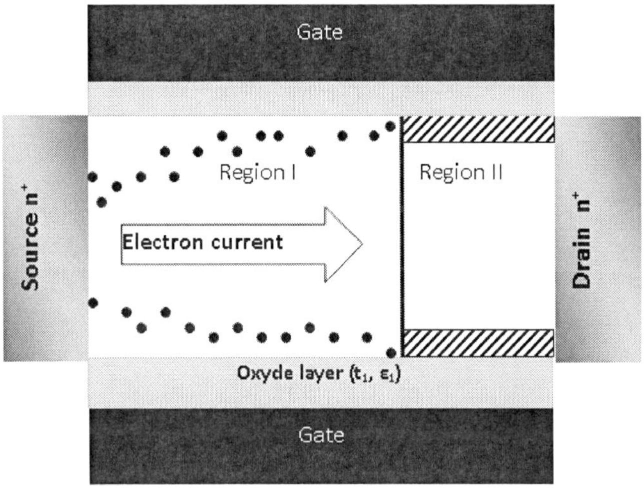

Figure 3. Creation of a damaged region in the DG MOSFET structure because of hot-carrier degradation process

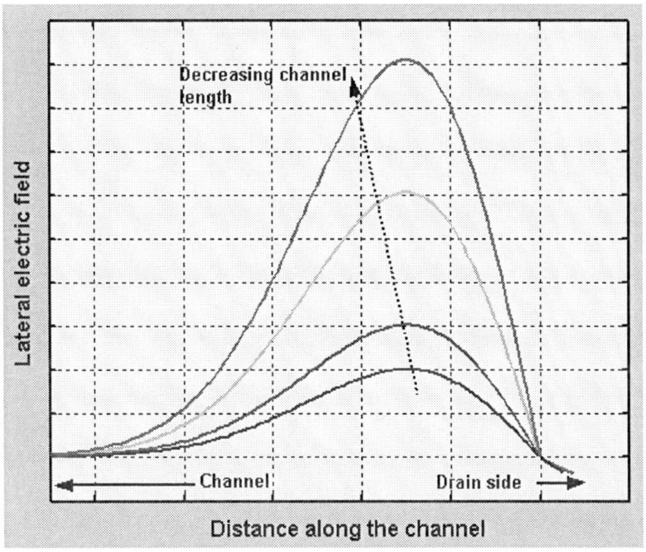

Figure 4. Maximum lateral electric field at the Si-SiO_2 interface along the channel in MOS transistors with decreasing channel lengths

2.2.2. Device Parameters degradation

The quantification of damages caused by hot-carriers can be measured basing on the shift in some pertinent device level parameters such as the threshold voltage, transconductance and drain current gathered by establishing characterization measurements before and after hot carrier injection. Despite that similar techniques are widely used for damages resulted from other stress conditions such as Fowler-Nordheim injection, and radiation exposure, but the hot-carriers degradation differs by its highly localized profile [26, 27]. The multitude advanced characterization techniques exploited in many research works provide us with a clear scheme of the nature of hot-carriers induced device degradation. However, the powerful simulation tools are becoming a promising solution not for dethroning the characterization experiments but to gain in conjunction with existing ones better insights into the physical processes responsible for hot-carrier induced parameters shifts.

3. Modeling of Multi-Gate Mosfets Including Hot-Carriers Effects

Multi-gates MOSFETs devices are considered currently as the workhorses of the next revolutionary miniaturization process, due to their good short-channel-effects immunity, rendering their usage a potential solution for future applications. However, when the channel length shrinks down, it results in more severe damages related to hot-carrier degradation effect, which is likely to become even more critical in coming-generation chips.

3.1. Hot-Carriers Effects in Subthreshold Regime

The subthreshold region of a MOSFET device operation can be seen as the smooth transition from weak to strong inversion. The importance of device current in such region is issued from the fact that it is a leakage current determining CMOS standby power. The device behaviour changes from square law to exponential when gate voltage approaches the threshold value, resulting in the domination of drift current. In the following section, we discuss the influence of hot-carriers effect on the main parameters characteristics of the subthreshold region for different prototypes of Multi-gate MOSFET structures, analytical models are developed and some simulation results are interpreted in order to deduce dominance criteria in term of immunity against hot-carriers degradation.

3.1.1. Surface potential

Consider the DG and GAA MOSFETs structures where a step-function interface charge distribution region is present near the drain side with length L_d=L-L_1, L represents the (DG/GAA) MOSFETs channel, L_1 is the fresh region length, our initial step in the determination procedure of the surface potential passes necessarily by the resolution of the 2-D Poisson equation governing the channel electrostatics potential Ψ for both structures:

$$(DG) \quad \frac{\partial^2\psi(x,y)}{\partial x^2}+\frac{\partial^2\psi(x,y)}{\partial y^2}=\frac{qN_A}{\varepsilon_{Si}} \tag{3}$$

$$(GAA) \quad \frac{\partial^2\psi(x,r)}{\partial r^2}+\frac{1}{r}\frac{\partial\psi(x,r)}{\partial r}+\frac{\partial^2\psi(x,r)}{\partial x^2}=\frac{qN_A}{\varepsilon_{Si}} \tag{4}$$

where ε_{si} is the silicon permittivity, q the electron charge and N_A the doping level of the channel.

The boundary conditions for Ψ are found for considered structures by satisfying continuity of both the normal component of the electric displacement at the (Si/SiO_2) interfaces, and the potential at the source/drain sides [28].

Following the assumption that the potential distribution along the silicon film has a parabolic profile given by the general form :

$$\psi(x, y) = a(x) + b(x)y + c(x)y^2 \tag{5}$$

Once, the symmetry conditions are exploited, the 2-D channel potential for both structures are obtained,

$$\psi(x, y) = \psi_s(x) + \frac{\varepsilon_{ox}}{\varepsilon_{Si}} \frac{\psi_s(x) - V_g^*}{t_{ox}} y - \frac{\varepsilon_{ox}}{\varepsilon_{Si}} \frac{\psi_s(x) - V_g^*}{t_{ox} t_{Si}} y^2 \tag{6}$$

$$\psi(x, r) = \left(1 + \frac{C_{ox} t_{Si}}{4\varepsilon_{Si}}\right)\psi_s(x) - \frac{C_{ox}}{4} \frac{t_{Si}}{\varepsilon_{ox}} V_g^* + \frac{C_{ox}}{t_{Si}\varepsilon_{Si}} \left[V_g^* - \psi_S(x)\right] r^2 \tag{7}$$

Here ε_{ox} is the oxide permittivity, t_{ox} and t_{si} are the thicknesses of the oxide layer and the silicon channel, respectively. C_{ox} denotes the oxide capacitance having expression depending on the considered device, and V^*_g represents the normalized gate voltage.

Upon substituting (6) in (3) and (7) in (4), we obtain the differential equation that deals only with surface potential Ψ_s as follows:

$$\frac{d^2\psi_s(x)}{dx^2} - \frac{1}{\lambda^2_{(DG/GAA)}} \psi_s(x) = D_{(DG/GAA)} \tag{8}$$

with

$$\lambda_{DG} = \sqrt{\frac{\varepsilon_{si} \cdot t_{ox} \cdot t_{si}}{2\varepsilon_{ox}}} \quad \text{and} \quad D_{DG} = \frac{qN_A}{\varepsilon_{si}} - \frac{1}{\lambda^2_{DG}} V_g^*$$

$$\lambda_{GAA} = \sqrt{\frac{\varepsilon_{si} \cdot t_{ox} \cdot t_{si}}{4\varepsilon_{ox}}} \quad \text{and} \quad D_{GAA} = \frac{qN_A}{\varepsilon_{si}} - \frac{1}{\lambda^2_{GAA}} V_g^*$$

At this level of modelling, we have to discriminate between damaged and fresh regions by introducing additional term and boundary condition that express the hot-carrier degradation effect. In fresh region ($0 \le x \le L_1$) the equation to be solved is,

$$\frac{d^2\psi_s(x)}{dx^2} - \frac{1}{\lambda^2_{(DG/GAA)}} \psi_s(x) = D_{(DG/GAA)} \tag{9}$$

Subject to the boundary conditions

$$\psi(0,y)=V_{bi}$$

$$\psi(L_1,y)=V_p$$

V_{bi} is the junction voltage between the source/drain and intrinsic silicon and V_p is the boundary potential among both regions

The resolution of such differential equation leads us to the surface potential form:

$$\psi_{s1}(x)=-\lambda^2_{(DG/GAA)}D_{1(DG/GAA)}+\frac{\phi_{D1(DG/GAA)}\sinh\left(\frac{x}{\lambda_{(DG/GAA)}}\right)-\phi_{s1(DG/GAA)}\sinh\left(\frac{x-L_1}{\lambda_{(DG/GAA)}}\right)}{\sinh\left(\frac{L_1}{\lambda_{(DG/GAA)}}\right)} \tag{10}$$

with $\phi_{D1(DG/GAA)}=V_{p(DG/GAA)}+\lambda^2_{(DG/GAA)}D_{1(DG/GAA)}$ and $\phi_{s1(DG/GAA)}=V_{bi}+\lambda^2_{(DG/GAA)}D_{1(DG/GAA)}$

In damaged region ($L_1\le x\le L$), by following the same methodology as above for the following problem,

$$\frac{d^2\psi_{s2}(x)}{dx^2}-\frac{1}{\lambda^2_{(DG/GAA)}}\psi_{s2}(x)=D_{2(DG/GAA)} \tag{11}$$

where $D_{2(DG/GAA)}=D_{1(DG/GAA)}-\left(\frac{qN_f}{C_{ox}}\right)\left(\frac{1}{t_{ox}^2}\right)$, and N_f represents the hot-carrier induced localized interface charge density per square area [29].

Subject to the boundary conditions

$$\psi(L_1,y)=V_p$$

$$\psi(L,y)=V_{bi}+V_{ds}$$

where V_{ds} is the drain to source voltage.

Hence, the solution in the second region is,

$$\psi_{s2}(x)=-\lambda^2_{(DG/GAA)}D_{2(DG/GAA)}+\frac{\phi_{s2(DG/GAA)}\sinh\left(\frac{x-L_1}{\lambda_{(DG/GAA)}}\right)-\phi_{D2(DG/GAA)}\sinh\left(\frac{x-L}{\lambda_{(DG/GAA)}}\right)}{\sinh\left(\frac{L-L_1}{\lambda_{(DG/GAA)}}\right)} \tag{12}$$

with $\phi_{D2(DG/GAA)}=V_{p(DG/GAA)}+\lambda^2_{(DG/GAA)}D_{2(DG/GAA)}$ and $\phi_{s2(DG/GAA)}=V_{bi}+\lambda^2_{(DG/GAA)}D_{2(DG/GAA)}$

$V_{p(DG/GAA)}$ is the potential developed at the boundary of damaged and non-damaged regions and using the continuity equation of the electric field at $x = L_1$, the value of this potential can be derived.

A more general framework have been proposed for the GCGS DG MOSFET structure with hot-carrier effect, which has been introduced in fresh form in [30]. The whole channel is divided into 3 regions and the associated surface potential is obtained separately, potentials at boundaries of different regions are developed using the continuity equation of the electric field at $x = L_1$ and $x=L_1+L_2$.

The variation of surface potential along the channel for GCGS and conventional DG MOSFETs for nanoscale dimensions indicates two interesting features; first, the incorporation of GCGS design introduces a shift in the surface potential profile along the channel compared to the conventional case. Second, the surface potential for both structures with traps deviates considerably from that of the fresh device where the potential surface in the damaged region is increased when the interface charges N_f are present.

3.1.2. Threshold voltage

For nanoscale multi-gate MOSFETs, threshold voltage is a crucial factor reflecting the device sensitivity. In the case of a fresh device, the minimum point at which the potential reaches its minimum value at low drain voltage is approximately at the middle of the channel, but when hot-carriers are introduced, the location of minimum point is altered and more rigorous calculation is needed. Since the determination of the minimum potential is required for the threshold voltage modelling, the problem reduces to the resolution of the following basic equation; the procedure is repeated for each region of the channel,

$$\frac{\partial \psi_{s1,s2}(x)}{\partial x} = 0 \tag{13}$$

Without great ingenuity, locations and values within regions for both structures are deduced, it is to note that the minimum surface potential is always obtained in the first region irrespective of the value of the hot-carrier induced localized interface charge density. Then the threshold voltage can be calculated from the equation [31],

$$\psi_{s\min} = \psi_{s1\min} = -\lambda^2_{(DG/GAA)} D_{1(DG/GAA)} + \frac{\phi_{D1(DG/GAA)} \sinh\left(\frac{x_{\min 1}}{\lambda_{(DG/GAA)}}\right) - \phi_{s1(DG/GAA)} \sinh\left(\frac{x_{\min 1} - L_1}{\lambda_{(DG/GAA)}}\right)}{\sinh\left(\frac{L_1}{\lambda_{(DG/GAA)}}\right)} = 2\phi_B \tag{14}$$

where Φ_B is the bulk potential of silicon body.

The asymptotic behavior of the threshold voltage is characterized by some limiting cases [32], for very short channel lengths (L < 40 nm), the solution of the equation at low drain–source voltage is given by,

$$V_{th(DG/GAA)} = \frac{\left(2\phi_B + \frac{qN_A\lambda^2_{(DG/GAA)}}{\varepsilon_{Si}}\right) - 2\left(V_{bi} + \frac{qN_A\lambda^2_{(DG/GAA)}}{\varepsilon_{Si}}\right)\frac{\sinh\left(L/2\lambda_{(DG/GAA)}\right)}{\sinh\left(L/\lambda_{(DG/GAA)}\right)}}{1 - 2\frac{\sinh\left(L/2\lambda_{(DG/GAA)}\right)}{\sinh\left(L/\lambda_{(DG/GAA)}\right)}} \tag{15}$$

For long channel lengths (L>40nm and $L>>\lambda_{(DG/GAA)}$), after some mathematical manipulations, we can define a new simple compact threshold voltage model appropriate for long channel multi-gate MOSFETs as,

$$V_{th(DG/GAA)} = 2\phi_B + \frac{qN_A\lambda^2_{(DG/GAA)}}{\varepsilon_{Si}} \tag{16}$$

In the case of a Gcgs Dg Mosfet structure, an accurate model is obtained after some mathematical simplifications equation 14,

$$\sqrt{\left(-\phi_{D1}^2 - \phi_{s1}^2 + 2\phi_{D1}\phi_{s1}\cosh\left(\frac{L_1}{\lambda}\right)\right)} + \left(\lambda^2 D_1 + 2\phi_B\right)\sinh\left(\frac{L_1}{\lambda}\right) = 0 \tag{17}$$

The above equation can be easily transformed to a second degree algebraic one dealing with the threshold voltage explicitly, which can be deduced analytically using the familiar quadratic formula and basing on some physical consideration to eliminate the unaccepted solution [33]. Figure 5 compares the threshold voltage for the conventional and GCGS DG MOSFETs taking into account the effect of hot-carrier degradation. The threshold voltage increases rapidly with the channel length to attain the ideal value for L>30 nm, and we note that the contribution of hot-carrier effect is essentially discriminator for different curves in the nanoscale domain.

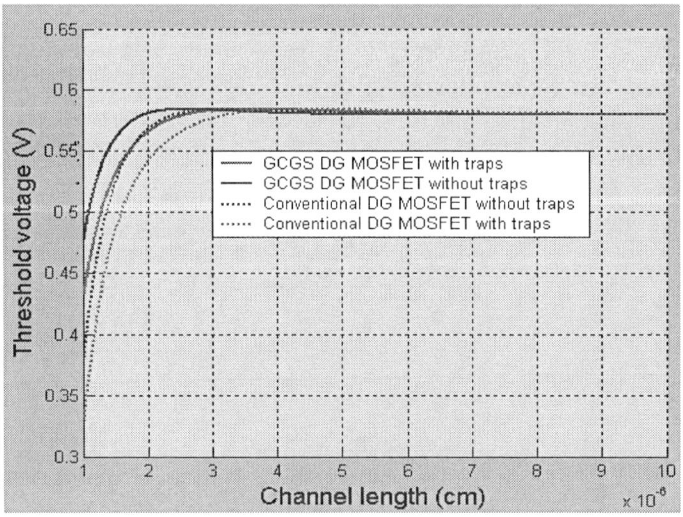

Figure 5. Calculated threshold voltage for conventional and GCGS DG MOSFETs with and without traps (L_1=L/2, L_d=L/3, t_1=1 nm, t_2=1.5 nm, t_{si}=10 nm, $N_{AH}=10^{17}$ cm^{-3}, $N_{AL}=10^{15}$ cm^{-3} and $N_f=10^{15}$ cm^{-2})

3.1.3. Subthreshold current

Theoretically, in subthreshold region no channel exists between the source and the drain, resulting in a disappearance of the drain current. However, in real devices, behavior is different, drain current follows an exponential decay, the low electron concentration results in low electric field along the channel and hence the subthreshold current is mainly due to diffusion of carriers. The calculation of such current can be derived by adopting in a similar manner the methodology proposed for 4-T and 3-T DG MOSFET [34], including the electron density at virtual cathode assumed to have a Boltzmann distribution profile,

$$J_n(y) = qD_n \frac{n_i^2 e^{\frac{\psi_{\min}(y)}{kT}}}{N_A L}\left(1 - e^{\frac{V_{ds}}{kT}}\right) \tag{18}$$

Here D_n denotes the diffusion constant, and n_i is the intrinsic silicon density.

The subthreshold current including interfacial traps effects is then obtained by integrating (18) along the silicon film to get,

$$I_{sub} = I_{sub1} + I_{sub2} \tag{19}$$

with I_{sub1} and I_{sub2} are the subthreshold currents associated with both regions respectively which are defined by,

$$I_{sub1} = \frac{2k_1 kT}{E_s}\left(e^{\frac{\psi_{\min}}{kT}} - e^{\frac{\psi_{\min}^s}{kT}}\right) \tag{20}$$

$$I_{sub2} = \frac{2k_2 kT}{E_s}\left(e^{\frac{\psi_{\min}}{kT}} - e^{\frac{\psi_{\min}^s}{kT}}\right) \tag{21}$$

with

$$k_1 = \left(\frac{q\mu_n W V_t n_i^2}{L_1 N_A}\right) \times \left(1 - e^{-\frac{V_P}{kT}}\right)$$
$$k_2 = \left(\frac{q\mu_n W V_t n_i^2}{(L - L_1) N_A}\right) \times \left(1 - e^{-\frac{V_{ds} - V_P}{kT}}\right)$$
$$E_s = 2\left(\psi_{\min} - \psi_{\min}^s\right)/t_{si}$$
$$\psi_{\min}^s = \psi(x_{\min}, 0)$$
$$\psi_{\min} = \psi(x_{\min}, t_{si}/2)$$

If we consider the proposed structure GCGS DF MOSFET with hot-carrier effect, a more detailed model can be obtained by taking into account that the integration along the channel should be done with respect to the three regions boundaries and following the same precedent procedure.

3.1.4. Subthreshold swing

The subthreshold swing can be seen as the gate voltage required to reduce the current from its on to an acceptable off value. It is an important parameter characteristic of the subthreshold regime. For a device to have good turn-on characteristics, the swing factor should be as small as possible. Clearly, it is a convenient measure of the turnoff characteristics of a Multi-gate MOSFET. This gate voltage swing, is by definition the change in the gate voltage required to reduce subthreshold current by one decade [35],

$$S = \frac{\partial V_{gs}}{\partial \log I_{ds}} = kT \ln(10) \times \left[\frac{\int_{0}^{t_{si}/2} e^{\psi_{\min}/kT} \left(\frac{\partial \psi_{\min}}{\partial V_{gs}} \right) dy}{\int_{0}^{t_{si}/2} e^{\psi_{\min}/kT} dy} \right]^{-1} \tag{22}$$

An approximate solution for this integral was proved in [36] to be of the form,

$$S = \frac{kT}{q} \ln(10) \times \left[\frac{\partial \psi_{s\min}}{\partial V_{gs}} \right]^{-1} \tag{23}$$

Hence, assuming the minimum potential situated in the first region, a closed form expression for subthreshold swing with the hot-carriers degradation effects can be written as,

$$S_{(DG/GAA)} = \frac{kT}{q} \ln(10) \times \left[1 + \frac{\left(\alpha_{(DG/GAA)} + \beta_{(DG/GAA)}\right) - \sinh\left((L - L_1)/\lambda_{(DG/GAA)}\right) - \sinh\left(L/\lambda_{(DG/GAA)}\right)\gamma_{(DG/GAA)} - \sinh\left(L_1/\lambda_{(DG/GAA)}\right) - \sinh\left((x_{\min} - L_1)/\lambda_{(DG/GAA)}\right)}{\sinh\left(L_1/\lambda_{(DG/GAA)}\right)} \right]^{-1} \tag{24}$$

with

$$\alpha_{(DG/GAA)} = \sinh\left(\frac{L_1}{\lambda_{(DG/GAA)}}\right) \cosh\left(\frac{(L - L_1)}{\lambda_{(DG/GAA)}}\right)$$

$$\beta_{(DG/GAA)} = \cosh\left(\frac{L_1}{\lambda_{(DG/GAA)}}\right) \sinh\left(\frac{(L - L_1)}{\lambda_{(DG/GAA)}}\right)$$

$$\gamma_{(DG/GAA)} = \frac{\sinh\left(x_{\min}/\lambda_{(DG/GAA)}\right)}{\sinh\left(L/\lambda_{(DG/GAA)}\right)}$$

where x_{min} is the location of minimum potential obtained by resolving equation (13).

For the GCGS DG MOSFET structure with the hot-carriers degradation effects, the subthreshold swing can be written [37],

$$S = \frac{kT}{q} \ln(10) \times [1 + \xi]^{-1} \tag{25}$$

with

$$\xi = \frac{a + b\left(\dfrac{\sinh\left(\dfrac{L - L_1}{\lambda}\right) + \sinh\left(\dfrac{L_1}{\lambda}\right)}{\sinh\left(\dfrac{L}{\lambda}\right)}\right)}{2\sinh\left(\dfrac{L_1}{\lambda}\right) \times \left[-\phi_{s1}^2 - \phi_{D1}^2 + 2\phi_{s1}\phi_{D1}\cosh\left(\dfrac{L_1}{\lambda}\right)\right]^{\frac{1}{2}}}$$

where

$$a = 2\phi_{s1} - 2\phi_{D1}\cosh\left(\frac{L_1}{\lambda}\right)$$

$$b - 2\phi_{D1} - 2\phi_{s1}\cosh\left(\frac{L_1}{\lambda}\right)$$

A more elementary model can be deduced with some approximations and simplifications as,

$$S = \frac{kT}{q}\ln(10)\left[1 + \frac{\sinh\left(\dfrac{x_{\min 1} - L_1}{\lambda}\right) - \sinh\left(\dfrac{x_{\min 1}}{\lambda}\right) - \sigma}{\sinh\left(\dfrac{L_1}{\lambda}\right)}\right]^{-1} \tag{26}$$

with

$$\sigma = \frac{(a_1 + a_2 + a_3)\sinh\left(\dfrac{x_{\min 1}}{\lambda}\right)}{\sinh\left(\dfrac{L_1 + L_2}{\lambda}\right)\sinh\left(\dfrac{L - L_1}{\lambda}\right) - \sinh\left(\dfrac{L_1}{\lambda}\right)\sinh\left(\dfrac{L - L_1 - L_2}{\lambda}\right)}$$

$$a_1 = \sinh\left(\frac{L_2}{\lambda}\right)\sinh\left(\frac{L - L_1}{\lambda}\right) \times \left(1 - \cos\left(\frac{L_1}{\lambda}\right)\right)$$

$$a_2 = \sinh\left(\frac{L_1}{\lambda}\right) \times \left(1 - \cosh\left(\frac{L_2}{\lambda}\right)\right) \times \left(\sinh\left(\frac{L - L_1}{\lambda}\right) + \sinh\left(\frac{L - L_1 - L_2}{\lambda}\right)\right)$$

$$a_3 = \sinh\left(\frac{L_2}{\lambda}\right) \times \sinh\left(\frac{L_1}{\lambda}\right) \times \left(1 - \cosh\left(\frac{L - L_1 - L_2}{\lambda}\right)\right)$$

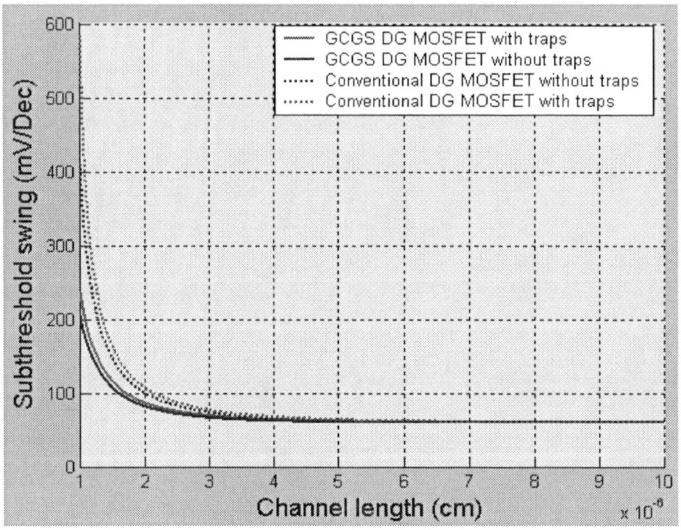

Figure 6. Calculated subthreshold swing for conventional and GCGS DG MOSFETs with and without traps (L_1=L/2, L_d=L/3, t_1=1 nm, t_2=1.5 nm, t_{si}=10 nm, $N_{AH}=10^{17}cm^{-3}$, $N_{AL}=10^{15}$ cm^{-3} and $N_f=10^{15}$ cm^{-2})

Figure 6 plots the subthreshold swing as a function of channel length calculated from our compact model, the comparison indicates that the GCGS DG MOSFET provides higher swing factor compared to conventional structure in both cases with or without traps.

3.2. Hot-Carriers Effects in Saturation Regime

The derivation of the direct current equations of multi-gate MOSFET devices is one of the main objectives in compact modelling. Obtained models should be easy to use and of low complexity in order to avoid NP-hard problems which admit in practice only approximate solutions [38]. Moreover, the impact of hot-carriers degradation becoming increasingly important should be flexibly integrated into the whole model, so that fundamental conclusions about how the architecture and the biasing conditions influence the operation of the device in linear and saturation regime can be extracted.

3.2.1. Drain current model

When the device is operating in or near the saturation region, the electrons moving from the source to the drain experience a significant velocity increase caused by the high horizontal electric field. Such situation is basically responsible about the hot-carrier injection into the gate oxide near the drain side, the localization of hot-carrier injection into the gate oxide leads to the localization of the resulting oxide damage in the form of interface trap generation. Although it is quite clear that the hot-carrier-induced oxide charge is localized very close to the drain, the shape of the charge distribution profile is not known exactly and many shapes have been proposed [39, 40, 41].

In our case, for the derivation of model equations, we consider the localized oxide-interface charge arising from hot-carrier injection to have a uniform charge profile

$$Q_{it}(y)=\begin{cases}0 & 0\le y\le L_1\\ Q_f=qN_f & L_1\le y\le L\end{cases} \tag{27}$$

where L_1 is the length of undamaged channel region.

As a result of the uniform charge profile, two device parameters are basically altered, first, the flat-band voltage is augmented by an additional term in damaged region reflecting the hot-carrier degradation effect [42].

$$V_{FB}(y)=\begin{cases}V_{FB1}=\Phi_{MS}-\dfrac{Q_{ox}}{C_{ox}} & 0\le y\le L_1\\ V_{FB2}=\Phi_{MS}-\dfrac{Q_{ox}}{C_{ox}}-\dfrac{Q_f}{C_{ox}} & L_1\le y\le L\end{cases} \tag{28}$$

where Φ_{MS} represents the work function difference, Q_{ox} represents the constant positive oxide-interface charge density independent of hot-carrier injection.

The second affected parameter concerns the channel electron mobility, where we have a reduction law represented by Mathiessen's rule in the damaged region [43],

$$\mu_n(y)=\begin{cases}\mu_{n1}=\mu_0 & 0\le y\le L_1\\ \mu_{n2}=\dfrac{\mu_0\mu_{ox}}{\mu_0+\mu_{ox}} & L_1\le y\le L\end{cases} \tag{29}$$

with μ_0 the bulk mobility, and $\mu_{ox}=1000\left(\dfrac{3\times10^3}{N_{it}}\right)\times\left(\dfrac{T}{80}\right)$ is the oxide-interface charge scattering mobility. The mobility model in device simulation is an important parameter that greatly affects the calculation results [44, 45].

Once we have all these ingredients in hand, the compact model of drain current can be elaborated in the linear region. The drain current, defined as a positive quantity, is obtained integrating the total current density over the cross-sectional area of the channel. For the fresh region, we have:

$$I_{ds}\int_0^{L_1}dy=-\int_0^{W}\int_0^{V_p}\mu_n(y)Q_n(y)dV_{ch}dz \tag{30}$$

where W is the device width, and V_{ch} is the channel voltage.

The mobile electron charge along the channel region is expressed using the charge-sheet approximation [46],

$$Q_n(y)=-C_{ox}\left[V_{gs}-V_{ch}(y)-V_{FB}(y)-2|\Phi_P|\right]+\left[2\varepsilon_{Si}qN_A\left(2|\Phi_P|+V_{ch}(y)-V_{bi}\right)\right]^{\frac{1}{2}} \tag{31}$$

Here V_{gs} represents the gate voltage, and Φ_p is silicon barrier potential.

Integration of equation 30 is straightforward and consideration of the symmetry of the structure, lead for the first region to the result,

$$I_{ds} = \frac{2W\mu_{n1}C_{ox}}{L_1}\left\{\left(V_{gs} - V_{FB1} - 2|\Phi_P|\right)V_P - \frac{V_P^2}{2} - \frac{2}{3}\frac{\sqrt{2\varepsilon_{Si}qN_A}}{C_{ox}}\left[\left(2|\Phi_P| - V_{bi} + V_P\right)^{\frac{3}{2}} - \left(2|\Phi_P| - V_{bi}\right)^{\frac{3}{2}}\right]\right\} \tag{32}$$

By repeating the above calculations for the damaged region with appropriate modifications we arrive at

$$I_{ds} = \frac{2W\mu_{n2}C_{ox}}{(L - L_1)}\left\{\left(V_{gs} - V_{FB2} - 2|\Phi_P|\right)\left(V_{ds} - V_P\right) - \frac{\left(V_{ds}^2 - V_P^2\right)}{2} - \frac{2}{3}\frac{\sqrt{2\varepsilon_{Si}qN_A}}{C_{ox}}\left[\left(2|\Phi_P| - V_{bi} + V_{ds}\right)^{\frac{3}{2}} - \left(2|\Phi_P| - V_{bi} + V_P\right)^{\frac{3}{2}}\right]\right\} \tag{33}$$

Since a damaged DG MOSFET can be considered as two elementary transistors having different properties and connected in series, we can get the boundary voltage value by equating both expressions of drain current then resolving the nonlinear resulted equation numerically, so that the compact model can be well defined.

In the linear region, it is assumed that an inversion layer exists along the channel from the source to drain, which is true beyond the threshold and for small drain voltages. However, when the drain voltage reaches a certain value, the inversion charge at the drain end drops to zero. That is, the channel is pinched-off near the drain end, the drain voltage that results in the disappearance of the channel is referred to as the saturation voltage, and the corresponding current is approximately constant. The expression of the saturation voltage is obtained by either one of these two conditions mathematically equivalent [47],

$$Q_n(L) = 0 \tag{34}$$

$$\frac{\partial I_{ds}}{\partial V_{ds}} = 0 \tag{35}$$

Combining equations 31 and 34 yields,

$$V_{dsat} = V_{gs} - V_{FB2} - 2|\Phi_P| + \frac{\varepsilon_{Si}qN_A}{C_{ox}^2}\left\{1 - \sqrt{1 + \frac{2C_{ox}^2}{\varepsilon_{Si}qN_A}\left(V_{gs} - V_{FB2} - V_{bi}\right)}\right\} \tag{36}$$

The obtained I_{ds}-V_{gs} characteristics is given in Figure 7, it is easy to note the reduction in the drain current which becomes more apparent when the gate voltage is increased to higher values. The DG MOSFET including a high-k layer brings prominent advantages in terms of immunity against the hot-carrier degradation effect compared to conventional DG MOSFET. As expected, the threshold voltage increases with the amount of oxide interface charge.

3.2.2. Small-signal model parameters

Small-signal parameters particularly g_m and g_{ds} are required basically for analog circuit design [48]. In addition, they are also used for linearizing nonlinear drain current models.

These parameters can easily be derived from the device drain current expressions, which means the differentiability of the drain current equations with respect to all terminal voltages.

The transconductance g_m can be defined as the ratio of change in the drain current to the change in the gate voltage. Using equation (32) we can deduce a closed form for the considered device transconductance,

$$g_m = \left.\frac{\partial I_{ds}}{\partial V_{gs}}\right|_{V_{ds}} = \frac{2W\mu_{n1}C_{ox}}{L_1}\left\{V_P + \left(V_{gs} - V_{FB1} - 2|\Phi_P|\right)\frac{\partial V_P}{\partial V_{gs}} - V_P\frac{\partial V_P}{\partial V_{gs}} - \frac{3r}{2}\left[\left(2|\Phi_P| - V_{bi} + V_P\right)^{\frac{1}{2}}\frac{\partial V_P}{\partial V_{gs}}\right]\right\} \quad (37)$$

with

$$r = \frac{\sqrt{2\varepsilon_{Si}qN_A}}{C_{ox}}$$

$$\frac{\partial V_P}{\partial V_{gs}} = \frac{\left(\frac{\mu_{n1}}{L_1} + \frac{\mu_{n2}}{L - L_1}\right)C_{ox}WV_P - \frac{W}{L - L_1}\mu_{n2}C_{ox}V_{ds}}{\frac{W\mu_{n2}C_{ox}}{L - L_1}\left[V_P - \left(V_{gs} - V_{FB2} - 2|\Phi_P|\right) + \frac{3r}{2}\left(2|\Phi_P| - V_{bi} + V_P\right)^{\frac{1}{2}}\right] - \frac{W\mu_{n1}C_{ox}}{L_1}\left[\left(V_{gs} - V_{FB1} - 2|\Phi_P|\right) - V_P - \frac{3r}{2}\left(2|\Phi_P| - I\right.\right.}$$

Based on obtained analytical expression of the transconductance, figure 8 shows the advantages of introducing a high-k layer on the oxide layer of the DG MOSFET structure when considering the existence of interface traps over the performance of conventional DG MOSFET. As a result of the drop in the biasing current of the transistor and the channel carrier mobility, the transconductance g_m also decreases [49]. From this figure, it can be clearly seen that the degradation simulated above threshold value is increasing with increased gate voltage.

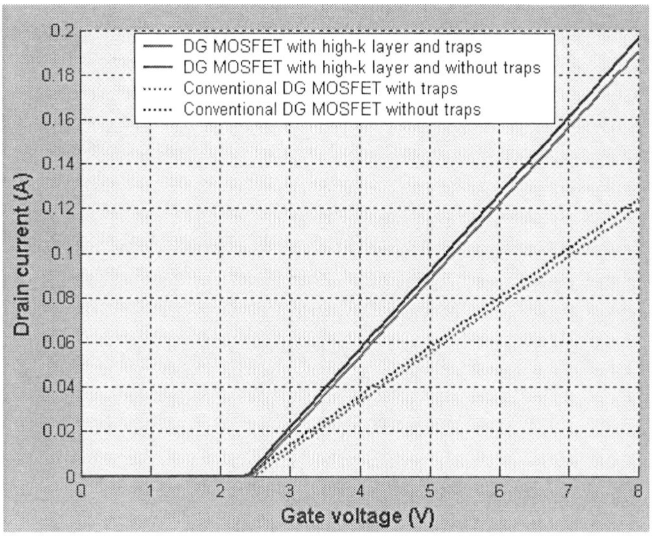

Figure 7. I_{ds}-V_{gs} characteristics for conventional and high-k layer DG MOSFETs with and without traps (L=100nm, L_d=L/3, W=200nm, t_{si}=20 nm, $N_A=1.45\times10^{10}cm^{-3}$, $N_f=10^{15}$ cm^{-2} , $N_{ox}=10^{10}$ cm^{-2})

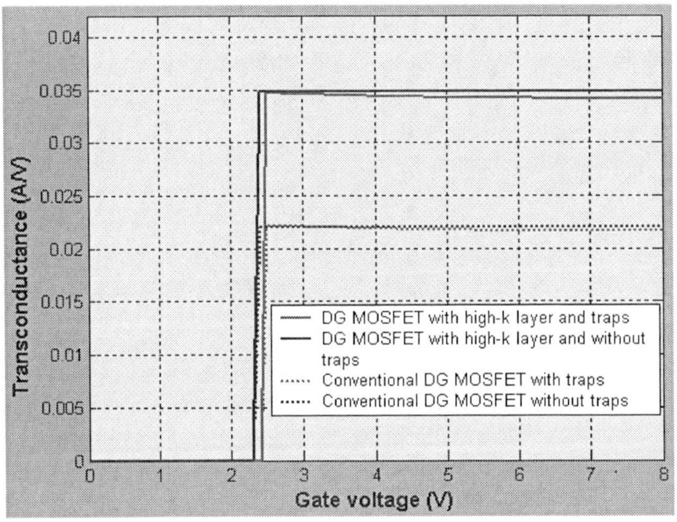

Figure 8. Transconductance variation for both conventional and high-k layer augmented DG MOSFET with and without traps (L=100nm, L_d=L/3, W=200nm, t_{si}=20 nm, $N_A=1.45\times10^{10}cm^{-3}$, $N_f=10^{15}$ cm^{-2}, $N_{ox}=10^{10}$ cm^{-2})

The ratio of change in the drain current to the change in the drain voltage is called output conductance and in a similar manner to above calculation we have,

$$g_{ds}=\left.\frac{\partial I_{ds}}{\partial V_{ds}}\right|_{V_{gs}}=\frac{2W\mu_{n1}C_{ox}}{L_1}\left\{\left(V_{gs}-V_{FB1}-2|\Phi_P|\right)\frac{\partial V_P}{\partial V_{ds}}-V_P\frac{\partial V_P}{\partial V_{ds}}-\frac{3r}{2}\left[\left(2|\Phi_P|-V_{bi}+V_P\right)^{\frac{1}{2}}\frac{\partial V_P}{\partial V_{ds}}\right]\right\} \quad (38)$$

with

$$\frac{\partial V_P}{\partial V_{ds}}=\frac{\frac{W\mu_{n2}C_{ox}}{L-L_1}\left[\left(V_{gs}-V_{FB2}-2|\Phi_P|-V_{ds}\right)-\frac{3r}{2}\left(2|\Phi_P|-V_{bi}+V_{ds}\right)^{\frac{1}{2}}\right]}{\frac{W\mu_{n1}C_{ox}}{L_1}\left[\left(V_{gs}-V_{FB1}-2|\Phi_P|\right)-V_P-\frac{3r}{2}\left(2|\Phi_P|-V_{bi}+V_P\right)^{\frac{1}{2}}\right]-\frac{W\mu_{n2}C_{ox}}{L-L_1}\left[V_P-\left(V_{gs}-V_{FB2}-2|\Phi_P|\right)+\frac{3r}{2}\left(2|\Phi_P|-V_{bi}+V_P\right)^{\frac{1}{2}}\right]}$$

4. Concluding Remarks

This chapter describes a compact modelling approach to study the performances of various multi-gate MOSFET designs. Based on the development of compact analytical expressions of the subthreshold and saturation parameters under different electrical and geometrical conditions, the complete behavior scheme of the device including hot-carrier effects can be established. The approach does not restrict the applicability only to fresh devices, but also offers a friendly framework which allows the incorporation of the short-channel and hot-carrier degradation effects to study the future nanoscale CMOS-based circuits. Besides providing information about the long term reliability behavior, the compact modeling arguments also help in making a priority order for different multi-gate prototypes when subject to additional constraints due to short-channel-effects. The elaborated analysis in this chapter is mainly relied on mathematical modeling and assumptions, taking the advantage of the low computational complexity of obtained results without the need of the metaheuristics optimization strategies.

REFERENCES

[1] Bentrcia, T; Tamrabet, Y; Chahdi, M. "A low complexity finite differences-neural network combined approach for the modelling of p-n junction electrical properties," *Journal of Electron Devices*, 2010, *7*, 203-210.

[2] Wang, J. "Computational intelligence in microelectronics manufacturing," *Computational Intelligence in Manufacturing Handbook*, CRC Press LLC, Boca Raton, USA, 2001.

[3] Bentrcia, T; Mouss, LH; Mouss, MD. "A reliability based approach for the design of hybrid expert systems in industrial diagnosis field," *2nd International Symposium on Innovation in Information and Communication Technology ISIICT'04*, 1,84-91, Amman, Jordan, Apr. 2004.

[4] Kuo, W; Zuo, M. *Optimal reliability modeling principles and applications*, John Wiley and Sons, NJ, USA, 2003.

[5] Merkle, C. "It's a small, small, small, small world," *MIT's Technology Review*, 100(2), 25-32, 1997.

[6] Pagey, MP. Characterization and modeling of hot-carrier degradation in sub-micron NMOSFETS, Msc Thesis, *Electrical Engineering Department*, Vanderbilt University, Nashville, Tennessee, USA, Aug. 2002.

[7] Pagey, MP. Hot-carrier reliability simulation in aggressively scaled MOS transistors, PhD Dissertation, *Electrical Engineering Department*, Vanderbilt University, Nashville, Tennessee, USA, Dec. 2003.

[8] Ciurea, ML; Iancu, V; Lazanu, S; Lepadatu, AM; Rusnac, E; Stavarache, I. "Defects in silicon: from bulk crystals to nanostructures," *Romanian Reports in Physics*, 2008, 60(3), 735-748.

[9] Renesas Technology Corp, Semiconductor Reliability Handbook, *Renesas Technology Corp*, Tokyo, Japan, 2006.

[10] Kuntman, A; Ardali, A; Kuntman, H; Kacar, F. "A Weibull distribution-based new approach to represent hot carrier degradation in threshold voltage of MOS transistors," *Solid-State Electronics*, 2004, 48, 217-223.

[11] Aschcroft, NW; Mermin, ND. *Solid state physics*, Harcourt College Publishers, FL, USA, 1976.

[12] Maroudas, D; Pantelides, ST. "Point defects in crystalline silicon, their migration and their relation to the amorphous phase," *Chemical Engineering Science*, 1994, 49(17), 3001-3014.

[13] Geng, H. *Semiconductor manufacturing Handbook*, McGraw-Hill, USA, 2005.

[14] Seebauer, EG; Kratzer, MC. Charged semiconductor defects structure, *thermodynamics and diffusion*, Springer-Verlag London, 2009.

[15] Chroneos, A; Kube, R; Bracht, H; Grimes, RW; Schwingenschlögl, U. "Vacancy–indium clusters in implanted germanium, " *Chemical Physics Letters*, 2010, 490, 38-40.

[16] Pintilie, I; Lindstroemb, G; Junkes, A; Fretwurst, E. "Radiation-induced point- and cluster-related defects with strong impact on damage properties of silicon detectors," *Nuclear Instruments and Methods in Physics Research A*, 2009, 611, 52-68.
[17] Krellner, C; Haas, S; Goldmann, C; Pernstich, KP; Gundlach, DJ; Batlogg, B. "Density of bulk trap states in organic semiconductor crystals: Discrete levels induced by oxygen in rubrene," *Physical Review B*, 2007, *75*, 245115.
[18] Schroder, DK. *Semiconductor material and device characterization*, Third Edition, John Wiley & Sons, NJ, USA, 2006.
[19] Lorenco, MA; Knights, AP; Homewood, KP; Gwilliam, RM; Simpson, PJ; Masher, P. "A comparative study of vacancies produced by proton implantation of silicon using positron annihilation and deep level transient spectroscopy," *Nuclear Instruments and Methods in Physics Research B*, 2001, 175-177, 300-304.
[20] Karwasz, GP; Zecca, A; Brusa, RS; Pliszka, D. "Application of positron annihilation techniques for semiconductor studies," *Journal of Alloys and Compounds*, 2004, 382, 244-251.
[21] Seo, MW; Kwak, DW; Cho, WS; Park, CJ; Kim, WS; Cho, HY. "Charge traps and interface traps in non-volatile memory device with Oxide-Nitride-Oxide structures," *Thin Solid Films*, 2008, 517, 245-247.
[22] Xiaolia, Y; Yi, S; Shulin, G; Jianmin, Z; Youdou, Z; Kenichi, S; Hiroki, I; Toshiro, H. "Effects of interface traps in silicon-quantum-dots-based memory structures," *Physica E*, 2000, 8, 189-193.
[23] Amat, E; Kauerauf, T; Degraeve, R; Rodríguez, R; Nafría, M; Aymerich, X; Groeseneken, G. "Channel hot-carrier degradation in pMOS and nMOS short channel transistors with high-k dielectric stack," *Microelectronic Engineering*, 2010, 87, 47-50.
[24] La Rosa, G; Rauch III, SE. "Channel hot carrier effects in n-MOSFET devices of advanced submicron CMOS technologies," *Microelectronics Reliability*, 2007, 47, 552-558.
[25] Jang, SJ; Ka, DH; Yu, CG; Cho, WJ; Park, JT. "Hot-carrier effects as a function of silicon film thickness in nanometer-scale SOI pMOSFETs," *Solid-State Electronics*, 2008, 52, 824-829.
[26] Landheer, D; Tao, Y; Xu, DX; Sproule, GI; Buchanan, DA. "Defects generated by Fowler-Nordheim injection in silicon dioxide films produced by plasma-enhanced chemical-vapour deposition with nitrous oxide and silane," *J. Appl. Phys.*, 1995, 78(3), 1818-1823.
[27] Siemieniec, R; Südkamp, W; Lutz, J. "Determination of parameters of radiation induced traps in silicon," *Solid-State Electronics*, 2002, 46, 891-901.
[28] Abdelhamid, HM. Compact Modeling of Multiple Gate MOS Devices, PhD dissertation, Department of Electronic, *Electrical and Automatic Engineering*, University of Rovira i Virgili Tarragona-Spain, 2007.
[29] Singh, AK. "An analytical study of hot-carrier degradation effects in submicron MOS devices," Euro . *J .Applied Physics*, 2008, 42, 87-94.
[30] Djeffal, F; Meguellati, M; Benhaya, A. "A two-dimensional analytical analysis of subthreshold behavior to study the scaling capability of nanoscale graded channel gate stack DG MOSFETs," *Physica E*, 2009, 41, 1872-1877.

[31] Ho, CS. Kuo-Yin Huang, Ming Tang, Juin J. Liou, "An analytical threshold voltage model of nMOSFETs with hot-carrier induced interface charge effect," *Microelectronics Reliability*, 2005, 45, 1144-1149.

[32] Djeffal, F; Ghoggali, Z; Dibi, Z; Lakhdar, N. "Analytical analysis of nanoscale multiple gate MOSFETs including effects of hot-carrier induced interface charges," *Microelectronics Reliability*, 2009, 49, 377-381.

[33] Bentrcia, T; Djeffal, F; Abdi, MA; Chahdi, M; Boukhennoufa, N. "An accurate two dimensional threshold voltage model for nanoscale GCGS DG MOSFET including traps effects," *3rd International Conference on Signals*, Circuits and Systems SCS'09, 1, 273-277, Djerba, Tunisia, Nov. 2009.

[34] Dey, A; Chakravorty, A; DasGupta, N; DasGupta, A. "Analytical model of subthreshold current and slope for asymetric 4-T and 3-T double-gate MOSFETs," *IEEE Trans. Electron Devices*, 2008, 55, 3442-3445.

[35] Abdi, MA; Djeffal, F; Arar, Dl Hafiane, ML. "Numerical analysis of double gàte and gate all around MOSFETs with bulk trap states," *J Mater Sci: Mater Electron*, 2008, 19, S248.

[36] Young, KK. "Analysis of conduction in fully depleted SOI MOSFETs," *IEEE Trans Electron Devices*, 1989, 36, 504-506.

[37] Bentrcia, T; Djeffal, F; Arar, D. "An analytical two dimensional subthreshold current model for nanoscale GCGS DG MOSFET including interfacial traps effects," *2nd International Conference on Electronic Systems CISE'2009*, 1,173-177. Batna, Algeria, Oct. 2009.

[38] Tovey, CA. "Tutorial on computational complexity," *Interfaces*, 2002, 32(3),30 -61.

[39] Marrakh, R; Bouhdada, A. "Modeling of interface defect distribution for an n-MOSFETs under hot-carrier stressing," *Active and Passive Elec., Comp*, 2000, 23, 137-144.

[40] Marrakh, R; Bouhdada, A. "Modeling of the I–V characteristics for LDD-nMOSFETs in relation with defects induced by hot-carrier injection," *Active and Passive Elec. Comp*, 2003, 26(4), 197-204.

[41] Sheu, CJ; Jang, SL. "Modeling of electron gate current and post-stress drain current of p-type silicon-on-insulator MOSFETs," *Solide-State Electronics*, 2003, 47,705-711.

[42] Gurfinke, M; Kim, J; Potbhare, S; Xiong, HD; Cheung, KP; Suehle, J; Bernstein, JB; Shapira, Y; Lelis, AJ; Habersat, D; Goldsman, N. *Characterization of interface and bulk oxide traps in SiC MOSFETs with epitaxialy grown and implanted channels*, 2007, IIRW FINAL REPORT, 111-113.

[43] Galup-Montoro, C; Schneider, MC. *Mosfet modeling for circuit analysis and design*, World Scientific Publishing, Singapore, 2007.

[44] Djeffal, F; Lakhdar, N; Meguellati, M; Benhaya, A. "Particle swarm optimization versus genetic algorithms to study the electron mobility in wurtzite GaN-based devices," *Solid-State Electronics*, 2009, 53, 988-992.

[45] Djeffal, F; Arar, D; Lakhdar, N; Bendib, T; Dibi, Z; Chahdi, M. "An approach based on particle swarm computation to study the electron mobility in wurtzite GaN," *Microelectronics Journal*, 2009, 40, 357-359.

[46] Leblebici, Y; Kang, SM. "Modelling of nMOS transistors for simulation of hot-carrier-induced device and circuit degradation," *IEEE Transactions on Computer-Aided-Design*, 1992, 11(2), 235-246.

[47] Arora, N. MOSFET modeling for VLSI simulation theory and practice, *International Series on Advances in Solid State Electronics and Technology*, World Scientific Publishing, Singapore, 2007.

[48] Zhou, X; Chiah, SB; Lim, KY. "A compact deep-submicron MOSFET gds model including hot-electron and thermoelectric effects," *Solid-State Electronics*, 2004, 48, 2125-2131, 2004.

[49] Naseh, S; Jamal Deen, M; Chen, C. "Hot-carrier reliability of submicron nMOSFETs and integrated nMOS low noise amplifiers," *Microelectronics Reliability*, 2006, 46, 201-212.

In: CMOS Technology
Editor: Min-jun Kwon, pp.159-173
ISBN: 978-1-61761-325-8

Chapter 5

MOSFET MODELING: RELIABILITY AND VALIDATION FOR ANALOG/RF IC DESIGN

Thomas Noulis[1,2*]
[1]HELIC Inc., 101 Montgomery street, suite 1950, San Francisco, CA 94104, (866-99-HELIC) - RnD Center 2, GR55535 Thessaloniki, Greece
[2]Electronics Lab., Physics Dept., Aristotle University of Thessaloniki, Aristotle Univ. Campus, 54124 Thessaloniki, Greece

ABSTRACT

In the majority of high performance analog signal processors, power efficiency, speed, noise and dynamic range should be traded against one another to find an optimum device bias condition for a given application. Due to these constraints IC designers are forced to operate MOSFETs in moderate inversion, where the model accuracy is questionable. Considering that the model will always limit the simulation accuracy, it is of great importance the designer to be aware of the capabilities – limitations of the applied CMOS model. Concerning the modeling of a MOS device, the model should be simple enough that the results produced in a simulation to be reliable. The model should also be simple in order the simulation time to be minimal and the process for parameter extraction can be easily performed. A balance between the model simplicity and accuracy needs to be attained. Regarding Analog/RF IC design, inconsistencies have been observed mainly in relation to high frequency behavior, linearity-distortion estimation and noise performance. In this work, MOSFET model reliability issues are addressed in relation to the used MOSFET model, the operating mode, the frequency band of interest (baseband or wideband, RF) and the related applications.

Keywords: MOSFET, modelling and simulation, analog/RF IC design. BSIM, PSP, HiSIM, EKV

---

* Corresponding author : email: tnoul@physics.auth.gr

1. Introduction

Semiconductor technology rapid evolution due to the never-ending demand for better performance results to a steady CMOS feature downsizing by about a factor of 2 per two generations (for silicon CMOS technology, the evolution is measured in generations of 3 years). At present, CMOS manufacturing has progressed to the 45nm technology node resulting to CMOS ICs performance rise packing several 100 million transistors on a chip and operating with clock rates well into GHz region. Greatly important factor in this development are the fabrication process complexity levels and physics mechanisms of deep submicron FETs. These mechanisms should be accurately described and implemented into process modelling and circuit design EDA tools, in order the IC designers to be able to utilize-control their IC systems during the whole design flow.

In particular, concerning the modelling of a MOS device, the model should be simple enough in order the results to be reliable and the related simulation time to be practical. In other words, the model should also be simple in order the simulation time to be minimal and the process for parameter extraction can be easily performed. A balance between the model simplicity and accuracy needs to be attained. While in the active modelling of digital IC designs and generally digital design applications, the time evaluating the model equations is short enough to minimize the overall simulation time and the balance had favored model simplicity sacrificing accuracy in the device model, that's not the case for Analog/RF IC design. The used MOSFET model in Analog/RF applications should be reliable in all operating regions and capable to provide accurate MOSFET simulation, regarding different performance parameters such as noise, linearity –dynamic range, power consumption and small and large signal behaviour. Most of the well-known problems with conventional I-V models for MOSFETs were related to discontinuities in the conductance at the boundary between linear and saturation regions or at the boundary between weak and strong inversion conditions. These problems have been solved quite successfully by utilizing a unified modelling of the drain currents in all regions of operating biases. However, the other attention has been given to maintaining continuity in the charge and current models exclusively in the forward or reverse regions.

In this work, the modelling of the MOSFET transistor is addressed regarding its reliability for Analog/RF design. The most popular- advanced models are briefly presented and particular validation issues are described in relation to specific simulation inaccuracies.

2. Analog/RF MOSFET Models Overview, Accuracy Requirements and Validation

Regarding MOSFET modelling, the threshold-voltage-based BSIM model from Berkeley University – California, from its inception in 1995, until 2005, had been the only standard MOSFET model of the Compact Model Council (CMC) which operates within the framework of the Government Electronics and Information Technology Association (GEIA). Consensus has been reached by analog designers and EDA developers of compact MOSFET models that the surface-potential-based approach offers unique advantages for advanced circuit simulations and that was finally reflected by the adoption of the surface-potential-based PSP model as the next-generation standard MOSFET model. The particular model developed in the framework of the scientific cooperation between Arizona State University and NXP Semiconductors Research [1]. Its competitive model is HiSIM (Hiroshima-university STARC IGFET Model), a drift-diffusion-based advanced MOSFET model for circuit simulation with easy parameter extraction. Specifically, HiSIM 2.Y.Z is a surface-Potential-Based RF MOSFET Model for Circuit Simulation. Another well known model is finally EKV, a a mathematical model of metal-oxide semiconductor field-effect transistors (MOSFET) which is intended for circuit simulation and analog circuit design [3]. It was developed by C. C. Enz, F. Krummenacher and E. A. Vittoz (hence the initials EKV) around 1995 based in part on work they had done in the 1980s [2]. Unlike simpler models like the Quadratic Model, the EKV Model is accurate even when the MOSFET is operating in the subthreshold region. In addition, it models many of the specialized effects seen in submicrometre CMOS IC design.

The validation of the used MOSFET model is addressed regarding their capability for Analog/RF design. Extended analysis is presented mainly regarding the BSIM and PSP models, which are practically the "old" and the "new" CMC model standards. Related subjects are also addressed concerning the HiSIM and the EKV MOSFET models. In addition a brief presentation of the RF/Analog MOSFET modelling necessities and requirements is provided.

A. RF MOS Transistor Models and Benchmark Testing

Compared with the MOSFET modelling for digital and low frequency LF analog design applications, the HF modelling imposes more challenges. Specific further requirements should be maintained in RF modelling [5]. Some of these requirements are validated with particular simulation based benchmark tests.

i. RF modelling accuracy requirements

In particular, the model should accurately estimate bias dependence of small signal parameters at HF, and accurately predict nonlinear behaviour (distortion – linearity performance). The noise behaviour of the device should be accurately simulated since in HF designs (such as LNAs, PAs) additional noise sources contribute. The model should also include non quasi static (NQS) effects in order to describe (in the time domain) the device behaviour at high frequencies. The gate resistance should be accurately modeled not only regarding its value but its distribution and the related noise contribution and correlation to the other noise sources. In addition the extrinsic drain/source resistors should be modelled as real external resistors. A bias dependent overlap capacitance is necessary to be included and all external components (if an RF sub-circuit is used, as it is in the majority of the process design kits (PDKs)), should be physics based and geometrically scalable).

ii. MOSFET model benchmark testing for analog/RF IC design

The majority of the model deficiencies are related to RF IC design. In typical analog applications deficiencies also existed but they were faced quite successfully from BSIM3V3 model version. Particular benchmark tests can be applied in order to estimate the accuracy of each model. For typical analog applications these tests are described in [6] and are consider as traditional continuity tests and 1/f and thermal noise tests. The basic BSIM model inaccuracy was its discontinuity around the Vds=0V value. In particular BSIM3 and BSIM4 could not pass the Gummel symmetry test [7], a discontinuity exits in this point resulting in inaccurate simulations for circuits operating with MOSFET devices with Vds zero DC value (Mixers, Dividers etc) and with respect to source–drain terminal interchange. A related inaccuracy regarding the parasitic capacitances calculation is also present. These and other minor inaccuracies do not exist in the new MOSFET model standard of CMC, the PSP model [8]. In particular, the simulation results shown in Figure 1 demonstrate that PSP accurately reproduces the actual test data. Higher order derivatives for a MOS transistor operated at zero drain bias are simulated without discontinuity.

Additional harmonic balance, reciprocity and modified in general Gummel symmetry tests were applied regarding the suitability of PSP [9] – the model is capable to accurately simulate the linearity performance of analog circuits for all dc bias conditions.

Another great issue regarding MOSFET modelling is the related charge partition scheme. Regarding a quasi-static assumption in the device saturation, the proper partitioning is 40/60 [10]. The other partition schemes (0/100 and 50/50) luck strong physical basis. However, the quasi-static approximation incorporated in all the partition schemes fails when the time scale is comparable to the transit delay time of the transistor. In particular, when a MOSFET is given a step ramp in the gate voltage, the output signal simulated using all three quasi-static models is incorrect at the initial part of the transient. Specifically, for the 40/60 partitioning, there is an unphysical initial rise in the output voltage resulting in a bump feature. The partition 0/100 and 50/50 become popular because this bump problem associated with the 40/60 charge partition scheme, is removed. However because they are still quasi-static models, the simulated results are still inaccurate regarding the initial part of the transient. The problematic calculation of the parasitic capacitance Cdg is main reason regarding the problematic behavior of the available charge partition schemes. In particular, the correct Cdg behavior is obtained using the 40/60 partition with a finite capacitance value at saturation and increases in the linear region. The 0/100 partition yields a zero Cdg when the MOS device operates in saturation. Because the Cdg is zero in saturation, the 0/100 partition is able to avoid the bump problem associated with the 40/60 charge partitioning. It is rather controversial that while the 40/60 partition is physically based and calculates the correct Cdg behavior, this scheme can lead to unphysical overall circuit results at the initial part of the transient. In addition, with wrong Cdg characteristics the other partitions becomes more numerically robust during a circuit simulation.

The related benchmark test that is applied regarding the NQS bump MOSFET model problem is the Killer NOR gate test depicted in Figure 2.

The X node voltage reaches an unphysical potential of −8 V at the falling edge of the input A using the QS mode. This is caused by the negative drain current in MP1 during transistor turn-on predicted. This current should be supplied from MP2, which, however, is biased in the sub-threshold region. To provide the current needed, the X node voltage should go very negative and, in this case, reaches around −8 V.

For RF modelling at the frequency domain, the 40/60 scheme should be used unless the concerned frequency region is nearby the device cut-off frequency. Above the cut-off frequency is better to switch to the 0/100 scheme in order to avoid the incorrect voltage gain dependence on frequency. For transient simulation studies, the 0/100 scheme should be applied in order the unphysical bump problem observed in the 40/60 scheme, not to appear. The 50/50 partition scheme has no fundamental benefits over the other partition schemes. The above issues are faced in PSP's incorporated algorithms-for high-frequency and fast transient simulations - a version of the PSP model can be used that enables the simulation of non-quasi-static (NQS) effects including parasitic and coupling components.

In relation to noise modelling and simulation, extended research has also been reported. Concerning 1/f noise, the models that are used in the majority of MOSFET models were reported and verified in [12] – [16]. Regarding channel thermal noise – no specific problem also exists. Therefore the related simulation accuracy is mainly determined by the noise parameter characterization. However, regarding the gate induced noise, noise estimation problems exit. In particular, in the majority of the PDKs, an RF sub-circuit is used and the related gate resistance provides a noise source not accurately correlated with the rest MOSFET noise sources resulting in inaccurate noise simulations for RFICs in the GHz region.

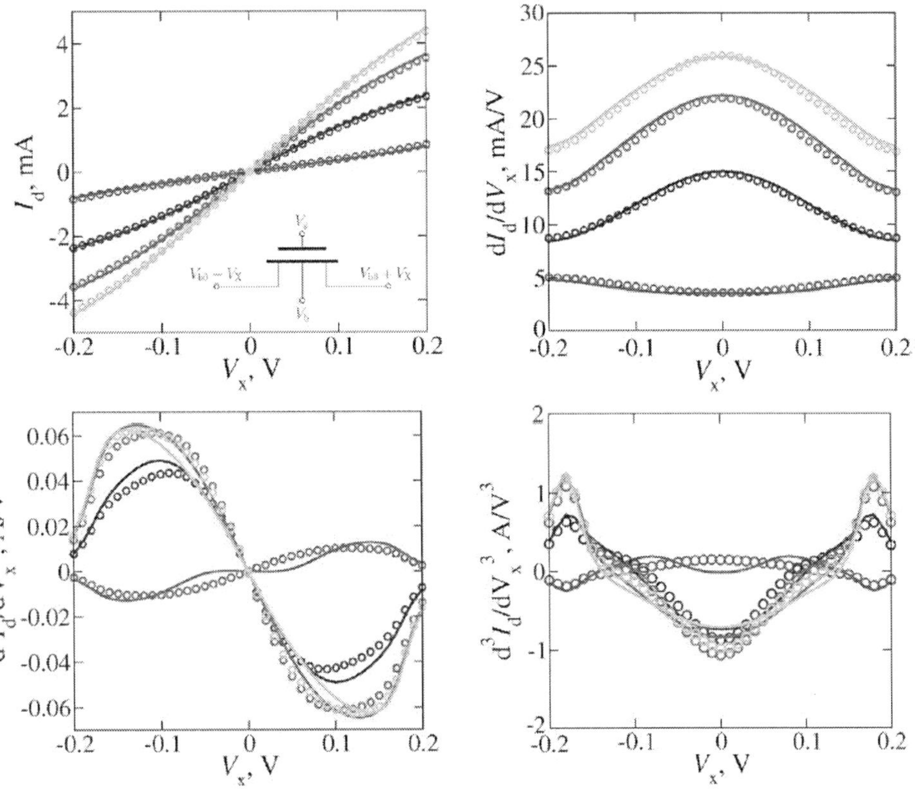

Figure 1. Drain current and derivatives for Jazz Semiconductor 0.18-μm technology n-channel 10/0.18 μm MOSFET; Vg varies from 0.6 to 1.8 V in steps of 0.4 V. Symbols represent experimental data, while solid lines refer to PSP; Vb = 0 V; and Vbo = 0 V. Inset shows the circuit diagram for simulation (taken from [9] with permission ©2009 IEEE)

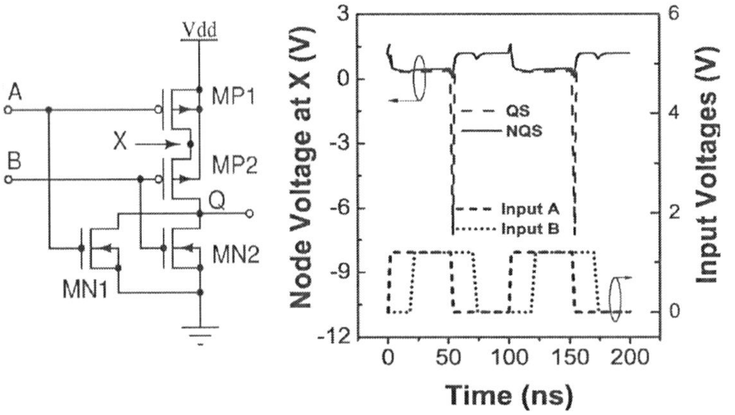

Figure 2. Schematic diagram of the killer NOR gate and comparison of the simulated voltage at node X of the killer NOR gate using the QS and NQS models. (taken from [1] with permission ©2006 IEEE)

B. PSP: Brief Overview of the New CMC Standard Compact MOS Model [8]

State-of-the-art models such as BSIM3 and BSIM4 models, are based on threshold voltage formulations, and are called threshold-voltage-based models. These models fail to satisfy requirements for advanced modeling. This deficiency has presently resulted in a wide consensus in the compact modeling community that traditional threshold-voltage-based models have reached the limit of their usefulness and need to be replaced with more advanced models based on surface potential formulations referred as surface-potential-based or inversion-charge-based models [8].

The PSP model uses a physics-based modelling for all regions of operation (including the moderate inversion and the accumulation) and avoids approximations beyond those already inherent in the charge-sheet models. Since the inversion charge is directly related to the intrinsic drain current, this formulation has certain elegance and, unlike Vth-based models, directly and mostly physically evaluates device characteristics in the weak inversion (sub-threshold) region. This flexibility of the qi-based formulation is limited by the fact that qi is not available for the model formulation in the accumulation region (RF applications) and in the source-drain overlap regions. Additionally, the inversion charge density is not a natural variable for the formulation of the noise model, gate current partition in the channel region, and makes it more difficult to formulate the large-signal NQS version of the model. As a result, the ψ_s-based formulation is the only viable foundation for the next compact MOSFET models generation. In fact, as the ψs-based formulation incorporates both Vth-based and qi-based methods as special cases that follow from the general approach under additional assumptions. To serve as a gateway to advanced MOS IC design for new generations of CMOS technology, PSP contains an accurate description of the major physical effects for the characteristics of scaled MOSFETs. These include Coulomb scattering - non-universality of the mobility model, HALO doping, vertical profile non-uniformity, quantum-mechanical effects, polysilicon depletion region, non-singular velocity saturation model, gate tunneling and gate-induced drain and source leakage (GIDL and GISL) currents, all MOSFET noise sources, an advanced junction model, and an STI-induced stress model. Additionally, the model formulation enables all these features be implemented without prohibitive increase in model complexity.

The structure of the PSP model is shown in Figure 3. The core model includes computation of the surface potentials at the source and drain ends of the channel respectively, the intrinsic drain current, and the terminal charges. The extrinsic model includes computation of the overlap charges, gate tunnelling current, substrate impact ionization current, and spectral densities for the electrical fluctuations (noise). PSP model also contains two additional modules dealing with the NQS formulation and junctions. Regarding the junctions, the doping gradients and thus the electric fields in the source-bulk and drain-bulk junctions are quite high. These high electric fields lead to increased junction leakage due to various tunnelling effects. In the PSP model the source-drain junctions are modelled by the JUNCAP2 model [17]. This model includes Shockley-Read-Hall generation/recombination, trap-assisted tunnelling, and band-to-band tunnelling, leading to a relatively complicated expression for the resulting current. In the JUNCAP2 model, three components to both current and capacitance are distinguished for each junction: a bottom component, a gate-edge component, and an isolation-edge component. Each of these components is modelled independently and has its own set of model parameters. The JUNCAP2 model gives a junction currents good description, both in forward and reverse bias, and as a function of temperature. To calculate the current in a MOSFET at a given bias condition, the junction currents in all three components (bottom, gate-edge, and isolation edge) of both the source and drain junctions should be calculated. So, while the expression for the channel current is evaluated once, the expression for the junction current is evaluated six times. To alleviate this calculational burden, the “JUNCAP2 Express” option was developed and is briefly provided in Figure 4. In the bias-independent part of the model (only called in the initialization phase), the full expressions for all three components are used to calculate the coefficients of a very simple expression. Now, in the bias-dependent part, which is called repeatedly by the circuit simulator in all the bias- or time-steps of the circuit simulation, only this very simple expression needs to be evaluated [17].

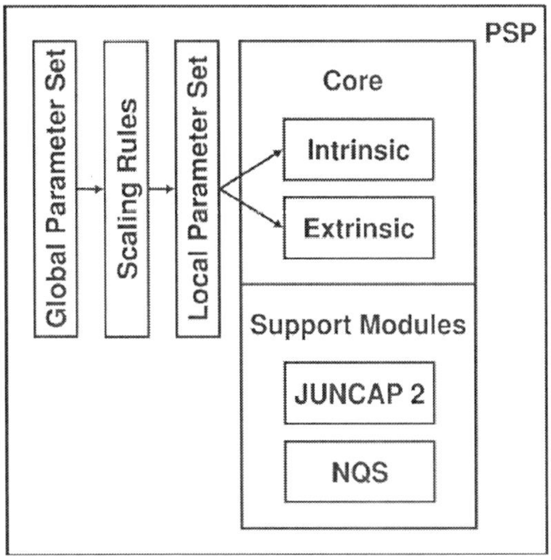

Figure 3. PSP model structure (taken from [8] with permission ©2006 IEEE)

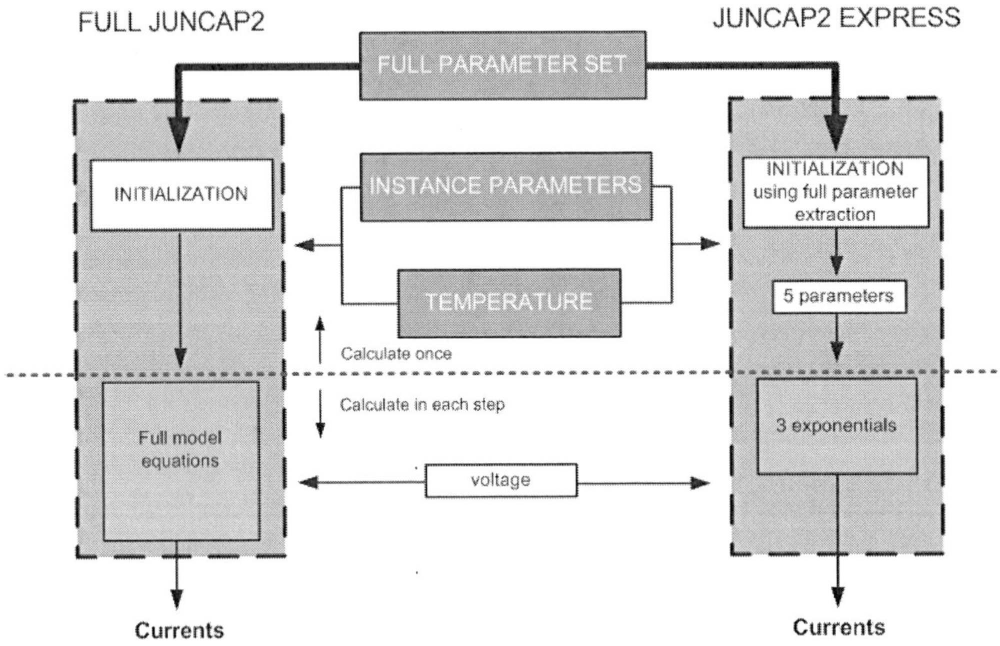

Figure 4. Illustration of the basic idea behind JUNCAP Express: in the frequently called voltage-dependent part of the model, the computationally expensive calculations are replaced by simple ones. This set of simple equations is mapped onto the full JUNCAP model in the initialization part of the model

Regarding the model parameters, global parameters include geometry dependencies, and before evaluating the MOSFET output characteristics, are converted in a small number of local parameters that are used in the core model. The use of local parameters defines the model parameter extraction can be performed by the local parameters extraction for each device geometry separately and then using scaling equations the global parameters are obtained for the relevant range of geometries [17].

Concluding, the mathematical structure of the PSP model takes advantage of the MOS device physics rapid progress accomplished during the last decade. In particular, both the intrinsic and extrinsic components of the model are essentially physics based and formulated in terms of the surface potential using. Traditional benchmark tests, as well as new requirements brought about by the extensive use of MOS technology for analog and RF applications, are automatically satisfied in view of the physical nature of the model. The PSP model has been placed in public domain and implemented in all major circuit simulators. It has been selected by the Compact Model Council as the new standard model for MOS transistors [18].

C. HiSiM and EKV MOSFET models brief overview

HiSIM (Hiroshima University Starc Igfet Model) is the name of a surface-potential-based compact model for advanced sub-100nm scale MOSFET, which has been developed more than a decade. HiSIM has evolved into one of the worlds leading MOSFET models for circuit simulation, practically applied by the semiconductor industry. More recently, the HiSIM approach of a consistently potential-based modelling has been successfully extended to a broad range of integrated devices which have a MOSFET core in common, including SOI-

MOSFET (HiSIM-SOI), high-voltage MOSFET (HiSIM-HV), Insulated-Gate-Bipolar Transistor (HiSIM-IGBT), thin-film transistor (HiSIM-TFT), MOS varactor (HiSIM-VAR) and Double-Gate MOSFET (HiSIM-DG). Due to its accuracy, it was the other competitive model in the CMC.

The EPFL-EKV MOSFET model is a scalable, compact simulation model built on fundamental physical properties of the MOS structure. The particular model is dedicated to the design and simulation of low-voltage, low-current analog, and mixed analog-digital circuits using submicron CMOS technologies. All aspects regarding the static, the quasistatic and non-quasistatic dynamic and noise models are all derived in a coherent way from a single characteristic, the normalized transconductance-to-current ratio. Symmetric normalized forward and reverse currents are used throughout these expressions. For quasistatic dynamic operation, both a charge-based model for the node charges and trans-capacitances, and a simpler capacitances model are available. The dynamic model, including the time constant for the non-quasistatic model, is described in symmetrical terms of the forward and reverse normalized currents. The charge formulation is further used to express effective mobility dependence of local field.

3. Analog/RF MOSFET Modelling General Approach ands Critical Parameters [5]

Common modelling approach for RFIC design in the majority of the commercially available PDKs (Process Design Kits), is to build sub-circuits based on the intrinsic MOSFET that has been modelled well for analog applications. These models accuracy depends on the sub-circuits development with the correct understanding of the device physics in HF operation - how to model the intrinsic devices and extrinsic parasitics HF behaviour and how to extract the sub-circuit parameters. Most RF modeling activities focus on the sub-circuit approach based on different compact MOSFET models that are developed for digital and low-frequency analog applications. With added parasitic components at the gate, at the source, at the drain, and at the substrate, these models can estimate the HF ac small-signal characteristics of short channel devices up to gigahertz range.

In the RF region, the models are still at a preliminary stage compared to the modelling work for digital and low-frequency analog applications. Efforts from industry and universities are required as the RF MOSFET models to be in a mature level, describing the AC characteristics more accurately, and in improving the prediction of noise characteristics, distortion and NQS behaviour.

The MOSFET division modelling approach into two portions - intrinsic part and extrinsic part – is provided in Figure 5. The extrinsic part consists of all the parasitic components, such as the gate resistance, gate/drain overlap capacitance, gate/bulk overlap capacitance, source series resistance, drain series resistance, source/bulk junction diode, drain/bulk junction diode, and substrate resistances. The intrinsic part is the core of the device without including those parasitics. The major parameters of the extrinsic part are presented and discussed below.

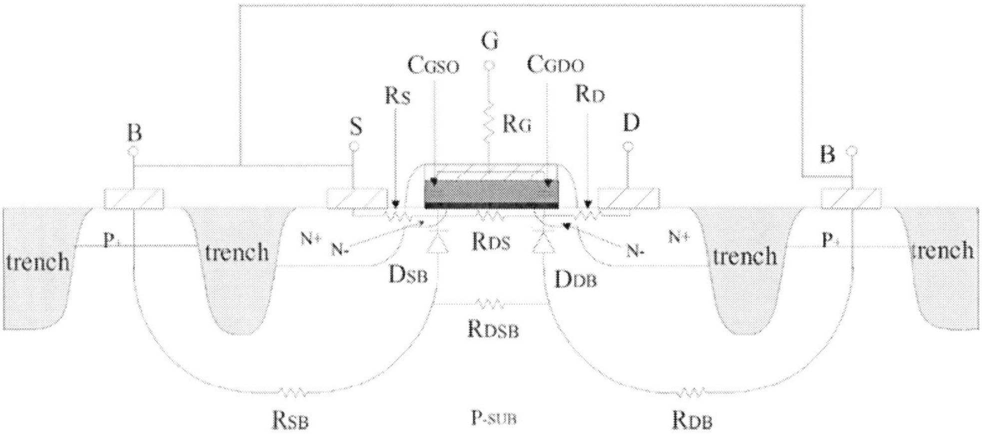

Figure 5. Illustration of different capacitance components in a MOSFET (taken from [5] with permission ©2005 IEEE)

Greatly critical parameter regarding the extrinsic MOSFET model configuration is the ***gate resistance***. The gate resistance impacts impedance matching to achieve maximum power transfer and modifies the noise figure of the transistor due to the thermal noise introduced by the gate resistance. Furthermore, the gate resistance also reduces the frequency at which the maximum available power gain of the device equals to 1, which is an important device parameter in RF circuit design. The gate resistance is in principle a bias-independent component at dc and low frequency, but may contain the contribution of an additional component with bias dependence at RF due to additional physical effects - the distributed transmission line effect on the gate and the distributed effect - NQS effect in the channel. The distributed transmission line effect becomes more severe as the gate width becomes wider at higher operation frequency. Therefore, multifinger devices are used in the circuit design with narrow gate width for each finger to reduce the influence of this effect.

The NQS effect (or distributed RC effect in the channel) is another effect that should be accounted for in modelling the HF behavior of a MOSFET. For the devices with NQS effects, additional bias and geometry dependences of the gate resistance are needed to account for the NQS effect. It has been reported that an additional resistive component in the gate should be added to represent the channel distributed RC effect [5]. In a MOSFET operating at high frequencies, the contribution to the effective gate resistance is not only from the physical gate electrode resistance but also from the distributed channel resistance, which can be "seen" by the signal applied to the gate - Figure 6. Therefore, the effective gate resistance consists of two parts, $R_G=R_{G,poly}+R_{G,nqs}$. $R_{G,poly}$ is the distributed gate electrode resistance from the polysilicon gate material and $R_{G,nqs}$ is the NQS distributed channel resistance seen from the gate (function of biases and geometry).

The source/drain resistances are bias dependent. However, a model without the bias dependence can work reasonably well in MOSFETs, since the LDD region in these devices (with advanced technologies of 180nm and below) has a very high doping concentration, resulting in a weak bias dependence of the R_D, R_S.

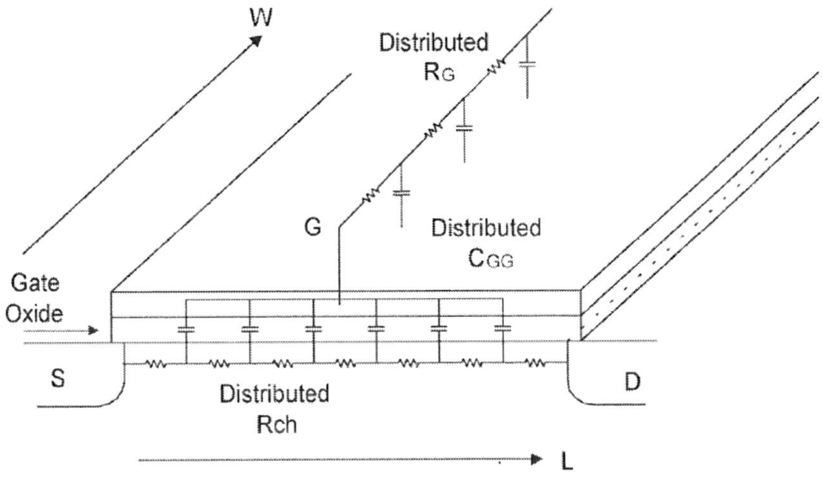

Figure 6. Equivalent gate resistance consists of the contributions from the istributed gate poly resistance and distributed channel resistance (taken from [5] with permission ©2005 IEEE)

The influence of the ***substrate resistance*** is usually ignored n compact models for digital and analog circuit simulation at low frequency. However, at high frequencies, the signal at the drain couples to the source and bulk terminals through the source/drain junction capacitance and the substrate resistance. The substrate resistance influences mainly the output characteristics, and can contribute as much as 20% or more of the total output admittance. It has been known that the substrate components become distributed at HF. Although it is always desirable to have a detailed distributed RC network to account for the contribution of the substrate components, it is too complex to be implemented in a compact model. A good compromise is to use a lumped RC network, if it is accurate in the required operation frequency range, to simulate the substrate components contribution. Simple equivalent circuit, for the substrate network, is provided in Figure 7.

It has been found that the bias dependence of the substrate resistances is actually very weak for devices with substrate ties isolated by shallow trench from the active region, and the above simple substrate resistance network is accurate up to 10GHz.

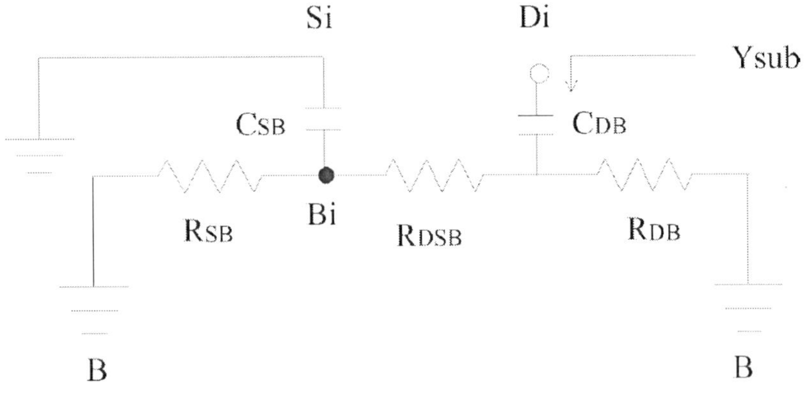

Figure 7. Equivalent circuit of the substrate network (taken from [5] with permission ©2005 IEEE)

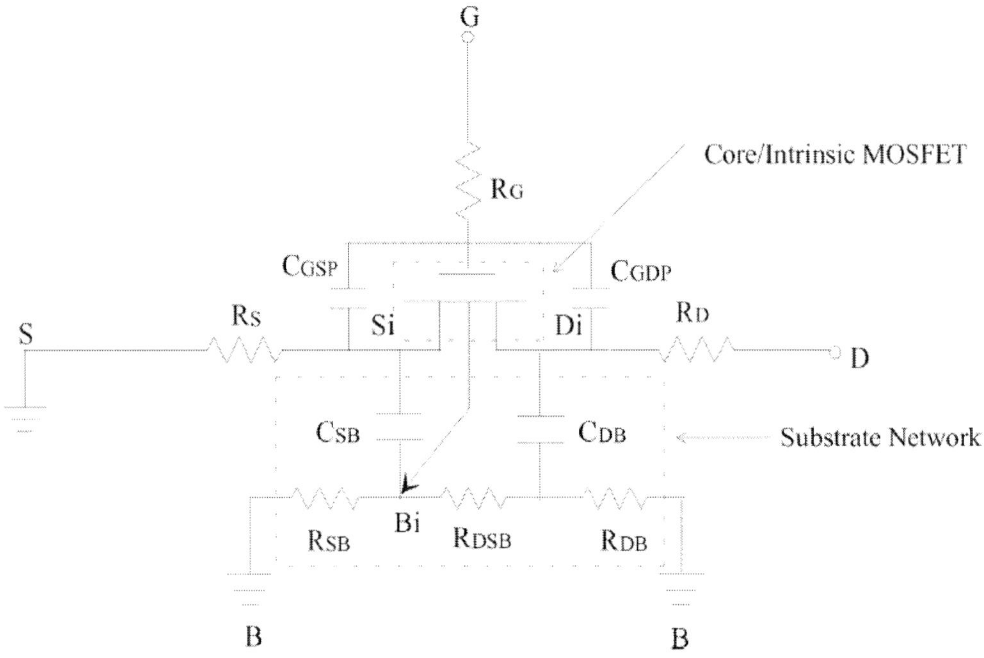

Figure 8. Sub-circuit RF MOS model with intrinsic and extrinsic components (taken from [5] with permission ©2005 IEEE)

Regarding the RF sub-circuit, with parasitic resistances at the drain, gate, source, and at substrate, a complete sub-circuit HF MOSFET model is provided in Figure 8. In the sub-circuit model, the characteristics of the intrinsic device is described by a MOS transistor compact model implemented in the circuit simulator, and all the extrinsic components have to be located outside the intrinsic device, so that the MOS transistor symbol in the sub-circuit only represents the intrinsic part of the device (It may also include the overlap capacitances at the source, drain and the bulk, depending on the intrinsic compact MOSFET model used). The source and drain series resistors have been incorporated outside the MOS intrinsic device for AC simulation, the gate resistance is added to the sub-circuit model - the substrate resistors are added for signal coupling estimation through the substrate, and two external diodes are added as to incorporate the influence of the substrate resistance at HF. The intrinsic substrate node should be connected at some point along the resistor R_{DSB}. However simulations show that connecting the intrinsic substrate to the source or the drain side has little influence on the simulated AC response. In some RF models, the intrinsic substrate is connected to the source side in order to save one node and one component for the sub-circuit model. Two external overlap capacitances, and as shown in Figure 8, with bias dependence can be added, but this is not always required, depending on the compact model used.

4. MOSFET Modelling in Relation to Analog / RF Circuits Critical Performance Parameters Estimation

It is greatly critical the Analog/RF designer to be aware of used MOSFET model capabities and accuracies or related inaccuracies. The basic problems concern the ***noise*** and the ***harmonic distortion*** performance – linearity. In particular, noise simulation is a key point of the design process. To evaluate circuit noise the designer uses one of the noise models available in the simulator. In particular SPICE proposes some specific built-in noise models for MOSFETs whereas others noise models, associated with more elaborated MOS transistor models like BSIM3 or BSIM4 which can also used in the SPICE simulator, are available. The selection of the noise model is very important in noise evaluation depending upon the type of MOSFET (n or p) and the operating mode. So any information or rule about the model to select, is very essential and useful to the designer.

For the channel noise models, the channel length modulation (CLM) effect is the major cause for the enhanced channel noise in short-channel MOSFETs. It begins to have an impact on the devices with channel lengths shorter than 0.5 m. On the other hand, the impact of the velocity saturation effect and the hot-electron effect in the gradual channel region seem to be negligible in the channel noise modelling of deep submicron MOSFETs. The channel noise is frequency independent and will increase when the channel length is reduced. This can be caused by the CLM effect and the noise contribution for substrate resistance through g_{mb}. On the other hand the gate induced noise and the correlated channel thermal and gate induced noise are proportional to f^2 and f, respectively, and both decrease when channel length is reduced because of C_{GS} decrease. From the simulated noise parameters with the noise sources implemented, it is concluded that the induced gate noise has strong impact on the noise performance of long-channel devices, especially for Noise Figure minimization. On the other hand, for short-channel devices, the enhanced channel noise will play the dominant role in the prediction of their noise performance. Further efforts are needed to model the induced gate noise, the correlation with channel thermal noise and its influence to the circuits at RF.

Concerning the MOSFET distortion behaviour - an RF model with accurate DC/AC fitting can predict the device distortion behaviour up to a very HF, enabling prediction of distortion behaviour with the standard modelling methodology. However, physical effects related to mobility degradation, velocity saturation, channel length modulation, and DIBL should be modelled carefully for accurate distortion prediction at high frequencies.

5. Summary

An overview of the available compact models has been performed mainly focusing on the new CMC model standard, the PSP model. Related basic benchmark test and the respective simulation inaccuracies were addressed in relation to analog/RF IC design applications. Critical performance parameters were analytically presented and the most commonly adopted RF MOSFET modelling approach was briefly presented. Critical parameter and accuracy information were provided, related to the available compact MOSFET model capabilities and the related accuracy in designing high performance Analog/RF IC systems.

REFERENCES

[1] Technical Note NXP-R-TN-2008/00299, "PSP 103.1" Users manual, In: W., XLi, G., Wu, G. D. J., Gildenblat, A. J., Smit, D. B. M. Scholten, & R. Klaassen, van Langevelde, Issued: 04/2009.

[2] Mitiko Miura-Marrausch et al, (2006). "HiSIM2: Advanced MOSFET model Valid for RF Circuit Simulation", *IEEE Tran. Electron Devices, vol. 53*, No. 9, September, 1994-2007

[3] Enz, C. C., Krummenacher, F. & Vittoz, E. A. (1995). "An Analytical MOS Transistor Model Valid in All Regions of Operation and Dedicated to Low-Voltage and Low-Current Applications", Analog Integrated Circuits and Signal Processing *Journal on Low-Voltage and Low-Power Design*, 8, 83-114, July.

[4] Enz, C. C., Krummenacher, F., Vittoz, E. A. (1987). "A CMOS Chopper Amplifier", *IEEE Journal of Solid-State Circuits 22, vol.3*, June, 335-342.

[5] Cheng, Y. et al., (2005). "MOSFET modelling for RF IC design", *IEEE Transactions on Electron Devices*, *vol. 52*, no. 7, July.

[6] Tsividis, Y. & Suyama, K. (1994). "MOSFET modeling for analog circuit CAD: Problems and prospects," IEEE *J. Solid State Circuits*, *vol. 29*, no. 3, 210-216, Mar..

[7] McAndrew, C., Gummel, H. & Singhal, K. (1995). "Benchmarks for compact MOSFET models," in Proc. *SEMATECH Compact Models Workshop*, Mar.

[8] Gildenblat, G., Li, X., Wu, W., Wang, H., Jha, A., van Langevelde, R., Smit, G. D. J., Scholten, A. J. & Klaassen, D. B. M. (2006). "PSP: An advanced surface-potential-based MOSFET model for circuit simulation," IEEE Trans. *Electron Devices*, *vol. 53*, no. 9, 1979-1993, Sep.

[9] Xin, Li et al. (2009). "Benchmark Tests for MOSFET Compact Models With Application to the PSP Model", *IEEE Tran. Electron Devices*, *vol. 56*, No. 2, February, 243-251.

[10] Liu, W. (2001). *MOSFET Models for SPICE simulation including BSIM3V3 and BSIM4,* John Wiley & Sons, United States.

[11] Wang, H. et. al, (2006). "A Unified Nonquasi-Static MOSFET model for large-signal and small-signal applications", *IEEE Tran. Electron Devices, Vol. 53*, No. 9, September.

[12] Noulis, T., Siskos, S. & Sarrabayrouse, G. (2007). "Analysis and selection criteria of BSIM4 flicker noise simulation models", *Intenational Journal of Circuit Theory and Applications, Vol. 36*, Issue 7, November, 813-823.

[13] Vandamme, E. P. & Vandamme, L. K. J. (2000). "Critical discussion on unified 1/f noise models of MOSFETs *IEEE Tran. Electron Devices*, *Vol. 47*, November, pp.2146–2152.

[14] Simoen, E. & Claeys, C. (1999). "On the flicker noise in submicron silicon MOSFETs" *Solid-State Electronics*, *vol. 43*, 865-882.

[15] Kolarova, R., Skotnicki, T. & Chroboczek, J. A. (2001). "Low frequency noise in thin gate oxide MOSFETs", *Microelectronics Reliability*, *Vol. 41*, (4), April, 579-585.

[16] Brederlow, R., Weber, W., Schmitt-Landsiedel, D. & Thewes, R. (1999). "Fluctuations of the low frequency noise of MOS transistors andtheir modeling in analog and RF-circuits", *International Electron Devices Meeting (IEDM),* 159-162.

[17] Marin, M., Akue Allogo, Y., De Murcia, M., Llinares, P. & Vildeuil, J. C. (2004). "Low frequency noise characterization in 0.13 μm p-MOSFETs. Impact of scaled-down 0.25, 0.18 and 0.13 μm technologies on 1/f noise", *Vol. 44(7)*, July, 1077-1085.
[18] Andries, J. Scholten et al., (2009). 'The New CMC Standard Compact MOS Model PSP: Advantages for RF Applications', *IEEE J. Solid State Circuits*, *Vol. 44*, No. 5, May, 1415-1424.
[19] *Compact Model Council Homepage*. [Online]. Available: www.eigroup.org/cmc.

In: CMOS Technology
Editor: Min-jun Kwon, pp.175-185

ISBN: 978-1-61761-325-8

Chapter 6

CATALOG OF VERSATILE QUADRATURE OSCILLATORS USING GROUNDED COMPONENTS

Sudhanshu Maheshwari*[*] *and Bhartendu Chaturvedi
Department of Electronics Engineering,
Z.H. College of Engg. and Technology, AMU, Aligarh, India

ABSTRACT

From their introduction in 1968 by Smith and Sedra and subsequent reformulation in 1970 by them, current conveyors have proved to be functionally flexible and versatile, rapidly gaining acceptance as both a theoretical and practical analog building block. Differential Voltage Current Conveyors are undoubtedly the most widely accepted operational devices in continuous time, current mode and voltage mode analog signal processing. This chapter is based on the CMOS implementation and application of Differential Voltage Current Conveyor (DVCC) and its' modified version, a relatively new active element suited for differential signal processing. In this chapter, a new circuit topology for realizing several second-order versatile (current mode as well as voltage mode) quadrature oscillator (VQO) is proposed. Each circuit employs three differential voltage current conveyors and all grounded passive components, ideal for IC implementation. The circuits exhibit non-interactive frequency control and low THD. Effects of non-idealities are also analyzed. PSPICE simulations using 0.5μ CMOS parameters confirm the validity and practical utility of the proposed circuits.

1. INTRODUCTION

In the last decade, various new current-mode (CM) active building blocks have received considerable attention owing to their larger dynamic range and wider bandwidth with respect to operational amplifier (op-amp) based circuits. As a result, current-mode active components have been increasingly used to realise high-speed and high-bandwidth circuits operating in

[*] Corresponding author: maheshwarispm@rediffmail.com

the current or the voltage mode. To name a few, current conveyors of various generations and type have been increasingly employed. Similarly many tunable active elements like current controlled current conveyors and more recently current controlled current differencing buffered amplifiers have also been used with success. Differential voltage current conveyor (DVCC) is another very interesting option especially due to its differential input nature, which allows better rejection of noise and common mode signals. The popularity of the active element is due to its easy CMOS compatibility and suitability for wide variety of analog signal processing applications [1-14].

The contents presented in the chapter are organized as sections: Section-2: Available Oscillators, Section-3: Proposed Circuits, Section-4: Non-ideality and Parasitic Study, Section-5: Simulation Results, Section-6: An Extended Application and Section-7 as Conclusion.

2. Available Oscillators

Numerous oscillator circuits using different types of current conveyors and passive components have been reported [4–14]. Whereas most of the works reported were based on voltage mode outputs [11-12], there are many circuits which generate current mode outputs [6, 8-10, 12-13]. Current mode sinusoidal oscillators are of special interest with the recent popularity of current-mode circuits. Moreover, oscillators with quadrature outputs (with 90° phase shift) find applications in communication and instrumentation systems. Several of the already reported circuits exhibit this property [9, 11, 13] whereas some circuits provide multiphase outputs [10]. Another important feature desired from current mode circuits is high output impedance for ease of cascading [2, 8], without any additional sensing elements in form of current followers. Furthermore, it is desirable for an oscillator circuit to show control over its oscillation frequency independent of the sustainability condition. This feature is also exhibited by many of the available circuits. Recently, Differential Voltage Current Conveyor (DVCC) has also become popular for realizing analog signal processing circuits [1-7]. An oscillator based on differential difference current conveyors with dual outputs was also published [7]. However, none of the circuits provided four current outputs, and the errors in simulated frequency of oscillation were relatively high. Similarly another circuit based on two differential voltage current conveyor and five grounded components enjoyed single element control property, and provided a current and a voltage output [12]. A very compact circuit based on single DVCC uses floating resistors and provided one current output [6]. Recently a compact quadrature oscillator with two DVCC and four grounded components was proposed, which needed extra stages for sensing current outputs [13]. Another recent current mode quadrature oscillator using two DVCC and four grounded passive components provides four high impedance quadrature current outputs and a number of voltage outputs. Owing to the smaller passive components count, the work does not exhibit non-interactive frequency control [2]. Very recently some third order oscillators using DVCC were also successfully developed, which show improved performances [3-4].

This paper presents a new circuit topology realizing versatile quadrature oscillators using DVCC and grounded passive components. A total of eight distinct circuits can be realized using the proposed topology. The proposed circuit exhibit non-interactive frequency control, provide four progressive quadrature current outputs and two quadrature voltage outputs. All these features make the new circuits a unique addition to the already available literature on the subject. A detailed non-ideal analysis is also included. Simulation results are given for verification purpose.

3. Proposed Circuits

3.1. Circuits' Description

The differential-voltage current conveyor (DVCC) is characterized by the following port relationship [1].

$$V_X = V_{Y1} - V_{Y2};\ I_{Y1} = I_{Y2} = 0;\ I_{Z+} = I_X;\ I_{Z-} = -I_X \tag{1}$$

The symbol and CMOS implementation of the proposed modified DVCC are shown in Figure 1, and is characterized by following relationship:

$$V_X = V_{Y1} - V_{Y2};\ I_{Y1} = I_{Y2} = 0;\ I_{Zi+} = k_1 I_X;\ I_{Zj-} = -k_2 I_X \tag{2}$$

where i = 1, 2 and j=1.

The new proposed circuit topology for versatile quadrature oscillator (VQO) is shown in Figure 2. The proposed quadrature oscillator circuit topology employs one modified DVCC (equation 2) and two DVCCs (eqn. 1).

In all eight circuits can be derived from Figure 2 by changing some of the terminals' connections (nodes A and B). Table 1 shows the connection details for all the eight circuits. The proposed quadrature oscillator circuits employ one modified DVCC (eqn. 2) and two DVCCs (equation 1). A modified DVCC incorporates two Z+ stages, one with k_1=1 and the other with k_1=3. Moreover, it also has a Z- stage with k_2=3 (eqn. 2). It is to be further noted that in all the proposed circuits DVCC3 is the modified one (Figure 1) whereas DVCC1 and DVCC2 are the ones with unity current gain (eqn. 1) [1]. However when implementing only two Z+ outputs (for DVCC1 and/or DVCC2, as required in some of the proposed circuits), transistors M15-M20 (Figure 1b) need not be implemented. Both Z+ stages are implemented by keeping the aspect ratios of the transistors M8, M9 same as transistor M7 and the aspect ratios of transistors M13, M14 to be same as transistor M12. As far as DVCC3 implementation is concerned the required gain (as given in eqn. 2) from X to Z terminal is realized by maintaining the aspect ratios of different transistors as defined in eqns. (3 – 4).

$$(W/L)_{7,8} = a;\ (W/L)_{12,13} = b \tag{3}$$

$$(W/L)_{9,15,16,17} = 3a;\ (W/L)_{14,18,19,20} = 3b \tag{4}$$

The modified DVCC uses only 20 transistors as shown in Figure 1b. The complete circuitry excluding passive components would need either 48 or 56 transistors; the one with 48 transistors (VQO-I) being the simplest of all in terms of circuit complexity.

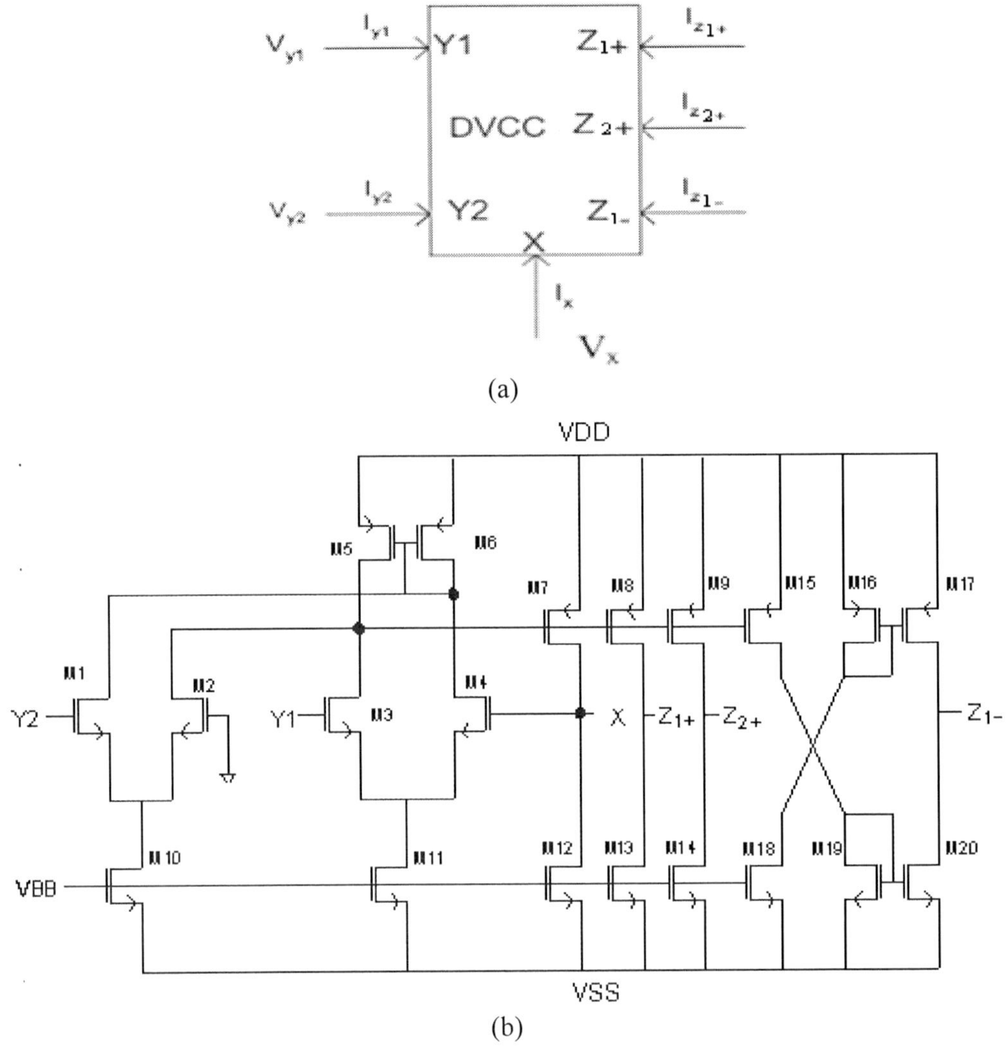

Figure 1. Modified DVCC: (a) symbol, (b) CMOS implementation

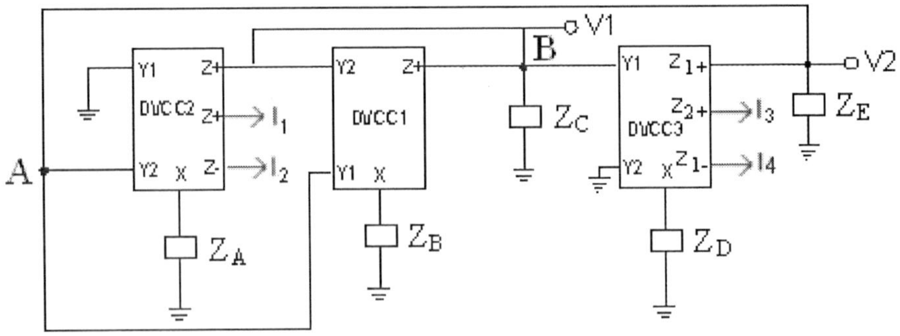

Figure 2. Proposed versatile quadrature oscillator (VQO) topology

Table 1. Eight possible quadrature oscillator circuit connection details

Circuits ⇩	Connections from DVCCs				Change in passive components				
	'A' connected to		'B' connected to		Z_A	Z_B	Z_C	Z_D	Z_E
	DVCC-1	DVCC-2	DVCC-1	DVCC-2					
VQO-I	Y_1	Y_2	Y_2 & Z_+	Z_+	R_2	C_1	R_1	R_3	C_2
VQO-II	Y_1	Y_1	Y_2 & Z_+	Z-	R_2	C_1	R_1	R_3	C_2
VQO-III	Y_2	Y_2	Y_1 & Z-	Z_+	R_2	C_1	R_1	R_3	C_2
VQO-IV	Y_2	Y_1	Y_1 & Z-	Z-	R_2	C_1	R_1	R_3	C_2
VQO-V	Y_1	Y_2	Y_2 & Z_+	Z_+	C_2	R_1	C_1	R_2	C_3
VQO-VI	Y_1	Y_1	Y_2 & Z_+	Z-	C_2	R_1	C_1	R_2	C_3
VQO-VII	Y_2	Y_2	Y_1 & Z-	Z_+	C_2	R_1	C_1	R_2	C_3
VQO-VIII	Y_2	Y_1	Y_1 & Z-	Z-	C_2	R_1	C_1	R_2	C_3

3.2. Circuit Analysis

The circuit of VQO-I is analyzed using eqns (1-2). The characteristic equation for the proposed circuit is found as.

$$s^2 + s\left(\frac{1}{C_1R_1} - \frac{1}{C_2R_3}\right) + \frac{1}{C_1C_2R_2R_3} = 0 \tag{5}$$

It is clear from equation (5) that the frequency of oscillation (FO) and the condition of oscillation (CO) are as follows:

$$\text{FO: } \omega_0 = \sqrt{\frac{1}{C_1C_2R_2R_3}} \;;\text{CO: } C_1R_1 \geq C_2R_3 \tag{6}$$

Equation (6) shows that the FO can be varied independent of CO by R_2. The four quadrature current outputs (I_1, I_2, I_3 and I_4) of the proposed circuit (VQO-I), depicted in Figure 2, are related as:

$$I_4 = -I_3,\ I_2 = -I_1 \tag{7}$$

$$I_3 = jkI_2,\ k = \omega C_2R_2 \tag{8}$$

For k= 1, all the four quadrature current outputs, which are given in equation (7, 8) have equal amplitude.

The two quadrature voltage outputs (V_1 and V_2) of the proposed circuit (VQO), depicted in Figure 4, are related as

$$V_1 = jkV_2,\ k = \omega C_2R_3 \tag{9}$$

For k= 1, both the two quadrature voltage outputs, which are given in equation (9) have equal amplitude.

Similar analysis can be performed for other VQOs listed in Table 1.

4. Non Idealities and Parasitic Study

4.1. Non Ideal Analysis

A non ideal DVCC is characterized by

$$V_X = \beta_1 V_{Y1} - \beta_2 V_{Y2}\,;\; I_{Y1} = I_{Y1} = 0\,;\; V_{Z+} = \alpha_1 I_X\,;\; V_{Z-} = -\alpha_2 I_X \quad (10)$$

where β_i (i = 1, 2) is the voltage transfer gain and α_i (i = 1, 2) is the current transfer gain of the ith DVCC. These transfer gains differ from unity by the voltage and current tracking errors of the DVCC. More specifically, $\alpha_i = (1 - \delta_i)$ and $\beta_i = (1 - \varepsilon_i)$, where δ_i is the current transfer error (tracking error) of the ith DVCC and ε_i is the voltage transfer error (tracking error) of the ith DVCC.

The VQO-I is analyzed using equation (10) and non ideal characteristic equations, frequency of oscillations (FO) and condition of oscillations (CO) are found as:

The characteristic equation for the VQO-I is given in equation (11)

$$s^2 + s\left(\frac{1}{\alpha_{11}\beta_{21}C_1R_1} - \frac{\alpha_{13}\beta_{11}\beta_{13}}{\beta_{21}C_2R_3}\right) + \frac{\alpha_{12}\alpha_{13}\beta_{12}\beta_{13}}{\alpha_{11}\beta_{21}C_1C_2R_2R_3} = 0 \quad (11)$$

The FO and CO for VQO-I is given in equations (12) and (13)

$$\text{FO: } \omega_0 = \sqrt{\frac{\alpha_{12}\alpha_{13}\beta_{12}\beta_{13}}{\alpha_{11}\beta_{21}C_1C_2R_2R_3}} \quad (12)$$

$$\text{CO: } \alpha_{11}\alpha_{13}\beta_{11}\beta_{13}\beta_{21}C_1R_1 \geq \beta_{21}C_2R_3 \quad (13)$$

where β_{11} and β_{21} are the voltage transfer gains of DVCC1 from the Y_1 and Y_2 terminals to the X terminal, respectively, β_{22} is the voltage transfer gains of DVCC2 from the Y_2 terminal to the X terminal, β_{12} is the voltage transfer gains of DVCC2 from the Y_1 terminal to the X terminal, β_{13} is the voltage transfer gains of DVCC3 from the Y_1 terminal to the X terminal and α_{11} is the current transfer gain of DVCC1 from X to Z_+ terminal, α_{12} is the current transfer gain of DVCC2 from the X to Z_+ terminal and α_{13} is the current transfer gain of DVCC3 from the X to Z_+ terminal.

The active and passive sensitivities of VQO-I circuit are low, as given in equation (14).

$$\left|S^{\omega_o}_{\alpha_i}\right| = \left|S^{\omega_o}_{\beta_i}\right| = \left|S^{\omega_o}_{C_i, R_i}\right| = \frac{1}{2} \quad (14)$$

4.2. Effect of DVCC Parasitics

Analogous to the second-generation current conveyor (CCII), the DVCC has a small parasitic resistance R_X at port X and high output impedance ($R_Z//C_Z$) at port Z. Where ever the X terminal of the DVCC is connected to a resistor, the parasitic resistance at the X terminal of the DVCC (R_X) can be absorbed as a part of the main resistance. However, the X-terminal of DVCC1 in VQO-I is terminated by a capacitor; hence R_{Xi} (i=1, 2 for DVCC1 and DVCC2 respectively) appears in series with them. A reanalysis of the circuits yields the modified frequency as "ω_o / p", where $p \approx (1+R_{Xi}/R_1)^{1/2}$. As the value of R_{Xi} is much smaller then the external resistor (R_1) so the effect of second component of 'p' will be negligible and 'p' would be approximately equal to 1, thus only slightly lowering the frequency of oscillators. The effects of the capacitors at port Y and Z of the DVCCs are also negligible because these capacitors are quite small (and process dependent) as compared to the external capacitors. However the effective values of the capacitors after parasitics' inclusion is given below

$$C_1\,(\text{effective}) = C_1+C_{X1},\; C_2\,(\text{effective}) = C_2+C_{Y2}+C_{Y1}+C_{Z3} \tag{15}$$

$$C_1\,(\text{effective}) = C_1+C_{Y2}+C_{Z2}+C_{Z1},\; C_2\,(\text{effective}) = C_2+C_{X2},\; C_3\,(\text{effective}) = C_3+C_{Y2}+C_{Y1}+C_{Z3} \tag{16}$$

From equation (15) & equation (16) it is clear that the parasitic capacitances appear in shunt with external capacitors in VQO-I, thus ensuring a possibility of pre-distorting the designed values. Therefore it is to be concluded that the circuits are not adversely affected by the parasitic capacitances and X-terminal resistances. This would be further confirmed in the following section

5. Simulation Results

PSPICE simulations were performed using the CMOS realization of DVCC with 0.5μ level 3 MOSFET parameters. The aspect ratios is given in Table 2. The supply voltages used were ±2.5V and V_{BB}=-1.6V. The VQO-I circuit was designed with: $C_1 = C_2 = 30$pF, $R_1 = 10.5$ KΩ, $R_2 = 1.5$ KΩ, $R_3 = 10$ KΩ. The simulated frequency of oscillation of quadrature oscillator is 1.37MHz, which is the same as the designed value. The output waveforms for the four quadrature current outputs are shown in Figure 3. The power consumption was found to be 7mW. The T.H.D of the proposed circuit at all outputs is with in 2.5%, which is low. The waveforms of two quadrature voltage outputs are shown in Figure 4.

Next, another set of results are shown where the two quadrature outputs have equal amplitudes as well. The designed values used were $C_1 = C_2 = 30$pF, $R_1 = 4.2$kΩ, $R_2 = 4$kΩ, and $R_3 = 4.3$kΩ. The designed frequency of oscillation of quadrature oscillator is 1.25MHz. The waveforms and their Fourier spectrum are shown in Figure 5 and 6 respectively. The value of capacitors used (30pF) are consistent with the IC feasibility. The FO is found to vary from 0.53MHz to 1.67MHz for R_2, variation from 10KΩ to 1KΩ respectively. The result of non-interactive FO control is given in Figure 7.

Table 2. Aspect ratio used

Transistors	W(μm)	L(μm)
M1, M2, M3, M4	1.6	1
M5, M6	8	1
M10, M11	29	1
M7, M8	20	1
M12, M13	90	1
M9, M15, M16, M17	60	1
M14, M18, M19, M20	270	1

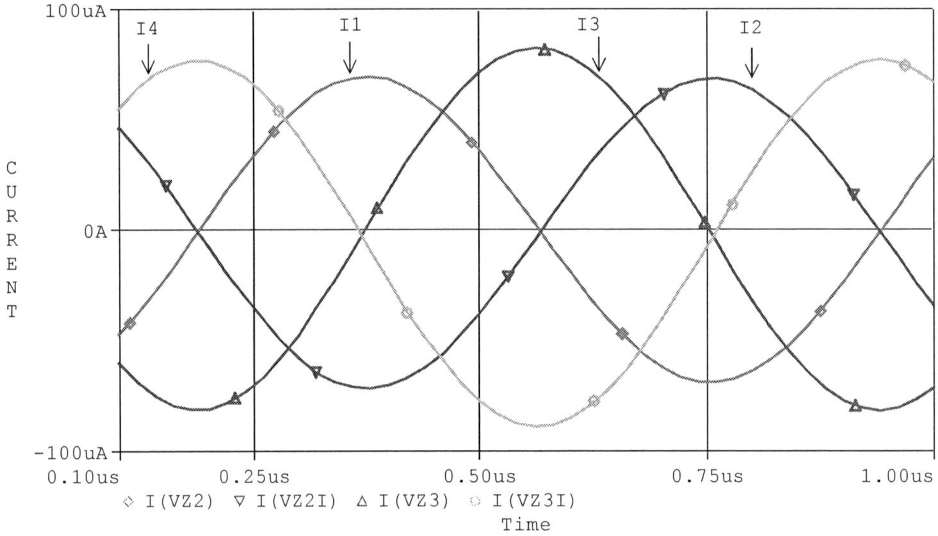

Figure 3. Four quadrature current outputs of VQO-I circuit at 1.37MHz

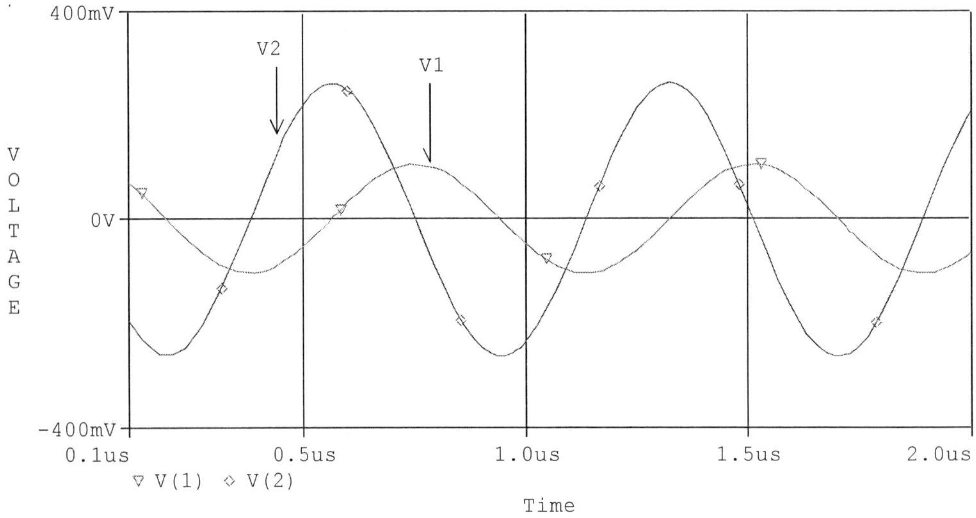

Figure 4. Two quadrature voltage outputs of VQO-I circuit at 1.37MHz

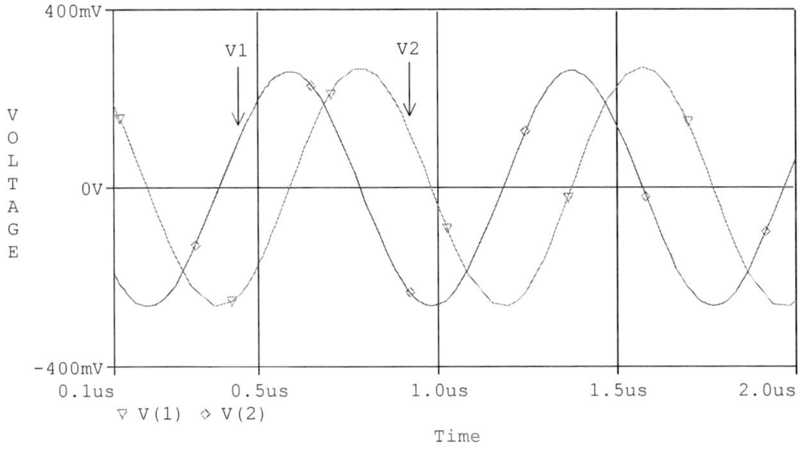

Figure 5. Two quadrature voltage outputs with equal amplitude at 1.25MHz

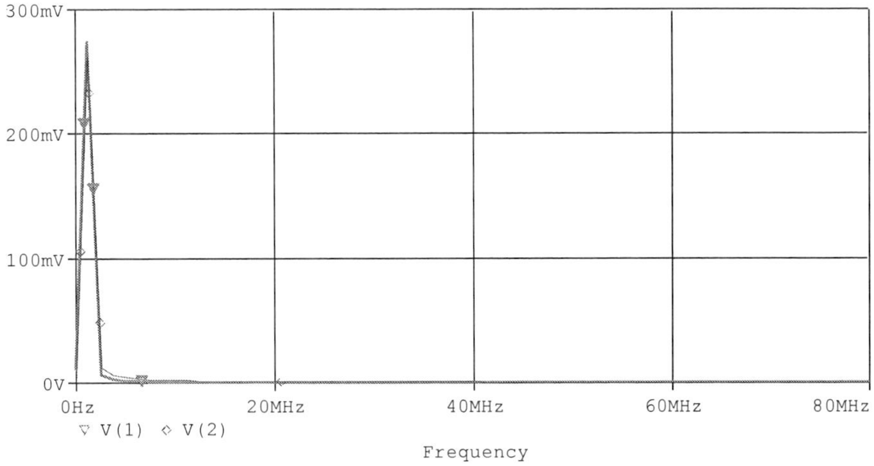

Figure 6. Fourier spectrum of voltage outputs with equal amplitude

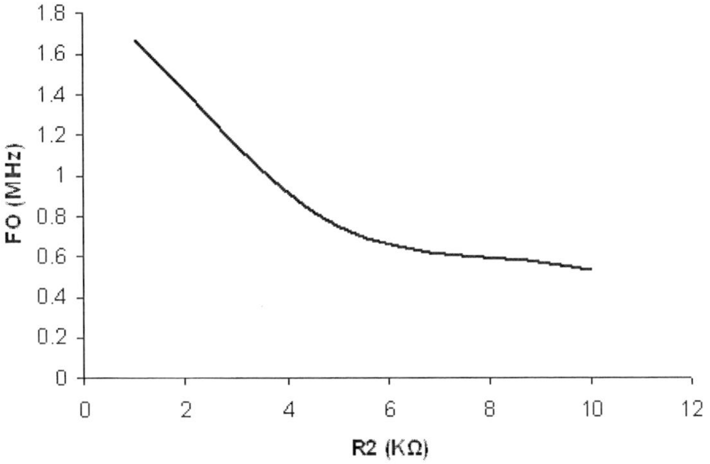

Figure 7. FO variation with R_2 for R_1=10.5KΩ, R_3=10KΩ and C_1=C_2=30pF

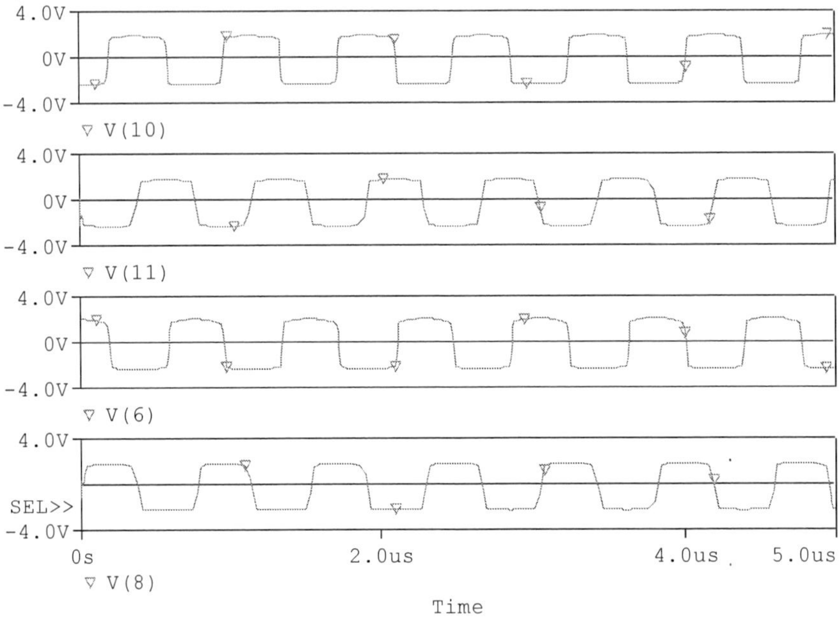

Figure 8. 4-phase clock wave shapes

5. An Extended Application

Here, a new application of the proposed quadrature oscillators in clock generation is given. A four phase clock is generated by using given VQO-I (Figure 2 & Table 1). For generation of four phase clocks, four voltages' outputs are taken instead of four current outputs (as shown in Figure 2) at the Z terminals of DVCC (1) and DVCC (3). As DVCC (1) and DVCC (3) both had Z+ and Z− output stages, so as to generate four phases. The four-phase clock voltage outputs are at a progressive phase shift of 90°. Note that the Z terminals exhibit a high output resistance, sufficient to saturate the DVCCs. The output levels depend on the supply voltage (V_{DD} and V_{SS}). No additional resistors are being used, thus the new scheme is compatible with monolithic implementation. The results of four phase clock are shown in Figure 8 and very well justify this new application. The outputs can also be made unidirectional by employing appropriate additional circuitry.

Conclusion

A new second-order versatile quadrature oscillator circuit topology, realizing eight distinct circuits, each with four quadrature current outputs and two quadrature voltage outputs with low THD has been proposed. The circuit topology uses three DVCCs and grounded passive components. The circuits also enjoy high impedance current output. An extended application in four-phase clock is further given. PSPICE simulations using 0.5μ CMOS parameters support the validity and practical utility of the proposed circuit.

REFERENCES

[1] Elwan, H. O. & Soliman, A. M. (1997). "Novel CMOS differential voltage current conveyor and its applications", IEE Proc., *Circuits Devices Syst., vol. 144*, 195-200.

[2] Maheshwari, S. (2008). High output imedance current-mode all-pass sections with two grounded passive components, IET: *Circuits Devices and Systems, 2*, 234-242.

[3] Maheshwari, S. (2008). High performance voltage mode multifunction filter with minimum component count, *WSEAS Transactions on Electronics, Vol. 5*, no. 6, 244-249.

[4] Maheshwari, S. (2009). Analog signal processing using a new circuit topology, IET: Circuits *Devices and Systems, vol. 3*, no. 3, 106-115.

[5] Maheshwari, S. (2009). Quadrature oscillator using grounded components with current and vltage outputs, IET: *Circuits Devices and Systems, vol. 3*, no. 4, 153-160.

[6] Gupta, S. S. & Senani, R. (2000). "Grounded-capacitor current mode SRCO, novel application of DVCCC." *Electron. Lett., vol. 36*, no. 3, 195-196.

[7] Gupta, S. S. & Senani, R. (2001). "CMOS Differential Difference Current Conveyors and Their Applications", *IEE Proc Circuits Devices Systems, vol. 148*, 335-336.

[8] Maheshwari, S. & Khan, I. A. (2005). "Current controlled third order quadrature oscillator", *IEE Proc. Circuits Devices and System, vol. 152*, 605-607.

[9] Keskin, A. U. & Biolek, D. (2006). "Current mode quadrature oscillator using Current Differencing Transconductance Amplifier (CDTA)", *IEE Proc.-Circuits Devices Systems, vol. 153*.

[10] Khan, I. A. & Hasan, S. (2006). "Current mode four phase quadrature oscillator using CCIIs", *J. of Active and Passive Electronic Devices, vol. 1*, 273-279.

[11] Maheshwari, S. (2007). "High input impedence VM-APSs with grounded passive elements", *IET Circuits Devices Systems, vol. 1*, 72-78.

[12] Kumar, V., Keskin, A. & Pal, K. (2007). "DVCC Based Single Element Controlled Oscillators Using All Grounded Components and Simultaneous Current-Voltage Mode Outputs", *Frequenz, vol. 5-6*, 141-144.

[13] Maheshwari, S. (2007). "Novel Cascadable Current Mode First Order All-Pass Sections", Int. *J of Electronics, vol. 94*, 995-1003.

[14] Soliman, A. M. (2009). "Generation of oscillators based on grounded capacitors current conveyors with minimum passive components", *JCSC, vol. 18(5)*, 857-873.

In: CMOS Technology
Editor: Min-jun Kwon, pp.187-219
ISBN: 978-1-61761-325-8

Chapter 7

FEEDTHROUGH: AN ENERGY EFFICIENT CMOS LOGIC FAMILY FOR ARITHMETIC CIRCUITS

*Victor Navarro–Botello, Juan A. Montiel–Nelson, and † Saeid Nooshabadi**
Institute for Applied Microelectronics
University of Las Palmas de Gran Canaria
E35017 Canary Islands, Spain
† Department of Information and Communications
Gwangju Institute of Science and Technology
500–712, Republic of Korea

Abstract

In this chapter we present a comprehensive and up-to-date study on the feedthrough logic (FTL) concept for designing high performance arithmetic circuits in CMOS technology.

The FTL logic family, for high speed and low power CMOS applications, was introduced by the authors in the recent past. FTL works successfully on the domino concept with the added feature that gates commence evaluation even before their input signals are valid. This is accomplished by means of initial quasi transition of all the FTL cells to a high gain point. This fact results in very fast evaluation time in the computational blocks for the final evaluation when inputs arrive. Furthermore, the well known problems of domino logic, such as the need for output inverters and charge redistribution, are completely eliminated, thus reducing the chip area and delay, and improving the performance.

The FTL is well suited to arithmetic circuits where the critical path is made of a large cascade of inverting gates. Furthermore, FTL based circuits perform better in high fanout and high switching frequencies due to both lower delay and dynamic power consumption.

Experimental results, from the chip measurements, demonstrated superior performance of the FTL ripple carry adders (RCA) when compared with the dynamic domino and traditional CMOS logic styles. Our 14–bit low power implementation performs faster, (2.6 times smaller propagation time delay, and 1.85 times higher maximum frequency), and provides a better energy efficiency (3.11 times or 67.9%), when

*E-mail address: vnavarro,montiel@iuma.ulpgc.es, saeid@gist.ac.kr

compared with the dynamic domino style. On the other hand, an 18–bit high speed FTL deign, working at maximum frequency, outperforms the dynamic domino logic in terms of the propagation delay (19.5 times less), maximum frequency (12.1 times more), and energy efficiency per bit (29.97 times or 96.7% better). Moreover, the same 18–bit high speed FTL adder outperforms other high performance adders, such as multilevel CSAs, in terms of both, energy efficiency (1.72 times) and propagation time delay (1.78 times).

However, FTL is very sensitive to the device mismatch, and the variations in the capacitive loads in the the manufacturing process. We show how the the sensitivity of the FTL based design can be improved through very clever design techniques.

This chapter also discusses the capabilities of the FTL logic in practical applications, and how to extend the use of this logic to larger word–length arithmetic circuits.

Keywords: feedthrough logic, CMOS digital integrated circuits, digital arithmetic, high speed integrated circuits, low power design, CMOS arithmetic circuits

AMS Subject Classification: 93C62, 94C10, 06E30, 11A25.
PACS 05.45-a, 52.35.Mw, 96.50.Fm.

1. Introduction

Reduction in the energy dissipation, in CMOS integrated circuits, while maintaining the high performance, has been the topic of intense research in the recent past. The proposed design techniques trade power for performance in the delay critical sections of the circuit [1]–[5]. To achieve this goal the mix of dynamic and static circuit styles [2], use of dual supply voltages [3]–[5], and dual V_T transistors [5], have been proposed.

To improve the performance of arithmetic circuits, a new logic family called *feedthrough logic* (FTL) was proposed by Montiel–Nelson and Nooshabadi for the integrated circuits in GaAs technology [6, 7]. FTL works on the domino concept for dynamic circuits, with the added feature that gates commence evaluation even before their inputs are valid. This fact results in very fast evaluation time in the computational blocks. Furthermore, the well known problems associated with the domino logic — such as the limitation of non–inverting only logic, charge redistribution and the need for output inverters — are completely eliminated [8], thus reducing the chip area and delay, and improving the performance.

FTL shows a high design flexibility; it can be used in domino–like cascaded stages, differential style, multiple output logic with iterative networks. It also can be pipelined with fast dynamic latches [6].

The FTL principle of operation was presented in [9]. Unlike other dynamic logic families, FTL resets the output nodes to low when the clock signal (ϕ) is set high. When the clock signal goes low, cascaded gates rise to their switching threshold value of V_{TH}, performing a partial transition to a high gain point. During this low phase of ϕ, cascaded stages evaluate their inputs in a domino–like fashion, with the output nodes only making a partial transition from the V_{TH} point to high or low logic levels.

The main challenge in the design of FTL based circuits is, however, maintaining the V_{TH} stability for long cascaded circuit structures, which is the key factor in the fast logic evaluation and high performance of FTL based circuits.

The FTL concept was successfully introduced by the authors in CMOS technologies for high performance and low power arithmetic circuits [10]. This research was also extendend to high speed circuits, and the adder sensitivity against variations in the capacitive load, temperature, power supply, process corner and noise coupling was analyzed for the high speed and low power FTL families [11]. Our pre–layout results in [11, 12] showed substantial performance improvement of FTL with respect to standard static fully complementary CMOS, pseudo–NMOS and dynamic domino logic design styles.

This chapter provides a comprehensive insight into the FTL design style, which includes the previously published work and extensive new material not available elsewhere. Specifically the reader will find the following up-to-date material on FTL.

- It provides an in–depth analysis and evaluation of the performance of the basic cells, and comparison with the equivalent standard static CMOS logic. The presented analysis and evaluation develops further insight into the working of the proposed FTL structures that can be utilized to optimize FTL based circuits.
- It presents the physical implementation of a set of Ripple Carry Adders (RCA) FTL adders. We compare their features with a corresponding set in the dynamic domino CMOS logic, based in post–layout simulation results. Useful implementation details are given to facilitate the use of FTL design style in other arithmetic circuits.
- The analysis of the performance variation of the implemented RCA cells against process variations and device mismatching due to the manufacturing process is presented, through a set of Monte–Carlo simulations.
- Chip measurements results are parented and discussed for the RCA cells. We show how our 14–bit low power FTL adder improves the energy efficiency (3.11 times or 67.9%), enhances the maximum achievable clock frequency (1.85 times), and provides for a smaller evaluation time (2.6 times), when compared with the dynamic domino RCA. Furthermore, our 18–bit high speed FTL RCA outperforms the dynamic domino logic in terms of the maximum clock frequency (12.1 times), propagation time delay (19.5 times), and energy efficiency (29.97 times or 96.7%) when operating at the maximum clock frequency.
- Differential RCA cells are introduced to the reader. Pre–layout and sensitivity analysis sections have been presented with the corresponding simulation results.

The material in this Chapter is organized as follows. Section 2 explains in details the principle of operation of FTL and provides a simple, but effective, way for estimating of the delay performance of FTL and standard static CMOS gates. Section 3 presents various FTL circuit structures and provides the simulation results for a simple inverter gate implemented in those structures. In this section simulation results are analyzed and practical applications of FTL are identified. The design and analysis of high performance and low power adders in various FTL structures are presented in Section 4. The section also presents the differential and non–differential RCA cells, as well as the the pre–layout simulation results. The key issues addressed in the design of adders with long propagation chain is of the particular interest for the development of other high performance FTL structures. The sensitivity analysis of the designed adders with respect to word–length size and technology process parameters are discussed in Section 5. Section 6 presents the implementation of the non–differential RCA cells using the $0.13\ \mu$m 1.2 V / 3.3 V 1P8M Logic High Speed Process from UMC. Our post–layout results demonstrate that the FTL concept can be utilized to increase the pipeline logic depth, to beyond ten gate stages used in the current deep–submicron CMOS technologies [13, 14]. Monte–Carlo analysis of the physical implementation of RCA structures in both dynamic domino and FTL is presented in Section 7. Section 8 sums up the measurement results of the FTL adders implemented on a test–chip, and Section 9 discusses and develops the context for the use of FTL structures. Finally, the main conclusions are presented in Section 10.

2. FTL Principle of Operation

The basic structure of a FTL gate is shown in Fig. 1(a). It consists of an NMOS logic network (*NMOS block*), an NMOS transistor (T_r) for resetting the output node to low logic level, together with a pull up PMOS load transistor (T_p). T_r and T_p are controlled by the clock signal (ϕ).

During the high phase of ϕ (reset phase), the FTL output node is pulled to ground (GND) through T_r. When ϕ goes low (evaluation phase), T_r is turned off, and the output node conditionally evaluates to either high or low logic levels. If the logic network evaluates to high, node *out* is pulled up toward V_{cc} (inverting logic), otherwise, it will remain low. Note that, the difference between the FTL and the other dynamic logic families, such as domino CMOS, that are based on charge storage, is that it does not need output inverters to restore the polarity or to avoid discharging of the output node. An FTL gate can freely evaluate to high or low logic levels during the evaluation phase.

Consider a long chain of inverters as shown in Fig. 1(b). Let us assume that all output nodes of the inverting chain are completely discharged as result of a previous reset phase. Also assume that all the gates in the chain are subjected to identical load and drive conditions. Now, let us observe the three intermediate nodes N_{N-1}, N_N, and N_{N+1} in the chain.

When the clock signal falls, the input node of the $N+1$ stage (N_N) is at the ground level, causing the output node (N_{N+1}) to begin charging. But the input node (N_N) is itself the output node of the previous stage, which have its own input node (N_{N-1}) at low potential. So, because of identical drive and capacitive load condition at each of the intermediate stages, both output nodes (N_N and N_{N+1}) will begin rising together. On the other hand, because of the inverting nature of the gates, as input nodes at the intermediate stages reach the voltage threshold of the gate (V_{TH}, point where $V_o = V_i$ in the DC transfer curve) the output nodes will stop rising and will remain at V_{TH} (Fig. 2).

However, due to eventual propagation of the input of the inverter chain across the intermediate stages, will force the input node of the N stage (N_{N-1}) to high or low logic levels, which in turn will cause its output node (N_N) to evaluate to the opposite logic level. All the intermediate stages behave in the identical manner. They start evaluation by first reaching the V_{TH} voltage, and then make the final transition to the final logic level.

So, we conclude that all the gates in the chain are subjected to a simultaneous initial pre–evaluation stage to an intermediate voltage (V_{TH}), before going through the final evaluation

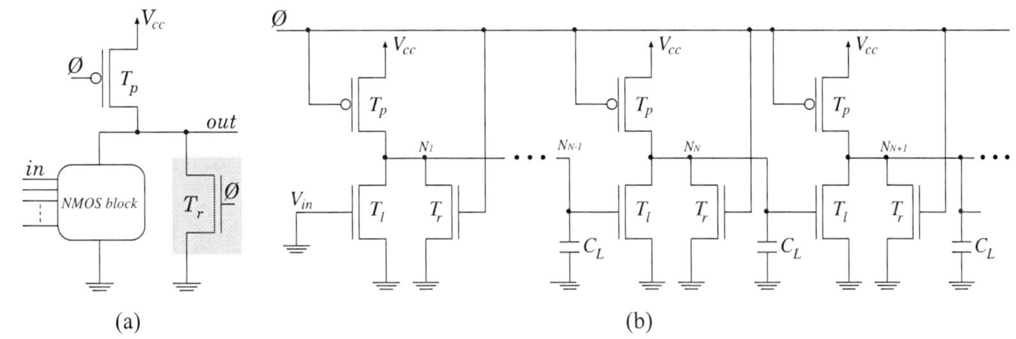

Figure 1. (a) Basic structure of FTL gate, and (b) Long chain of FTL inverters: transistor level circuit diagram.

stage. This pre–evaluation not only reduces the swing in the final transition, but also speeds it up. The speed up is due the fact that at the V_{TH} point, all gates in the inverting chain are operating in a high gain region. This feature distinguishes the FTL from other logic families. At V_{TH} point any small variation in the input nodes would cause a fast transition of the voltage at the output node. In all other logic families for the output node to begin transition, the inputs need to cross the threshold voltage. Furthermore, in FTL when the valid inputs to a gate are asserted, the gate output needs only make a partial transition from V_{TH} to V_{OH} or V_{OL}. The higher speed of FTL is due to the reduction in both low–to–high and high–to–low propagation time delays.

Next, we develop a very simple relation for the estimation of the propagation time delay (t_p), for the FTL and standard static CMOS inverters, in the inverter chain structure of Fig. 1(b). The transistor level circuit diagram for a FTL inverter and its small signal equivalent circuit are shown in Fig. 3(a) and 3(b), respectively. To compute the propagation time delays, we can use the equivalent small signal circuit to obtain the various circuit parameters associated with node N, including the net current into the node (I_N), the node capacitive value (C_N) and the voltage transition behaviors. However, to simplify the expression for the delay estimation we use an equivalent average current ($\overline{I}_N$) and capacitive load ($\overline{C}_N$) at

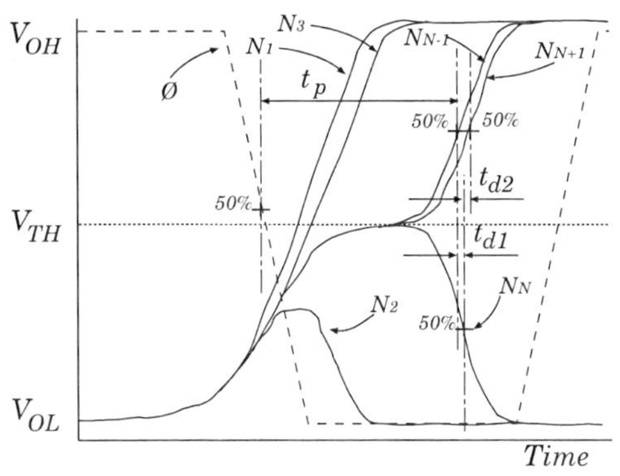

Figure 2. Plot of voltages at intermediate nodes (N_{N-1}, N_N and N_{N+1}) of a long chain of FTL inverters.

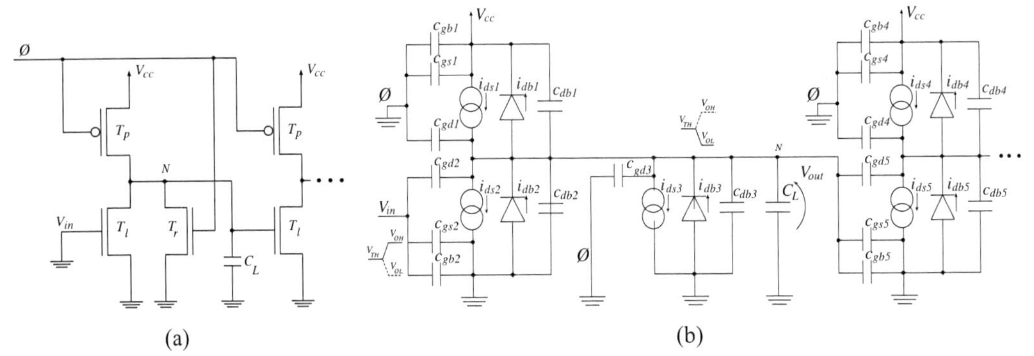

Figure 3. Analysis of FTL inverter: (a) circuit diagram, and (b) equivalent circuit for the transient analysis.

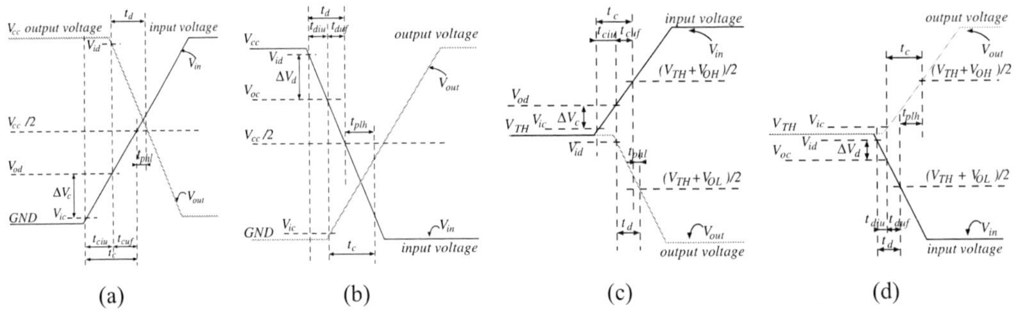

Figure 4. Simple model for gate propagation time delay. Output versus input voltage for static CMOS (a) high–to–low, (b) low–to–high; and FTL (c) high–to–low, (d) low–to–high transitions.

node N in both high–to–low and low–to–high transitions, for the region of interest as:

$$\Delta t = \overline{C}_N \frac{\Delta V}{\overline{I}_N} \tag{1}$$

Next, with the aid of Fig. 4, we develop a simple model for the gate propagation time delay in both the standard static CMOS and FTL. For the low–to–high transition, at the output of a gate in a cascaded chain, we identify two voltage levels: V_{id} and V_{oc}. Voltage level V_{id} denotes the input voltage when the input discharging current is 50% of its maximum value. V_{oc} denotes the input voltage at the point where the output charging current is 50%. Similar voltage points for the high–to–low transition are V_{ic} and V_{od}. Points V_{ic} and V_{od} are identified in Fig. 5 for the standard static CMOS and FTL inverters.

The values of V_{ic}, V_{od}, V_{id} and V_{oc} from simulation are −23.83 mV, 611.33 mV, 1172.88 mV and 415.57 mV for the standard static CMOS gate; and 690.10 mV, 780.90 mV, 625.85 mV and 554.48 mV for FTL. For the standard static CMOS, the identification of V_{ic}, V_{od}, V_{id} and V_{oc} points are made with respect to the starting point of minimum undershoot and maximum overshoot voltage points in the observed transient analysis.

Consider the high–to–low output transitions in Fig. 4(a) and 4(c). If an input transition takes $t_{c_{iu}}$ to charge from V_{ic} to V_{od} voltage, and then $t_{c_{uf}}$ to get to the final 50% transition

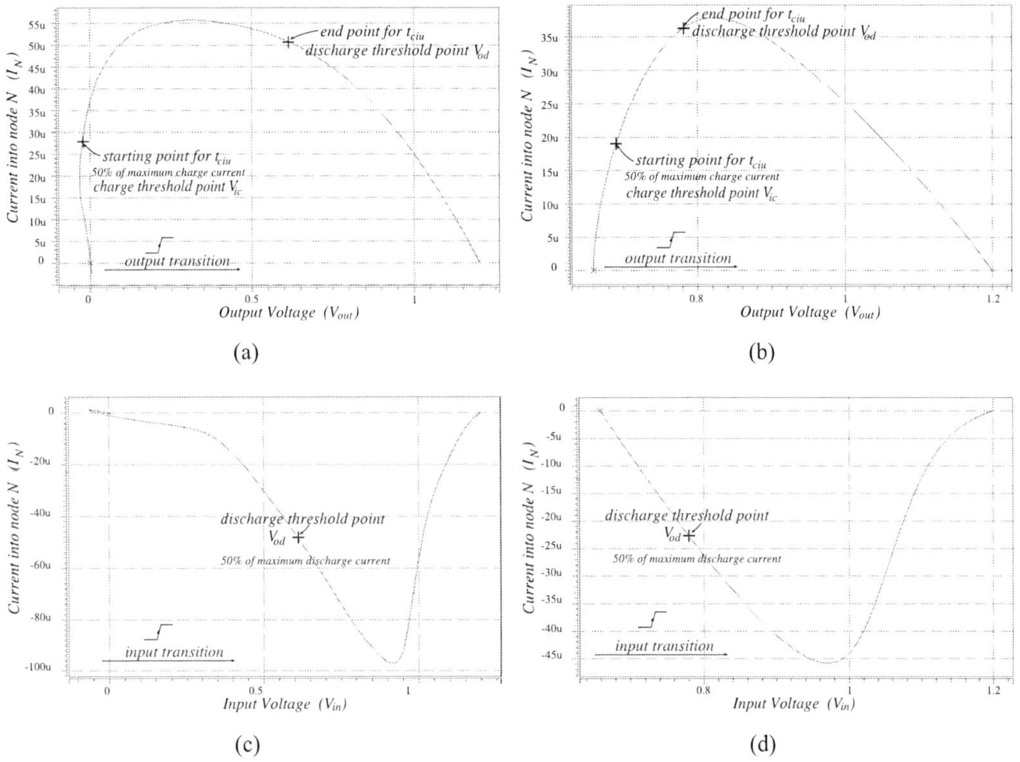

Figure 5. Plots of charge/discharge current for estimation of $t_{c_{iu}}$. Current into output node (I_N) versus output voltage (V_{out}) for low–to–high output transition for: (a) static CMOS and (b) FTL. Current into output node (I_N) versus input voltage (V_{in}) for high–to-low output transition: (c) static CMOS and (d) FTL. The technology is 0.13 μm 1.2 V / 3.3 V 1P8M Logic High Speed Process from UMC.

point, the total charge time (t_c) will be $t_c = t_{c_{iu}} + t_{c_{uf}}$. The same procedure is applied for obtaining the total discharge time (t_d) for the low–to–high output transitions. Assuming the chain of identical inverters cascaded together, where the role of input and output can be easily changed, from Fig. 4 we can develop the delay expressions:

$$t_{p_{hl}} = t_d - t_{c_{uf}} = t_{d_{iu}} + t_{d_{uf}} - t_{c_{uf}}$$

$$t_{p_{lh}} = t_c - t_{d_{uf}} = t_{c_{iu}} + t_{c_{uf}} - t_{d_{uf}}$$

$$t_p = \frac{t_{p_{hl}} + t_{p_{lh}}}{2} = \frac{t_{d_{iu}} + t_{c_{iu}}}{2} \tag{2}$$

Therefore, we obtain the average propagation time delay t_p in terms of the input charge (discharge) time from the initial voltage V_{ic} (V_{id}) to the final voltage V_{od} (V_{oc}). This simplifies the analysis as we only require linearization of current around a small transition range.

Next, by computing $t_{c_{iu}}$ and $t_{d_{iu}}$ using Equation 1, we demonstrate the superior performance of FTL over the standard static CMOS logic.

To compute the average current values flowing into node N, we average the current values at two points of interest in the transition. The average discharge current ($\overline{I}_{Nd}$) in the low–to–high output transition is computed from the linear approximation of the curves in Fig. 5(a) and 5(b) for the standard static CMOS and FTL inverters. We first use the curves in Fig. 5(c) and 5(d) to identify the V_{od} points. Similarly, we identify V_{ic} points in Fig. 5(a) and 5(b). Current end points, for the low–to–high output transition, are fixed at V_{ic} and V_{od} points. The same procedure is applied to identify the average charge current (I_{Nc}) values for the high–to–low output transition. Current end points, for the high–to–low output transition, are fixed at V_{id} and V_{oc} points. Note in Fig. 5(a) and 5(b) that for an FTL gate the current flows into the output node as soon as input transition begins. This is contrary to the behavior of the traditional CMOS logic where a large input swing is required before a noticeable current is observed flowing into the output node. This is the main advantage of FTL.

Average values for the charge and discharge currents are $\overline{I}_{Nc} = 39.332\ \mu$A and $\overline{I}_{Nd} = -72.515\ \mu$A for the standard static CMOS inverter, and $\overline{I}_{Nc} = 27.607\ \mu$A and $\overline{I}_{Nd} = -30.058\ \mu$A for the FTL inverter.

The voltage swing for the low–to–high transition ΔV_c is obtained as the difference of the final voltage V_{od} and the initial voltage V_{ic}. Similarly, voltage swing for the high–to–low transition ΔV_d is obtained from the final voltage V_{oc} and the initial voltage V_{id}. Computed voltage swings for the standard static CMOS are 635.10 mV and -757.21 mV for ΔV_c and ΔV_d, respectively. Voltage swings for FTL are 90.80 mV and -71.37 mV for ΔV_c and ΔV_d, respectively.

Finally, by inserting the computed values for the average current and voltage swing in Equations 1 and 2, we obtain,

$$t_p = \frac{\overline{C}_N}{2} \frac{\Delta V_c}{\overline{I}_{Nc}} + \frac{\overline{C}_N}{2} \frac{\Delta V_d}{\overline{I}_{Nd}}$$

$$t_{p_{CMOS}} = 13.2953 \times \overline{C}_{N_{CMOS}}(\text{fF})$$

$$t_{p_{FTL}} = 2.8317 \times \overline{C}_{N_{FTL}}(\text{fF})$$

We observe that propagation time delay in the FTL structure is 4.69 times less for the same average capacitance. This value is near to the results obtained from the simulation, which shows a speed up factor of 3.85. The reason for the error in the current estimation can be attributed to the averaging of current values from the current end points, which assumes a linear current behavior. In addition, average capacitive values for the high–to–low and low–to–high transitions are almost equal for both, standard static CMOS and FTL, based on the simulation.

3. CMOS–FTL evaluation

3.1. Proposed structures

To satisfy the trade–off between the performance and power consumption, two basic FTL structures were proposed in [11]. These structures are shown in Fig. 6. Two FTL structures are derived from the pseudo–NMOS logic family for high speed applications, (Figs. 6(a) and 6(b)) and the standard static fully complementary CMOS logic family for low power

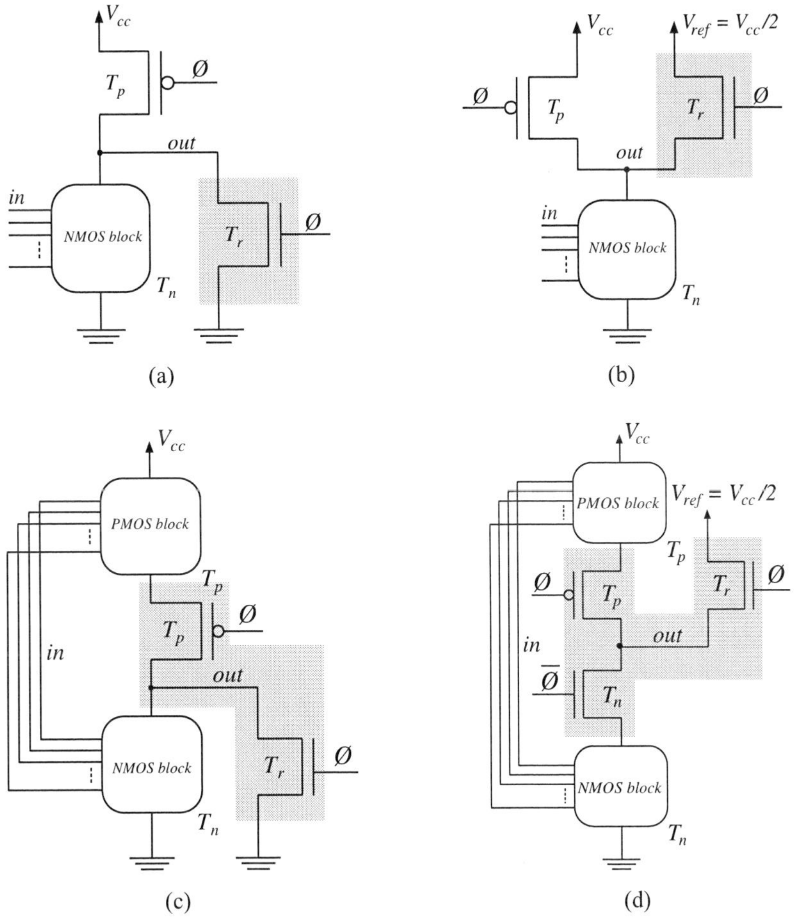

Figure 6. Proposed FTL structures: (a) high speed structure with $V_{ref} = 0$, (b) high speed structure with $V_{ref} = V_{cc}/2$, (c) low power structure with $V_{ref} = 0$ and (d) low power structure with $V_{ref} = V_{cc}/2$.

applications, (Figs. 6(c) and 6(d)). Each of these basic structures can operate in two configurations with reference voltage set to 0 ($V_{ref} = 0$) or $V_{cc}/2$ ($V_{ref} = V_{cc}/2$). With reference to Fig. 2, we observe that the initial stages have a different behavior from the other stages, as they never have enough time to reach to the V_{TH} point. However, with $V_{ref} = V_{cc}/2$, dissimilarity in the initial voltages is removed. As will be seen in Section 4, the low power structure with $V_{ref} = V_{cc}/2$, will provide advantage in the shorter word–length adder circuits. Considering the complexity associated with providing for the extra supply voltage $V_{cc}/2$, clock phase $\overline{\phi}$ and transistor T_n, low power structures with $V_{ref} = V_{cc}/2$ configuration are not suitable for long word–length adder circuits. On the other hand, high speed structures with $V_{ref} = V_{cc}/2$ can provide much higher performance in long word–length adder circuits.

The high speed structures have static power consumption, in the evaluation stage, due to the short–circuit current from the power supply through T_p and NMOS evaluation block. In reset phase, there is no static power consumption for $V_{ref} = 0$. However, circuit topology with $V_{ref} = V_{cc}/2$ exhibits static consumption because current flows to GND through T_r (reset transistor) and the NMOS block.

Low power structures remove the static power consumption, in both reset and evaluation phases, by means of T_p and T_n transistors as shown in Fig. 6(c) and 6(d). Note that the low power structure with $V_{ref} = 0$ uses a single clock phase. As no current flow is possible to GND during the evaluation phase, T_n transistor from Fig. 6(d) is removed. The main disadvantage of low power structures is that they have a higher input load than high speed ones, and, therefore, are slower. But because of reduced static power consumption, low power structures outperform the standard static CMOS structure, even at low operating frequencies.

On the other hand, high leakage current in the deep-submicron regime is becoming a significant contributor to power dissipation of CMOS circuits as threshold voltage, channel length, and gate oxide thickness are reduced. In order to minimize the leakage power dissipation, several circuit techniques have been proposed such as multi-threshold voltage CMOS (MTCMOS) using sleep transistor, variable threshold voltage CMOS (VTCMOS) using variable substrate bias voltage, or input pattern control [15,16]. Note that FTL design conduces to reduced leakage current as it share some of the common design techniques of low power design. The clocked transistors T_p and T_n of FTL structures are sleep transistors, therefore, reducing the overall subthreshold current due to stack effect. Moreover, these transistors can be implemented using high V_T devices, while the logic block transistors can be implemented using low V_T devices, as a variant of MTCMOS. Furthermore, $V_{ref} = 0$ structures also contribute to the reduction in the leakage current during the sleep mode. Only high speed FTL structure using $V_{ref} = V_{cc}/2$ is susceptible to high leakage currents for nanoscale CMOS circuits.

3.2. Analysis Results for the Basic Cells

To compare the proposed FTL structures against the standard static CMOS structures, a basic inverter cell in a long chain of inverters (Fig. 1(b)) is simulated. We used a $0.13\ \mu$m CMOS process from UMC, using the parameters for typical process corner at $25°$C, with 10 fF capacitive load in all output nodes. Simulation results of propagation delays of the inverter chains for FTL, dynamic domino and static CMOS logic were presented in [11,12]. However here we develop further insight into the operation of FTL by using the circuit parameters, as presented in Section 2, and characterizing the performance of the inverter in terms of capacitive load and frequency of operation.

For a typical inverter in a long chain of inverters, the values of output voltage swing ΔV, average current flowing into node N, $\overline{I}_N$, average capacitance $\overline{C}_N$ at node N, and propagation time delay t_p for low–to–high and high–to–low transitions are obtained. These circuit parameters, as well as average values of t_p and the speed up (t_p ratio) with respect to the standard static CMOS, are shown in Table 1.

As seen from the data in Table 1 the low power structure with $V_{ref} = V_{cc}/2$ (*LP06*) does not provide a significative speed up advantage. We will show that this assertion is not true for short logic depth chains. Also, both high speed structures (*HS0* and *HS06*) seems to provide identical performance. Although this is true for the simple inverter gate in a long chain, as will be seen in the next section, the results are different for other practical circuits,

Table 1. Estimation of Circuit Parameters at Node N for the Standard static CMOS and FTL Structures.

Logic family	ΔV (mV)		$\overline{I}_N$ (μA)		$\overline{C}_N$ (fF)		delay (ps)		t_p	t_p
	charge	discharge	charge	discharge	charge	discharge	t_{plh}	t_{phl}	(ps)	ratio
CMOS	635.10	−757.31	39.332	−72.515	1.446	1.368	23.354	14.292	18.823	1.00
LP0	88.42	−42.59	14.611	−22.913	1.655	1.641	10.013	3.050	6.531	2.88
LP06	104.78	−94.18	12.321	−18.570	1.825	1.727	15.523	8.761	12.142	1.55
HS0	90.80	−71.37	27.607	−30.058	1.331	1.336	4.378	3.172	3.775	4.98
HS06	90.80	−71.37	27.607	−30.058	1.331	1.336	4.378	3.172	3.775	4.98

such as long word–length adders.

To provide further insight into the operation of FTL structures, we have simulated the propagation time delay versus the fanout for an inverter gate in a 21–stage inverter chain configuration. Each inverter in the chain is identically loaded. The simulation results are shown in Fig. 7(a) and 7(b) for the 1st and 21st stages, respectively. Different stages are shown because the behavior of FTL inverters vary from the first stages to the intermediate ones. Delay at 21st stage is representative of the tend of the FTL inverter delay for long enough chains. However, it must be noted that the delay parameter is almost constant from the 3rd stage onwards. The labeling convention in Fig. 7 follows the syntax "[*structure*][*vref*]_[*channel_length*]"; where *structure* refers to either high speed (*HS*) or low power (*LP*), *vref* refers to either 0 (0) or $V_{cc}/2$ (06), and the *channel_length* identifies two channel lengths of 0.12 μm (012) and 0.35 μm (035) for the high speed FTL structures. The longer channel lengths are used to reduce the static power consumption in the high speed structures.

Note in Fig. 7(a), for the 1st stage, the FTL structures provide no propagation time delay advantage except in the high speed version with $V_{ref} = V_{cc}/2$ and minimum channel length (*HS06_012*). This is because for this stage there is insufficient time for the rising of the output to V_{TH}, and consequently, a full transition must be performed. The transition, in absence of pre–evaluation to V_{TH} point, is similar to the CMOS or pseudo–nMOS case for the low power and high speed FTL, respectively. Just a slightly higher delay is obtained for the FTL due to presence of T_p transistors. Note also that the *LP06* structure significantly improves the delay performance at the first stage, so it can be considered as a suitable alternative for short depth chains. Fig. 7(b), on the other hand, shows that all FTL structures outperform the standard static CMOS for the 21st stage. Similar simulation results are obtained from the 3rd stage onwards because from this stage on, the pre–evaluation voltage reaches V_{TH}, where we get the advantage of the FTL principle.

Delay–power product versus frequency at the 3rd and 21st stages of cascaded inverters are shown in Fig. 7(c) and 7(d). We removed points in which the operating frequency exceeds the maximum operating frequency of the inverter for the corresponding stage number, because of higher evaluation or pre–charge times than half of the period of clock signal (ϕ). Note that pre–charge time is the limiting factor in the *HS06* inverters.

Power consumption, and consequently delay–power product vary with the stage number, because V_{TH} voltage is maintained at the node for a longer period. This is evidenced in

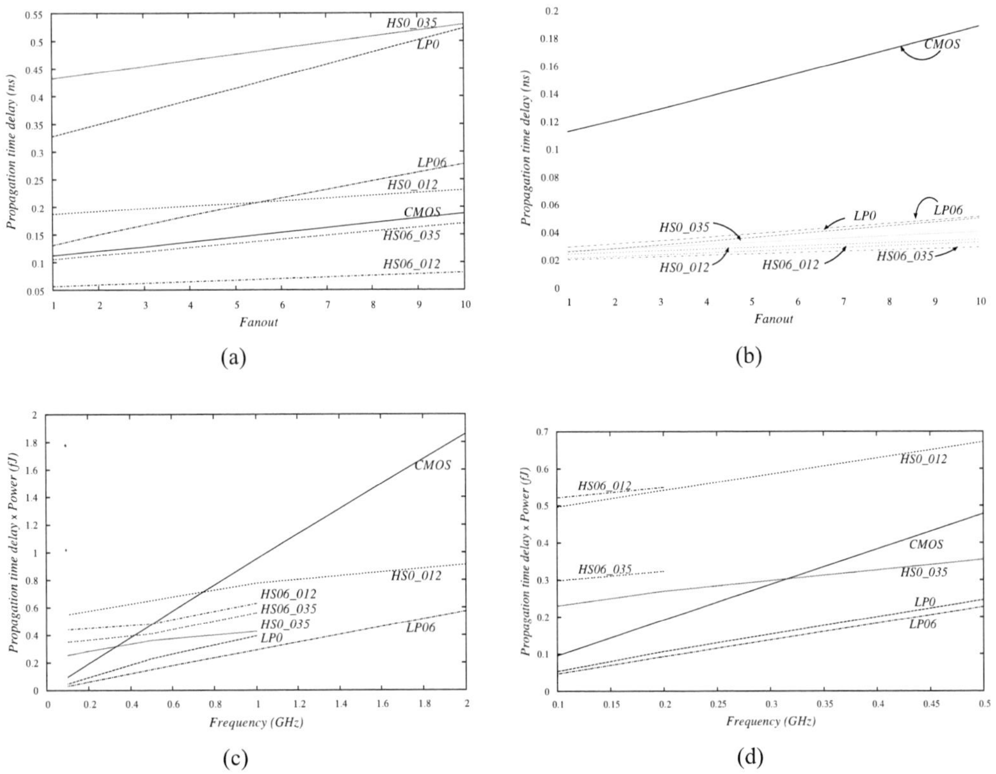

Figure 7. Results of static CMOS and FTL inverters: (a) propagation time delay (delay) versus fanout for the 1st stage, (b) delay versus fanout for the 21st stage, (c) delay–power product versus frequency for the 3rd stage, and (d) delay–power product versus frequency for the 21st stage.

the higher slope of the delay–power plots in Fig 7(d) when compared with those of Fig 7(c). The *LP0* structure achieves the lower delay–power product variation for the long cascaded chains, with just 14.3% higher dynamic energy dissipation. On the other hand, the highest dynamic energy variation is achieved by the *HS0_012* inverter, where the dynamic energy dissipation is increased three times from 3rd to 21st stage. Although this inverter has a higher dynamic energy variation across the stages, it exhibits a dynamic energy dissipation that is 10 times less than the CMOS inverter, at the 3rd stage. The rest of FTL inverter structures show higher dynamic energy variation, ranging from 40% to 75%, when going from the 3rd to the 21st stage. Note that the low power FTL inverters easily outperforms the delay–power product of CMOS, because of lower the static and dynamic dissipations. High speed FTL, on the other hand, is a good choice for high speed applications if high frequency of operation is considered to compensate for the high static energy dissipation. Furthermore, channel length can be used in these structures to trade speed for power performance. The simulated power–delay product of the 1st stage is higher in FTL than in the standard static CMOS. So to take advantage of FTL we must ensure that the number of stages are large enough to compensate for the disadvantage of the first stage.

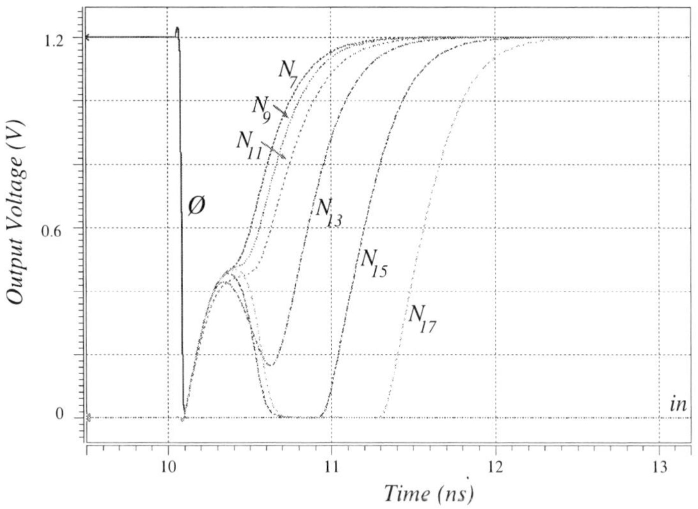

Figure 8. Effect of dissimilar load in the behavior of FTL inverter chain (LP0 structure).

However, for FTL to work in circuits with large number of stages, special care must be taken to avoid dissimilar capacitive loads in all the intermediate stages. This will ensure that all nodes rise together to the threshold voltage V_{TH}. This issue will be further discussed in the design and implementation of RCA cells.

The effect of dissimilar capacitive loads at various intermediate stages, on the performance behavior of the chain, is shown in Fig. 8 for the low power structure with $V_{ref} = 0$. In this example, an excess capacitive load of 1 fF is added to the output of the 11th stage. This excess load delays the rise to V_{TH} at stage 11, which in turn causes delay in the logic evaluation in the subsequent stages, causing the FTL inverter to behave as a CMOS dynamic logic NOR gate. A propagation time delay of $181ps$ is obtained at stage 21 for the low power structure, this is 59% higher than the delay in the standard static CMOS structure due to T_p transistor.

Simulation results show that FTL structures with $V_{ref} = V_{cc}/2$ are less sensitive to dissimilar capacitive load effect at the initial nodes, but tend to exhibit the same behavior in the final stages. On the other hand, high speed structures with $V_{ref} = V_{cc}/2$ do not reset all nodes to the same voltage level. The actual voltage levels at each stage depends on the input values during the previous evaluation cycle. However, as we will see later in the analysis of more practical circuits, the overall combined effect of two imbalances; intermediate capacitive load, and reset voltage values, has beneficial effect in increasing logic depth.

3.3. Applications

To sum up, FTL is well suited to applications where the critical path is made of a large cascade of inverting gates. Many arithmetic circuits such as adders, multipliers, FIR stages and other similar structures are clear candidates for implementation on FTL. Furthermore, FTL based circuits provide a clear advantage in high fanout and high switching frequencies due to both lower delay and dynamic power consumption. Lower delay variation versus load in FTL is specially suitable for applications, where improving the performance of deep pipeline stages is limited by the interconnect delay. In applications with lower power requirements, low power FTL structures can be employed, but at cost of lower speed.

The main challenge in employing FTL structures is to appropriately compensate for dissimilar capacitive loads for two identical nodes under the opposing charging and discharging conditions. Additional transistors are needed to compensate for this dissimilarity. Next section provides some examples of how this compensation can be provided for the long word–length adder structures. The additional transistors, however, will, marginally, increase the overall adder cell complexity.

4. Design Examples

In this section we present the design and optimization of FTL–based RCAs both differential and non differential adders [11, 12].

The structure of the basic sum and carry cells (both non differential and differential one) used in the design of long word–length adder circuits are similar to those in [6] and [7]. Due to the inverting property of the non differential RCA cells, and to maintain the correct polarity between them, the a and b inputs of even adder cells and the sum outputs of odd adder cells are buffered through inverters. Ignoring the transistors in the shaded area, these basic cells are presented in Fig. 9. Note that just $V_{ref} = 0$ structures are shown, but identical logic blocks are used for the $V_{ref} = V_{cc}/2$ cells.

The main problem of these structures is the dissimilar capacitive loads for different propagation paths. This capacitive load dissimilarity at the output of the carry cells, in a long chain, results in pre–evaluation voltages at the cell outputs that are different from V_{TH}, therefore, losing the advantage of the FTL. This effect is termed as *imbalance*, and henceforth, we call the *point of imbalance* to indicate the number of bits at the onset of the imbalance. The load capacitance dissimilarity results in loss of FTL advantage in the fourth bit of the adder in structures with $V_{ref} = 0$ when alternating the different propagation paths in the addder chain. However, structures with $V_{ref} = V_{cc}/2$ are less sensitive to dissimilar capacitive load effect, maintaining the FTL advantage to 16th adder bit. Similar behavior is observed for the differential and non differential adder cells.

To extend the FTL performance advantage to long word–length RCAs (up to 64 bits) a modified set of capacitive compensated structures for the adder cells have been employed. We compensate for dissimilar dynamic node capacitances by adding extra transistors, as shown in the shaded areas in Fig. 9. This solution increases area although power consumption does not change significantly, as power consumption is dominated by the routing capacities. Note that the added transistors do not alter the logic function of the cell, but improves the device symmetry for the different propagation paths (i.e, $a = 0, b = 1$; and $a = 1, b = 0$ cases). For example, the carry output node in the uncompensated non differential carry cell is loaded with the serial connection of a and b transistors. Unfortunately, the load of this serial connection does not remain constant for $a = 0, b = 1$; and $a = 1$, $b = 0$ configurations due to body effect, among others. But the corresponding symmetrically capacitive compensated cell does not suffer from this load dissimilarity, as the output node is loaded by two serial connections; one through a and b, and other one through b and a transistors. For the differential carry cells, we switched in the capacitive load of a

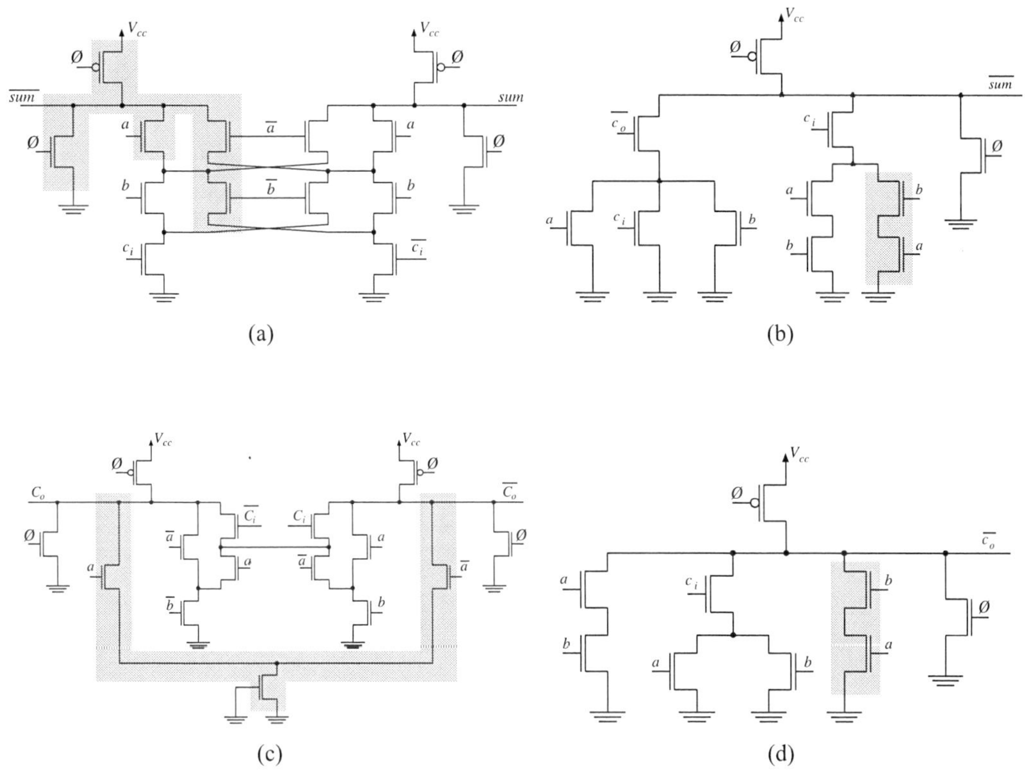

Figure 9. Ripple Carry Adder cells: (a) and (b) are the differential and non differential sum cells respectively, (c) and (d) are the carry cells.

cut–off device to one of the two branches of the differential structure to compensate for the dissimilarity.

From the capacitive compensated adder cells in Fig. 9, the differential version of the sum and carry cells (see Fig. 9(a) and 9(c)) are less sensitive to capacitive imbalance in the internal nodes, and therefore, most suitable for building higher performance long word–length adders. Although a differential sum cell is not functionally required, it provides a constant input load to the carry cell, improving the balance in the capacitive load.

Non differential structures have also been tested. The basic cells for the sum and carry functions are shown in Fig. 9(b) and 9(d). Non differential structures exhibit more sensitivity to dissimilar loading condition. To reduce the effect of dissimilar capacitive load, we insert non–inverting buffers between the output of the carry cells and the input of the sum cells.

Table 2 shows propagation time delay (t_p), power consumption, and figure of merit (FoM), Γ, expressed as delay–power product over frequency, and its normalized value, for the NP_DRCA domino CMOS [17], non differential high speed (*nodiff_hs*) and low power (*nodiff_lp*) structures; and capacitive compensated differential high speed (*diff_hs*) structure. Note that the FoM (Γ), is a measure of energy efficiency for a given clock frequency (or performance) [18]. Also note that we compared with a high performance compound (logic–logic) dynamic domino configuration [17], instead of a simple (logic–inverter). In

Table 2. Comparison Between the Dynamic Domino and FTL Adders in Terms of Performance (Frequency and Delay), Power, Γ, and their Ratios, and for the Differential and Non Differential Structures.

Logic family	Number of bits	f_{max} (MHz)	t_p (ns)	power (mW)	Γ (fJ/MHz)	Γ ratio	f_{max} ratio	t_p ratio
domino	40	77.059	12.060	0.133	20.752	1.000	1.000	1.000
domino	64	49.414	19.320	0.136	53.290	1.000	1.000	1.000
nodiff_hs0	40	390.396	2.172	4.053	22.549	0.920	5.066	5.552
*nodiff_hs0	64 (42)	188.164	4.937	5.901	154.829	0.344	3.808	3.913
nodiff_hs06	40	292.056	1.970	3.657	24.667	0.841	3.790	6.122
nodiff_hs06	64	195.122	3.021	5.950	92.122	0.578	3.949	6.395
nodiff_lp0	40	255.820	2.440	0.745	7.106	**2.920**	3.320	4.943
*nodiff_lp0	64 (42)	92.593	9.603	0.542	56.254	0.947	1.874	2.012
*nodiff_lp06	40 (28)	117.082	6.905	0.498	29.370	0.707	1.519	1.747
*nodiff_lp06	64 (28)	57.904	15.640	0.454	122.627	0.435	1.172	1.235
diff_hs0	40	487.329	1.577	6.634	21.468	0.967	6.324	7.647
diff_hs0	64	337.541	2.355	10.520	73.397	0.726	**6.831**	8.204
diff_hs06	40	340.832	1.366	4.634	18.572	1.117	4.423	8.829
diff_hs06	64	234.852	2.087	7.466	66.346	0.803	4.753	**9.257**

*Imbalanced adders: The numbers in parenthesis indicates the maximum number adder bits before the point of imbalance. The technology is 0.13 μm 1.2 V / 3.3 V 1P8M Logic High Speed Process from UMC.

Table 2 the relative performance of the FTL structures with respect to the domino logic is presented in terms of ratios of propagation time delay, achievable clock speeds, and Γ. The data in Table 2 are obtained by simulation at the highest achievable clock frequency for each of the 64–bit and 40–bit adders. The simulation parameters include a post–layout estimated minimum load of 10 fF on each node, and the typical process corner and 25°C. Note that the minimum achievable clock period (maximum frequency) is obtained by sum of the minimum evaluation and pre–discharge times required to assure no more than 5% degradation of the voltage levels.

From the data in Table 2, it can be observed that in the FTL structures, as the number of adder bits increase (before the point of imbalance), speed up improves with respect to the dynamic domino logic. However, this improvement comes at the cost of slight reduction in the normalized Γ ratio (column 7 in Table 2). The lower energy efficiency of long word–length adders in FTL, even in the absence of imbalance effect, is due to the fact that they stay at the V_{TH} point for a longer period, and deliver high current, and therefore, cause higher dynamic power consumption in the end stages than the middle ones. Table 2 shows that the low power structures are not suited for more than 40 bits, in the best case (*nodiff_lp0* adder). High speed adder structures, on the other hand, can be extended to a word–length of 64 bits without losing the FTL advantage (*nodiff_hs06* adder), but requires an additional $V_{ref} = V_{cc}/2$ voltage source. On the other hand, differential high speed adders (*diff_hs* adders) can operate up to 64–bits without neither experiencing the onset of the imbalance effect, nor needing any additional voltage source.

The highest speed up, with respect to the dynamic domino, in terms of frequency (6.8 times) and lower evaluation delay (8.2 times) is achieved with the 64–bit differential high speed adder structure with $V_{ref} = 0$. Its energy efficiency (column 7 in Table 2), on the other hand, is less than the dynamic domino logic by a factor of 0.726 (38% less).

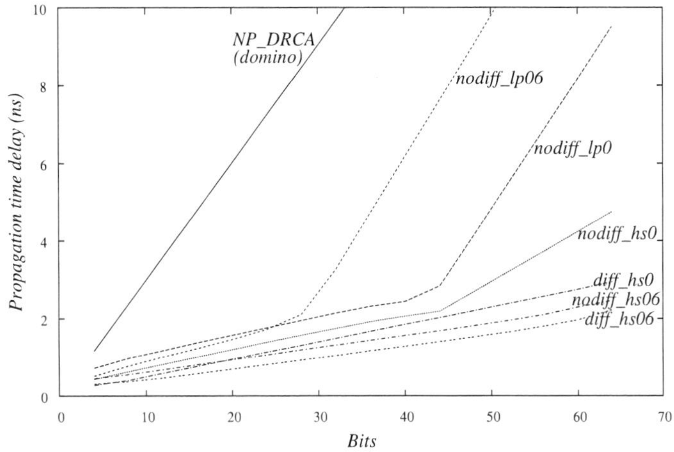

Figure 10. Sensitivity of the propagation time delay of dynamic domino, differential and non–differential FTL adders against number of bits.

For the 40–bit differential high speed adders with $V_{ref} = V_{cc}/2$ the energy efficiency improves with respect to dynamic domino logic by 10%, while achieving a higher frequency (4.4 times) and a lower evaluation delay time (8.8 times).

The maximum normalized Γ ratio is reached with non differential low power versions. In this FTL structure, for a 40–bit adder, the energy efficiency is 66% better (2.9 times) than the dynamic domino logic. Speed up factor, in terms of higher frequency (3.3 times), and lower evaluation time (4.9 times) is reached with respect to the dynamic domino.

A closer look at the data presented in Table 2, also, reveals that high speed structures with $V_{ref} = Vcc/2$ an average propagation time delay 13% smaller than structures with $V_{ref} = 0$. However, their maximum achievable clock frequencies are lower (39%). That is because the lower evaluation time comes at the expense of higher pre–discharge time.

It is also interesting to note that, low power FTL adders are limited by the early onset of the imbalance. However, before the point of imbalance these structures exhibit a remarkable energy efficiency with respect to domino dynamic logic.

5. Sensitivity Analysis

To prove the superior properties of the proposed FTL structures compared with the dynamic domino CMOS, we have characterized the performance of RCAs against the variations in number of bits and a set of technology parameter; capacitive load, temperature, power supply, process corner and noise coupling [11, 12]. The sensitivity analysis is important as the FTL structures are inherently sensitive to capacitive load dissimilarities. In this section we provide a comprehensive sensitivity analysis [11, 12] with results of differential high speed FTL adders and with an in–depth discussion of the results.

Fig. 10 shows the propagation time delay versus the number of adder bits for various adder structures and word–length configurations. We observe that, due to increased gate complexity, the effect of the capacitive imbalance is felt earlier in the low power structures

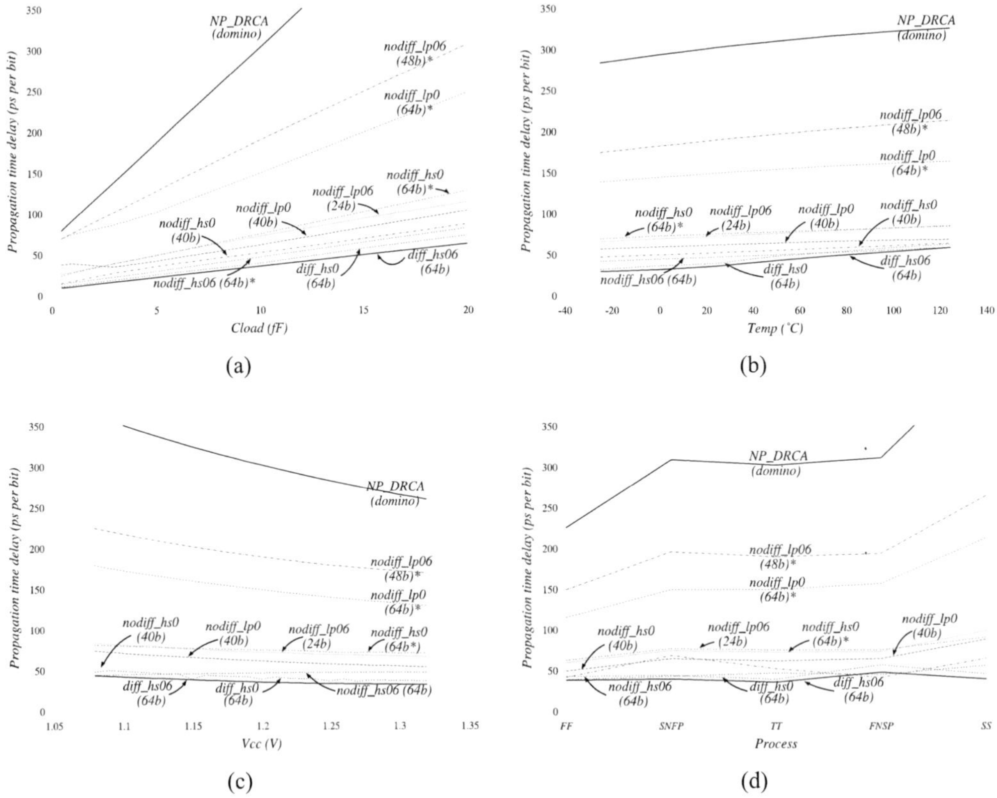

Figure 11. Sensitivity of the propagation time delay of dynamic domino, differential and non–differential FTL adders against (a) capacitive load, (b) temperature, (c) power supply, and (d) process corner. (*) mark indicates FTL structure has already pass the point of imbalance at the given number of bits.

(*nodiff_lp0*, *nodiff_lp06*). We also note that the differential structures perform better than the non differential ones. Further, high speed structures with $V_{ref} = V_{cc}/2$ provides the best results, providing a linear delay relation with respect to the word–length up to 64 bits. It can also be observed that in the FTL structures, before the point of imbalance, as the number of adder bits increases, the speed up improves with respect to the dynamic domino CMOS. However, beyond the point of imbalance, the rate of increase in the adder propagation time delay with respect to the number of adder bits in the low power FTL structure approaches that of the dynamic domino CMOS. However, the rate of increase for the high speed FTL structures is much lower.

Fig. 11(a)–(d) are plots of propagation time delay per bit versus the variations; in the inter–stage parasitic capacitive load (100% variation from the nominal extracted capacitance of 10 fF), operating temperature (variation from -25 to 125°C), power supply (10% variation from the nominal 1.2 V, ranging from 1.08 to 1.32 V), and process corner (variation from SS to FF corners), respectively. The plots show both, the balanced and imbalanced versions of the proposed FTL adders. The asterisk (*) mark on a curve indicates that the FTL structure has already pass the point of imbalance at the given number of bits.

As seen, the balanced versions of the FTL adders demonstrate lower sensitivity than the imbalanced ones, which in turn, show lower sensitivity than the dynamic domino CMOS for each of the considered parameter. The lower sensitivity of the balanced versions of the FTL structures to various parameters can be attributed to perfect matching of node capacitances. We have verified that by removing the capacitive compensating transistors from the adder cells in circuits of Fig. 9 and repeating the sensitivity analysis on resulting adder chain. Moreover, the differential FTL adders exhibit the lowest sensitivity for the capacitive load, and process corner variations. On the other hand, non differential high speed and low power FTL adders show the lowest sensitivity against power supply and temperature variations, respectivelly. It must also be highlighted that differential adders always obtain the best speed up ratios under all conditions.

The sensitivity of the balanced versions of the FTL adders versus the inter–stage capacitance (Fig. 11(a)) is 3.8 to 8.2 times less than the dynamic domino CMOS adders. Differential high speed FTL adders shows the highest improvement in sensitivity when compared with dynamic domino CMOS, making these structures ideal for high capacitive load conditions, i.e deep–submicron CMOS technologies where routing capacitances dominate.

For the temperature variations (Fig. 11(b)), the sensitivity of balanced versions of the FTL adders is 1.8 to 6.7 times less. Although typically the differential structures are less sensitive to temperature variations, lower sensitivity is detected in the non differential versions of FTL adder; where low power version shows slightly less sensitivity than the high speed version. This behavior can be attributed to better matching and symmetry of capacitive loads in the non differential adders. Consequently, non differential low power FTL adders are best suited when low jitter is required in high operating temperature ranges.

Balanced versions of non differential low power structures and differential high speed ones, are, approximately, 6 times less sensitive to the power supply variations than the dynamic domino CMOS. But note that the non differential high speed structures are, by one to two orders of magnitude, less sensitive (Fig. 11(c)). This characteristic of non differential high speed FTL adders make them ideally suited for low voltage applications, as we could reduce power consumption without incurring the additional cost of increasing the propagation time delay. Obviously, noise margins and reliability will impose a limit to lowering the power supply voltage.

Sensitivity of the balanced versions of the non differential FTL adder structures to process variations (Fig. 11(d)), is 3.9 to 8.7 times less than the dynamic domino CMOS adder. Differential FTL adders have even lower sensitivity to process variations (more than 1 order of magnitude with respect to dynamic domino).

Finally, the performance of non differential low power FTL adders with $V_{ref} = 0$ in the presence of a noise source that is capacitively coupled to all the nodes of the adder has been analyzed [11, 12]. We tried serveral values of coupling capacitances and, therefore, noise amplitudes. To create a maximum adverse at the output nodes, we synchronized the noise pulse width the commencement of the evaluation phase. The noise source has an average frequency and slew rate that is 10 times larger than the clock signal ϕ. A capacitive coupling as high as 3fF, and noise amplitude of more than 220mV, is required to slightly shift the point of imbalance in the FTL adder from bit 40 to bit 36. Higher capacitive coupling and noise amplitude are required for the high speed structures.

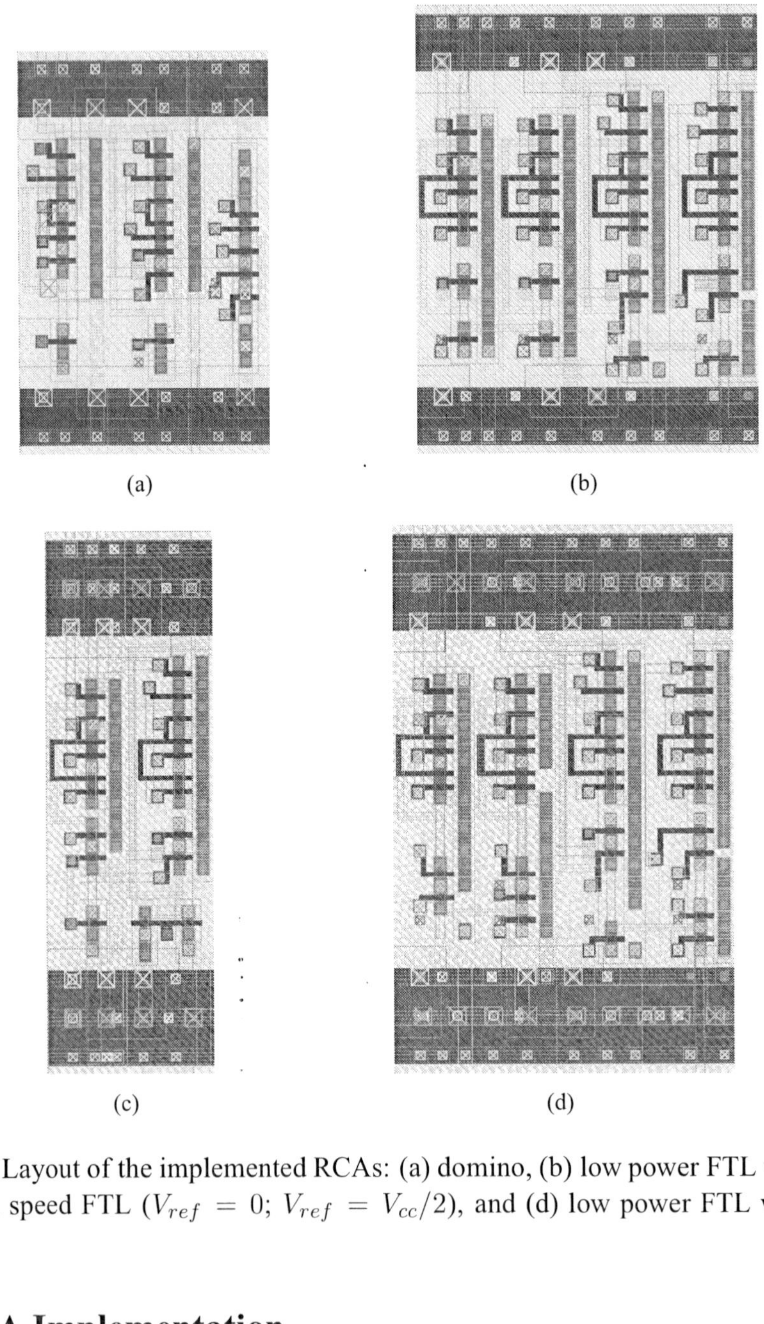

Figure 12. Layout of the implemented RCAs: (a) domino, (b) low power FTL with $V_{ref} = 0$, (c) high speed FTL ($V_{ref} = 0$; $V_{ref} = V_{cc}/2$), and (d) low power FTL with $V_{ref} = V_{cc}/2$.

6. RCA Implementation

To demonstrate the viability of FTL structures, as an alternative logic family, in practical applications, we laid out a set of FTL and dynamic domino CMOS RCAs. In order to obtain a fair comparison with the NP–domino RCA, we implemented the non differential versions of the FTL adders. Both, low power and high speed logic families for $V_{ref} = V_{cc}/2$ and $V_{ref} = 0$, were laid out using the 0.13 μm 1.2 V / 3.3 V 1P8M Logic High Speed Process from UMC.

The corresponding 1–bit RCA cell layouts, for the non differential FTL and dynamic domino logic families, are depicted in Fig. 12. We used the top metals for power planes and clock distribution. Also, we carefully laid out the FTL compensating transistors (shaded devices in Fig. 9) to minimize dissimilar capacitive loads for the adder cells. As FTL low power adders are more sensitive to capacitive imbalance, we carefully sized and laid out the design to minimize the effect of imbalance. The sizing of basic cells is similar to sizing in traditional CMOS logic or pseudo-nMOS for low power and high speed FTL, respectively. We optimized for delay parameter, and because of reduced fanout and capacitive loads, minimum dimension devices are used. But note that optimal sizing will depend on the fanout and routing capacitances. A key issue in the successful implementation of FTL adders is to ensure the symmetry in the circuit topology through the compensating transistors. Further symmetry in the layout may be is desirable, but as we will see, such fine grain symmetry is not essential even for the long word–length adders.

We designed 64–bit RCAs in each of the FTL structures, and through the post–layout simulations identified the bit position where the effect of imbalance is observed. We next, reduced the word–length for each FTL adder to the maximum number of bits before the onset of imbalance and simulated the corresponding adder circuit to characterize its performance. The adder structures were simulated with all extracted post–layout interconnect parasitics included, using a distributed RC model for the parasitic extraction.

Our post–layout simulation shows that it is possible to implement the balanced high speed and low power versions of the FTL structures with word–length extending to 40 and 20 bits, respectively, when $V_{ref} = V_{cc}/2$. This can be easily achieved without carrying high effort in the design of the layout (no complete layout symmetry, but high symmetry in the compensating transistors). For the case of $V_{ref} = 0$, however, the point of imbalance for high speed and low power structures would reduces to bit 32 and 16, respectively.

To improve the robustness of FTL structures against the capacitive imbalance in the physical implementation, we also did a symmetric layout of the of the adder cells, where each transistor and node were carefully placed to assure symmetry in the capacitive load. Strong symmetry requirements have been used to meet the design constraints of the common centroid technique [19]. Device splitting configured in parallel connection has been used to allow for complete symmetric layout in the transistors and routing. The symmetric designs allow for a factor from 1.2 to 1.5 increase in the number of bits before the onset of the imbalance. Although this increase in the number of bits is particularly noticeable in the low power structures, it comes at the cost of increased layout complexity, and therefore, higher routing capacitive load. This, incurs additional delay (1.4–2 times), layout area (2.3–2.7 times), and the power consumption (1.4–3.6 times), when compared with non symmetric layout cells.

Maximum number of bits, delay, power consumption, overall FoM, or energy efficiency rating (Γ), and the corresponding improvement ratios for extracted layouts are presented in Table 3. We simulated the FTL adders at both, maximum operating frequency and 100 MHz, and compared their performance with the equivalent adders in the dynamic domino logic. Note that performance parameters in Table 3 are expressed as per bit quantities. This facilities comparative evaluation of the performance of various adders differing in word–lengths and operating frequencies.

For each adder structure in Table 3 there are two more bit numbers (placed in parenthesis). These numbers correspond to the maximum numbers of bits for a clock period of

10 ns. The first number corresponds to an asymmetric clock, where the reset phase is set to the minimum time required, and the remaining time is used for the evaluation phase. The second number assumes that the clock signal is symmetric with 5 ns dedicated for both the reset and evaluation phases.

As seen in Table 3, all FTL adders perform better in terms of frequency, delay and energy efficiency in comparison with the domino logic. The best energy efficiency rating (a factor of 2.06 improvement) is obtained for the low power FTL adder, when compared with the dynamic domino logic. The clock frequency of this adder is 1.57 times higher than the domino logic. The high speed FTL adders allow for considerable improvement in the maximum operating frequency (4.5 times higher), while maintaining a superior energy efficiency (a factor of 1.64 better) in comparison with the dynamic domino logic. It is also remarkable that the performance of the low power FTL adder with $V_{ref} = 0$ does not depend on the clock frequency, because of the complete elimination of all the leakage and short–circuit currents.

Note that, even with a capacitive imbalance effect in full play, the high speed FTL adder can extend to an impressive 160 bits for 10 ns symmetric clock (99 bits for the asymmetric clock), which is near 3 (3.5 for assymetric clock) times longer than the dynamic domino adder. Under the similar clocking condition, the low power FTL structure, on the other hand, can accommodate no more than a few additional bits when compared with the dynamic domino logic. Its inferior performance is due to the early onset of the imbalance effect in the low power FTL structure. Also, note that high speed FTL structure with $V_{ref} = V_{cc}/2$ performs slightly better than the structure with $V_{ref} = 0$.

7. Monte–Carlo Analysis

To evaluate the performance of the FTL over the process variation and mismatch we run a series of Monte–Carlo simulations over the FTL *HS* adder structure, using $V_{ref} = 0$. A simplified distributed capacitance + coupling capacitance (C+CC) model for all inter and intra cell connections was used for the layout extraction, in order to speed up the simula-

Table 3. Results of the Implemented Adders, Taken From Post–Layout Simulations.

Logic family	Number of bits	freq (MHz)	t_p (ps/bit)	power (nW/MHz/bit)	$\Gamma \times 10^6$ (fJ/MHz/bit)	Γ ratio	f_{max} ratio	t_p ratio
domino	16 (54,28)	100.000 288.476	177.375	23.994 23.334	4255.891 4138.835	1.000 1.000	1.000	1.000
nodiff_lp0	16 (43,29)	100.000 452.552	64.063	32.300 31.902	2069.219 2043.747	**2.057** 2.025	1.569	2.769
nodiff_lp06	20 (42,31)	100.000 301.941	62.600	65.900 40.057	4125.340 2507.595	1.032 1.651	1.261	2.833
nodiff_hs0	32 (157,89)	100.000 737.251	31.750	396.875 107.621	12600.781 3416.969	0.338 1.211	**4.648**	5.587
nodiff_hs06	40 (160,99)	100.000 576.758	21.700	544.500 116.687	11815.650 2532.103	0.360 1.635	4.455	**8.174**

(x,y) in parenthesis are maximum number of bits for asymmetrical and symmetrical 10 ns period clock, respectively.

tions. We note that performing HSPICE simulation with this simplified C+CC model, the imbalance appears at the 36th bit instead of the 40th bit obtained with extraction with the complete distributed RC model. To obtain statistically significant Monte–Carlo simulation runs we used $\sigma = 3$ and 30 iterations, as recommended by the technology manufacturer and the HSPICE manual.

Waveform plots, for the low–to–high transition, for the 12th and 16th bit, are presented in Fig. 13. Waveform plots in Fig . 13(a) correspond in the Monte–Carlo analysis, with only process variation effect. We note that the delay variation is not significant, as was concluded in the sensitivity analysis in the previous section. The waveform plots of Fig. 13(b) include both process variation and device mismatch effects. Out of the total of 30 iterations 13 show the imbalance effect at the 16th bit, and 6 at the 12th bit.

Fig. 14 plots the propagation time delay versus the adder word–length. Results of Monte–Carlo iterations, with process variation and device mismatch, are shown for the high speed (*HS0*), the pseudo–nMOS (*nMOS*), and dynamic domino (*domino*) adders. Both high–to–low and low–to–high transitions are considered. We also simulated, as reference, the cases of with just process variation (*HS0(process)*), and typical corner (*HS0(typ)*), for the high speed FTL *HS0* adder. It can be seen that, before the onset of the imbalance, the delay of the *HS0* adder approaches its typical case, *HS0(typ)*. But once imbalance occurs, the delay variation increases and approaches the pseudo nMOS case (*nmos*). For the 12–bit word–length only few of *HS0* iterations exhibit delays that are extended to that of the nMOS cases. However, the number of such cases for the 16–bit word–length dramatically increases.

Finally, Fig. 15 shows the accumulated number of iterations with the imbalance in full play versus the number of adder bits, for high–to–low and low–to–high transitions. Two cases of process variation without and with device mismatch are considered. To see the effect of layout, we performed simulations on two circuits extracted from the compact layout (CL) and symmetric layout (SL) versions that were discussed in the previous section. It was observed that a) the point of imbalance for the compact layout version of the FTL *HS0* adder is no worse than the case of symmetric layout version, when effects of both process

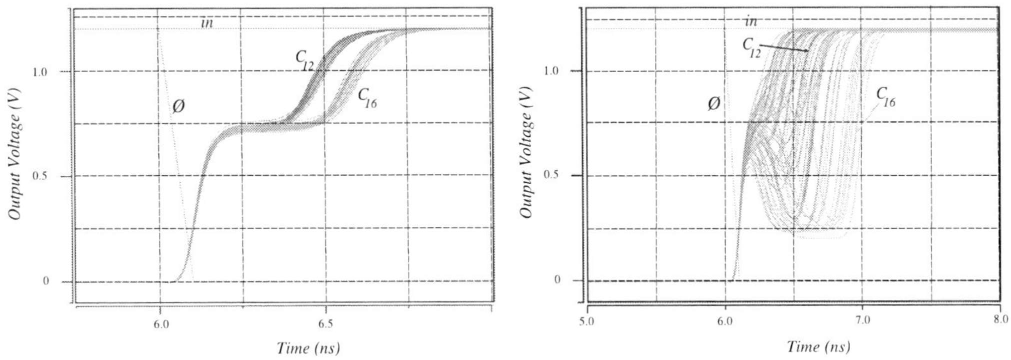

(a) (b)

Figure 13. Waveforms of Monte–Carlo simulations: (a) process corner, and (b) process corner and device mismatching.

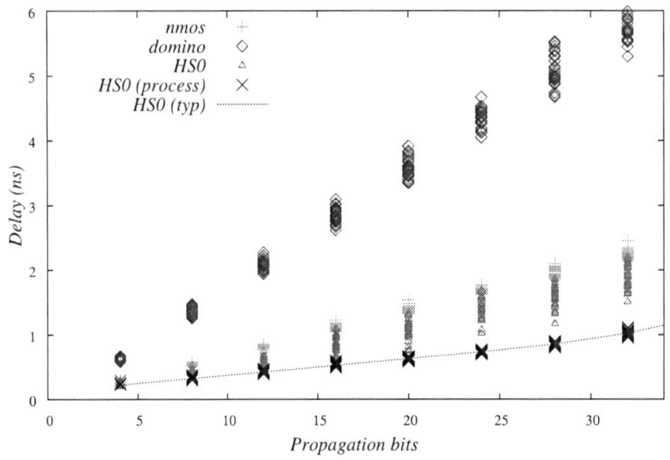

Figure 14. Propagation time delay vs. number of adder bits.

variation and device mismatch are taken into account. However, with just process variation considered (in the absence device mismatch) the imbalance point for the compact layout is much lower than the symmetric layout. This shows that in the presence of the process variation and mismatch, topological rather than the layout symmetry holds the key to the performance of FTL structures; b) the process variation, alone does not shift the imbalance point for the FTL *HS0* structure when compared to the typical case; and c) the limiting effect is the device mismatch, which causes significant imbalance at the 16th bit, although its effect is insignificant at the 12th bit. We can set the imbalance point to be near 14 bits if we allow for 50% of the cases to be at point of imbalance.

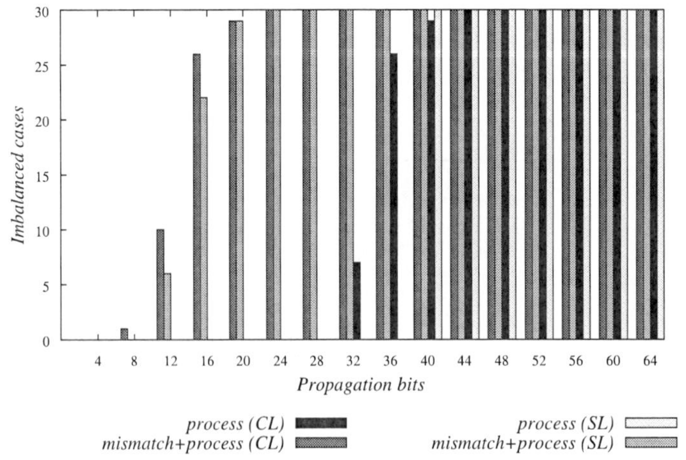

Figure 15. Imbalanced iterations vs number of adder bits, for compact layout (CL) and high–area symmetric layout (SL).

It can be concluded that, although the *HS0* FTL adder has low sensitivity to variations in the global parameters, such as process corner, it is highly sensitive to local parameters variations, such as device mismatch. Further work is required to improve robustness of FTL logic against the variations in the local parameters.

8. RCA Test Chip

To prove potentials of the FTL as a practical CMOS logic family, we implemented a set of of non differential FTL adders on a CMOS test–chip . The test–chip includes the layouts for a 64–bit RCA in the high speed FTL structure, which can be used with either $V_{ref} = 0$ (*HS0*, Fig. 6(a)) or $V_{ref} = V_{cc}/2$ (*HS06*, Fig. 6(b)), and two 32–bit RCAs in the low power structures, for use with $V_{ref} = 0$ (*LP0*, Fig. 6(c)) and $V_{ref} = V_{cc}/2$ (*LP06*, Fig. 6(d)), respectively. Note that we can also configure the low power structure *LP06* to use $V_{ref} = 0$. This configuration produces a modified version of the *LP0* structure (*LP06_0*). A die photograph of the FTL RCA test–chip is depicted in Fig. 16.

The test–chip includes additional circuitry for I/O and global signal buffers, clock tree buffers, multiplexors and serialization circuitry for the input and output data. Additional care has been taken in the physical implementation to achieve layout symmetry in the critical sections of the FTL adders. To reduce the clock skew, the buffered clock lines enter the FTL adder structures from four corners and are distributed across in a symmetrical fashion to each adder bit cell.

To measure the performance of the FTL test–chips, we designed a printed circuit board (PCB), connected to a high speed signal generator (HP8133A), and a high speed digitizing oscilloscope (HP54121) through 50 ohms transmission lines and low loss 50 ohms connector cables for the signal injection and monitoring of the carry input/output and the clock signals. With this set up we minimized the signal degradation and achieved accurate sampling (10ps) in the measurement. Our test set also includes power supplies (HM8142), and digital multimeters (HC3500T). The PCB is connected to a personal computer by means of the RS-232 serial port, to allow for easy access and configuration of the test–chip; *i.e.* the setup of the input words for the FTL adders, and the selection of an output from the FTL structures on the test–chip.

Figure 16. Die photograph of the manufactured FTL test chip.

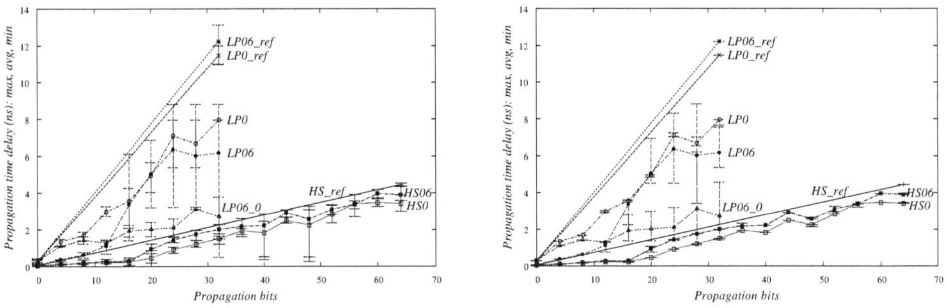

(a) (b)

Figure 17. Delay measurements: (a) chip-to-chip variations, (b) on-chip jitter.

We removed the contribution of the delay and power from the non–FTL circuitry through a series of calibrations. As part of the calibration, we also measured the total power consumption of the whole chip by forcing the FTL structures into permanent evaluation phase (clk=0; nclk=1). In this case the high speed and low power FTL RCAs behave similar to their corresponding CMOS and nMOS-like adders. This measurement is used as reference for comparisons. For calibration purposes the power dissipation was also measured when the FTL structures was forced into permanent reset phase (clk=1; nclk=0).

We performed a set of delay measurements by changing the word–lengths of the inputs into the FTL RCAs to observe the behavior of the delay as a function of number of propagation bits for all FTL structures and with $V_{ref} = 0$ and $V_{ref} = 0.6V$), using a clock with 50% duty cycle. For the reference purposes we also carried out a similar set of delay measurements for the cells configured in the permanent evaluation phase (clk=0; nclk=1). To calibrate for the delay through the test set up and the I/O buffers, we measured the delay by bypassing the carry input to the carry output.

For the test-each chip, we measured the average delays, taken over 1024 samples, and the maximum and minimum delays taken over over a 10 second time period, to account for jitters, or on-chip variations. We performed the measurements on a total of 7 chips. The average delay, and the maximum and minimum delay variations for the on–chip and chip–to–chip measurements, as a function of word–length are shown in Fig. 17.

From the test–chip measurement plots in Fig. 17 we observe that the maximum number of bits at the onset of imbalance is similar to the number obtained in the MonteCarlo analysis of Section 7. The imbalance effect appears at the 18th bit for the high speed *HS0* and *HS06* FTL adders, at the 14th bit for the low power *LP06* adder, and at the 10th bit for the *LP0* adder. A maximum difference of 4–bit is observed from the chip measurements and the MonteCarlo simulations for the *HS0* adder. This small difference can be easily explained due to the limited accuracy of the MonteCarlo device model and the simplified extraction model that was used to speed up the simulations. So, we can safely conclude that the early onset of the imbalance can be attributed to the transistor and capacitive mismatches in the manufacturing process. Note that the delay variation is small before the point of imbalance, but once the imbalance sets in, the chip–to–chip variation dramatically increases. This is because during the evaluation phase the imbalance can push the output nodes towards or away from the final logical value, depending on the characteristics of capacitive mismatch. The on–chip variation, on the other hand, is small except for the *LP06* structure, indicating that it is the most variation sensitive structure.

The calibrated power measurements on the test-chips, correspond to the sum of dynamic and static power consumption of all the FTL structures (64–bit *HS* adder, 32–bit *LP0* adder, 32–bit *LP06* adder) and clock buffers. In order to obtain an accurate estimation of power dissipation of the individual FTL structures before the points of imbalance, we correlated the total power consumption from the test–chip measurements with the power profiling from the post–layout simulations. To do that we performed a series of measurements on the test–chip set up in several configurations, in permanent evaluation phase, and two FTL operations modes with $V_{ref} = 0$ and $V_{ref} = 0.6V$. We also performed a series of power profiling through post–layout simulations of the corresponding adder circuits under the identical configurations. Through correlation, we obtained a set of static and dynamic correction scaling factors, to estimate power contribution of each the FTL adders on the test–chip, from their post–layout simulation values.

Table 4 presents the delay and energy performance of the FTL adders on the test–chips. We also estimated the performance of the domino dynamic logic adders, by applying the same correction scaling factor for the speed, static and dynamic power consumption as was applied to the post–layout simulation estimates of the reference adder cell (*LP0_ref*) configured in permanent evaluation phase (factors of 1.17, 0.94 and 1.44, respectively). We applied the same scaling factor as adder cells for *LP0_ref* and domino dynamic logic have very similar layout complexity, number of devices, and final post-layout performance.

We estimated the performance in terms of the measured static and dynamic and total power consumptions, and delay per bit, using both, a 100 MHz clock and the maximum achievable operating frequency. Note that in order to obtain a fair comparison (in terms of throughput), the dynamic power consumptions of the reference cells (*HS_ref*, *LP0_ref* and *LP06_ref*) correspond to the rate of change of the carry input, set to half of the clock frequency.

The first four columns of Table 4 identify the adder structure, the maximum number of bits before onset of imbalance, the static (P_s) power dissipation per bit, and dynamic energy consumption per clock period (E_d) per bit. The next two columns present the data for the initial delay (t_0) and the delay per bit (t_p). The initial delay is due to the CMOS/nMOS-like behavior of the first adder bit, where the FTL effect is not yet in play. Note that FTL adders with $V_{ref} = V_{cc}/2$ almost eliminate the initial delay because of pre-charging the output node to a voltage close to V_{TH}. All cells in the permanent evaluation phase (*HS_ref*, *LP0_ref*, and *LP06_ref*) are used to establish reference points for comparisons. As these structures behave similar to the corresponding CMOS/nMOS adders with the same capacitive load, it will allow us to observe the relative superior performance of FTL, compared to the standard static fully complementary CMOS/nMOS logic families.

Results of Table 4 indicate that the best performance is obtained by the 18–bit high speed FTL adder with $V_{ref} = 0V$, which improves the energy efficiency per bit by a factor of 16.6, (working at maximum operating frequency), reduces the propagation time delay by a factor of 6.38 and increases the maximum achievable frequency by a factor of 3.7, with respect to the corresponding nMOS-like structure (*HS_ref*). The corresponding improvements with respect to domino dynamic logic are a factor of 30, increase in the energy

Table 4. Estimated Performance of FTL Adders with Number of Bits Set at the Point of Imbalance, from Test Chip Measurements

Logic family	Bits	P_s (uW/bit)	E_d (nW/MHz/bit)	t_{p0} (ns)	t_p (ps/bit)	freq (MHz)	$\Gamma \times 10^6$ (fJ/MHz/bit)	Γ ratio	f_{max} ratio	t_p ratio
HS_ref	18	69.654	10.071	0.055	68.436	777.248 100.000	6822.14 48357.27	1.000 1.000	1.000	1.000
domino		0.027	58.942	0.167	208.855	238.612 100.000	12333.72 12366.24	0.553 3.910	0.307	0.328
HS0		38.296	25.089	0.055	10.723	2880.459 100.000	411.59 4375.45	**16.575** 11.052	**3.706**	**6.382**
HS06		42.502	17.070	0.014	17.149	1388.836 100.000	817.52 7581.32	8.345 6.378	1.787	3.991
LP0_ref	10	0.009	27.084	0.280	349.564	264.880 100.000	9478.97 9497.53	1.000 1.000	1.000	1.000
domino		0.027	58.942	0.167	208.855	396.816 100.000	12324.36 12366.24	0.769 0.768	1.498	1.674
LP0		0.013	41.500	0.280	190.411	403.530 100.000	7908.06 7926.36	**1.199** 1.198	**1.523**	**1.836**
LP06_ref	14	0.009	26.394	0.298	372.759	181.263 100.000	9856.81 9871.57	1.000 1.000	1.000	1.000
domino		0.027	58.942	0.167	208.855	298.020 100.000	12329.04 12366.24	0.799 0.798	1.644	1.785
LP06		2.948	45.885	0.069	86.835	385.764 100.000	4648.08 6544.63	2.121 1.506	2.128	4.293
LP06_0		0.022	48.341	0.298	81.753	550.597 100.000	3955.20 3969.68	**2.492** 2.487	**3.038**	**4.560**

efficiency, 19.5 times reduction in the propagation time delay and 12.1 times increase in the maximum operating frequency. Note that, in the high speed FTL structures energy efficiency degrades with the lowering of operating frequency. On the other hand, due to absence of static power dissipation in the dynamic domino logic, its energy efficiency remains unchanged with the frequency. At 100 MHz clock rate, the high speed *HS* FTL adder achieves a energy efficiency factor of only 2.83 over the domino dynamic logic.

On other hand, the energy efficiencies of the low power adders with $V_{ref} = 0V$ do not change with frequency. This is again due to absence of static power dissipation in the low power structures. However, the $V_{ref} = 0.6V$ structure shows significant degree of variation in the energy efficiency with the change in the frequency of operation due to presence of the static power dissipation with $V_{ref} = 0.6V$ power supply.

The best performance is obtained by the *LP06* adder structure using $V_{ref} = 0V$ (*LP06_0*). This adder exhibits 2.5 times better energy efficiency, 4.56 times reduction in the propagation time delay per bit and 3 times increase in the maximum frequency, when compared with the corresponding CMOS-like reference cell (*LP06_ref*). If we compare the performance results for this FTL adder with results for the dynamic domino logic, we obtain a 3.11 times better energy efficiency, 2.6 times reduction in the propagation time delay and 1.85 times increase in the maximum frequency. Note that the performance of *LP0* adder is marginally worse due to higher delay difference (near 3 times) from the post–layout simulations to the test–chip measurements. The loss of performance of *LP0* with respect to *LP06* is most likely due absence of topological symmetry that is present in the later (see Fig. 6(c) and Fig. 6(d)). Nevertheless, when compared with domino dynamic logic, even with similar performance level in terms of propagation time delay (1.10 ratio) and maximum frequency (1.02 ratio), *LP0* exhibits a remarkable superior energy efficiency by a factor of 1.56.

We compared the obtained results for the *HS0* FTL RCA cell with a high performance delay and low power version of a multilevel 32–bit carry skip adders (CSA) [20, 21], that outperforms some well–known high performance adders, such as the Brent–Kung, Ladner–Fisher, Koge–Stone, or Sklansky adders, in terms of delay and energy efficiency. To reduce the delay and power consumption of the CSA adder in [21], it is divided into variable-sized blocks that balance the inputs to the carry chain. Within blocks, highly optimized carry look-ahead logic, which computes block generate and block propagate signals, is used to further decrease delay. These adder architectures decrease power consumption by reducing the number of logic levels, glitches, and transistors.

Note that the CSA adders in [21] and [20] are implemented in 0.25μm and 0.13μm CMOS technologies, respectively. So, we used the improvement ratios in [20] and the performance estimations in [21] to obtain the estimated performance of the CSA adder in [21] in the equivalent 0.13μm CMOS technology. Table 5 sums up the performance parameters of CSA adders in [20, 21], and our *HS0* adder. Our *HS0* RCA cell outperforms the adders in [20, 21] in terms of propagation time delay (by a factor of 2.2 and 1.78), and energy efficiency (2.89 and 1.72 ratios), while obtaining similar power consumption (1.07 and 0.79 ratios).

Table 5. Comparison of HS0 FTL adder and other high performance adders.

Logic family	t_p (ps/bit)	E_t (nW/MHz/bit)	$\Gamma \times 10^6$ (fJ/MHz/bit)
[21]	23.53	**30.13**	708.96
[20]	29.06	40.94	1189.64
HS0	**13.17**	38.38	**411.59**

9. Discussion and Further Work

From the test–chip measurements we can make the following inferences:

- The symmetry in the circuit topology is the key to the higher performance of FTL structures. High level of symmetry is required in the layout of the compensating devices for proper operation, but is not needed in the rest of the layout.
- The FTL structures are not suited to be used as assorted logic. They are most suitable for regular structures like adder and multiplier cells.
- The high speed FTL structures offer the best tolerance with respect to the mismatch in the process parameters. This is due to lower complexity in the topology of the circuit.
- The high speed FTL structures offer the best energy efficiency for high clock frequencies, even though they suffer from high static power dissipation.

- The low power FTL structures should be used when power dissipation (and not the energy efficiency) is of primary concern.
- To safeguard against the imbalance effect caused by the capacitive and device mismatch, the FTL structures should be employed for small size arithmetic blocks like pipeline RCAs, carry save adders, and carry select adders, where word-length depth per pipeline stage is no more than 8 and 16 bits, for the low power and high speed FTL structures, respectively.

To facilitate design of FTL circuits, the following design guidelines are in order:

1. Candidate circuits for FTL are easily implemented as they form a chain of inverting cells. Moreover, the inverting cells should maintain a close logic symmetry.
2. It is important to identify the different propagation paths of the basic FTL inverting cell, and accurately estimate the difference between their load conditions. Additional transistors must be added to compensate for the dissimilarities that may have been found.
3. Buffering can help ensuring that the output load is constant, independent of the propagation path and the load cells.
4. Estimate the post–layout capacitances and optimize sizing to achieve the better performance in terms of the desired parameter; speed, power, energy, or sensitivity.
5. For the physical implementation, start with the compensating transistors. We also need to achieve high symmetry in the routing, because the symmetry in the signal paths is a key factor in the reliability of middle and long word–length chains. Strict symmetry of the logic through the compensation capacitances in the the basic cell is vital, but layout symmetry is not required unless device mismatch is no longer the limiting factor in the FTL implementation.
6. To minimize the device mismatch effect, place all devices along the same orientation. Otherwise, the device mismatches are higher in the current CMOS technologies [22]. It is also desirable to obtain maximum cell symmetry to minimize stress effects.
7. Buffering of both, the input and output nodes, with identical buffer cells is desirable to ensure the identical voltage conditions at the input nodes, and maximum isolation of the outputs to the variations in the capacitive load or fanin.

A major weakness of the FTL style over static, domino and np-CMOS designs is that it only suits regular arithmetic structures and not assorted gates. Although FTL no doubt shows excellent performance and reliability over conventional design styles, further work is required in the design and characterization of various FTL gates for the assortment of digital circuits. Moreover, the RCA is not a practical structure beyond 16 bits, as higher performance topologies are available, such as parallel prefix, Manchester carry chain adder (MCC), carry look ahead (CLA), or CSA. These structures require more than just a full adder cell.

For example, CSA is a clear candidate to be implemented in FTL. The basic operation of this topology can be translated to a cascaded operation of multiplexors (for the skip section) and RCA cells (for the adder blocks). Each operation can be separated; parallel computation of the carry output at each adder block, skipping of correct carry signals to the carry inputs by the multiplexors chain, and finally, correct computation of the sum words at each adder block. Fixed word–lengths of the adder blocks can be used for the initial design. Note that the results of our FTL *HS0* adder design are superior in terms of energy (1.72 times) and delay (1.78 times) per bit when compared with other high performance adders [20]. Therefore, a simple pipelined adder formed by two HS0 adder stages will exhibit even better performance than these high performance topologies. Consequently, high delay and energy efficiency improvement can be expected from CSA implementations in FTL style.

Other important drawback on the FTL design is the difference between the post–layout simulations and chip measurement results in the delay parameter. The limited accuracy of the models for the HSPICE simulator in the UMC technology dictates the accuracy of the Monte–Carlo analysis. To design reliable circuits with FTL, efforts have to be focused on the utilization of implemented design blocks in a larger circuit design. So, characterization of implemented circuits, based on chip measurement results, will be a key issue for the future development of FTL.

The design of asynchronous FTL circuits also seems as an interesting possibility. FTL operation inherently implies the management of three logic levels; high, low, and *still computing* (while nodes are kept at V_{TH}). This feature allows for identification of the *end–of–operation* (EOO) condition, that are a quite natural to asynchronous logic. A key factor in the design of EOO logic is to obtain fast enough detectors so that the delay overhead will not be significant when compared with the very high speed of FTL logic.

Finally, further work is required to improve the robustness of FTL against the device mismatch of the manufacturing process. Although pipelining is an effective way for extending the use of FTL to long word–length arithmetic circuits, some efforts must be focused on studying lower sensitive FTL structures and characterizing the sensitivity parameter against other design parameters (such as maximum short–circuit current, or DC gain at V_{TH}). FTL implementations of differential and cascode voltage switch logic (CVSL) adders are clear candidates to be analyzed.

10. Conclusion

In this paper, we proposed a new class of logic family for CMOS technology based on the *feedthrough evaluation* concept. FTL structures can be tailored for the high speed or low power circuits with long logic depth, such as high performance ripple carry adders.

FTL logic stages (differential and non differential) can be cascaded without using interstage inverters, reducing propagation time delay with respect to domino dynamic logic families in CMOS. Charge redistribution problems associated with dynamic families are also removed in the FTL structures.

We designed, and laid out a set of non differential RCAs to evaluate the performance of the FTL adder structures. The test–chip measurements of this set of adders show that propagation time delay of arithmetic circuits employing high speed FTL structures outperform the dynamic domino structure by a factor of 12.1. The energy efficiency per bit is improved by a factor of 30. Even high performance and delay efficient multilevel CSAs, which improves energy and delay of Bret–Kung and many other parallel adders, are outperformed by our high speed FTL adder in terms of both, energy efficiency (1.72 times) and delay (1.78 times). Low power FTL structures outperform the domino logic by a factor of 2.6 in terms of propagation delay time, while improving the energy efficiency per bit by a factor of 2.83.

References

[1] Krishnamurthy, R. K., Alvandpour, A., Mathew, S., Anders, M., De, V., & Borkar, S., "High–performance, low–power, and leakage–tolerance challenges for sub–70nm microprocessor circuits", in *Proc. 28th Euro. Solid-State Cir. Conf.*, Sep. 2002, pp. 315–321.

[2] Mathew, S., Anders, M., Krishnamurthy, R. K., & Borkar, S., "A 4 GHz 130 nm address generation unit with 32–bit sparse–tree adder core", in *IEEE VLSI Cir. Symp.*, Jun. 2002, pp. 126–127.

[3] Krishnamurthy, R. K., Hsu, S., Anders, M., Bloechel, B., Chatterjee, B., Sachdev, M., & Borkar, S., "Dual supply voltage clocking for 5 GHz 130 nm integer execution core", in *IEEE VLSI Cir. Symp.*, Jun. 2002, pp. 128–129.

[4] Mathew, S., Anders, M., Bloechel, B., Nguyen, .T, Krishnamurthy, R. K., & Borkar, S., "A 4GHz 300mW 64b integer execution ALU with dual supply voltages in 90nm CMOS", in *IEEE Int. Solid-State Cir. Conf.*, Feb. 2004, pp. 162–163.

[5] Vangal, S., Hoskote, Y., Somasekhar, D., Erraguntla, V., Howard, J., Ruhl, G., Veeramachaneni, V., Finan, D., Mathew, S., & Borkar, N., "A 5 GHz floating point multiply–accumulator in 90 nm dual Vt CMOS", in *IEEE Int. Solid-State Cir. Conf.*, Feb. 2003, pp. 334–335.

[6] Nooshabadi, S., & Montiel–Nelson, J. A., "Fast Feedthrough Logic: A High Performance Logic Family for GaAs", *IEEE Trans. Circ. Syst. I*, vol. 51, no. 11, pp. 2189–2203, Nov. 2004.

[7] Montiel–Nelson, J. A., Nooshabadi, S., Armas, V., Sarmiento, R., & Núñez, A., "High Speed GaAs Subsystem Design using Feed Through Logic", in *Proc. Design, Automation and Test in Euro. Conf. 1999 (DATE'99)*, Mar. 1999, pp. 509–513.

[8] Rabaey, J. M., Chandrakasan, A., & Nikolić, B., *Digital Integrated Circuits: A Design Perspective –2e.* Upper Saddle River, NJ: Prentice Hall, 2003.

[9] Montiel–Nelson, J. A., Nooshabadi, S., & Eshraghian, K., "Gallium Arsenide Based Fast Feed Through Logic (FTL)", in *Proc. IEEE Int. Symp. Cir. and Sys.*, Jun. 1997, pp. 1884–1887.

[10] Navarro–Botello, V., Montiel–Nelson, J. A., & Nooshabadi, S., "Low power and high performance arithmetic circuits in feedthorugh CMOS logic family for low power applications", *ASP J. Low Power Electronics*, vol. 2, no. 2, pp. 300–307, Aug. 2006.

[11] Navarro–Botello, V., Montiel–Nelson, J. A., & Nooshabadi, S., "Analysis of high performance fast feedthorugh logic families in CMOS", *IEEE Trans. Cir. & Syst. II*, vol. 54, no. 6, pp. 489–493, Jun. 2007.

[12] Navarro–Botello, V., Montiel–Nelson, J. A., & Nooshabadi, S., "High performance low power CMOS dynamic logic for arithmetic circuits", *Microelectron. J.*, vol. 38, no. 4–5, pp. 482–488, Apr. 2007.

[13] Oklobdzija, V. G., "Clocking and clocked storage elements in a multi–gigahertz environment", *IBM J. Res. and Dev.*, vol. 47, no. 5/6, pp. 567–583, Sept./Nov. 2003.

[14] Hrishikesh, M. S., Jouppi, N. P., Farkas, K. I., Burger, D., Keckler, S. W., & Shivakumar, P., "The Optimal Logic Depth Per Pipeline Stage is 6 to 8 FO4 Inverter Delays", in *Proc. Int. Symp. Comp. Arch.*, May 2002, pp. 14–24.

[15] Kim, K. K., Kim, Y. B., Choi, M., & Park, N., "Leakage Minimization Technique for Nanoscale CMOS VLSI", *IEEE Design and Test of Computers*, vol. 24, no. 4, pp. 322–330, July/Aug. 2007.

[16] Roy, K., Mukhopadhyay, S., & Mahmoodi-Meimand, H., "Leakage current mechanisms and leakage reduction techniques in deep-submicrometer CMOS circuits ", in *Proc. IEEE*, vol. 91, no. 2, pp. 305–327, Feb 2003.

[17] Fang, C.,Huang, C., Wang, J., & Yeh, C., "Fast and compact dynamic ripple carry adder design", in *Proc. IEEE Asia Pacific Conf. on ASIC*, Aug. 2002, pp. 25–28.

[18] Burd, T. D., & Brodersen, R., "Processor design for portable systems", *J. VLSI Sig. Proc. Sys.*, vol. 13, no. 2–3, pp. 203–221, Aug./Sep. 1996.

[19] Hastings, A., *The art of analog layout.* Upper Saddle River, NJ: Prentice Hall, 2001.

[20] Lin, Y. S., & Radhakrishnan, D., "Delay efficient 32–bit carry–skip", *VLSI Des.*, vol. 2008, no. 2, pp. 1–8, Jan. 2008.

[21] Chirca, K., Schulte, M., Glossner, J., Wang, H., Mamidi, S., Balzola, P., & V. Navarro–Botello, J. A. Montiel–Nelson, and S. Nooshabadi, Vassiliadis, S., "A Static Low–Power, High–Performance 32–bit Carry Skip Adder", in *Proc. IEEE of the Digital System Design, EUROMICRO Systems*, Aug. 2004, pp. 615–619.

[22] Pelgrom, M. J. M., Duinmajer, A. C. J., & Welbers, A. P. G, "Matching Properties of MOS Transistors", *IEEE J. Solid–State Circuits*, vol. 24, pp. 1433–1440, Oct. 1989.

In: CMOS Technology
Editor: Min-jun Kwon, pp.221-244

ISBN: 978-1-61761-325-8

Chapter 8

DISCUSSION ON 1/F NOISE IN CMOS TRANSISTORS: MODELLING-SIMULATION AND MEASUREMENT TECHNIQUES*

T. Noulisa†, S. Siskosa, L. Bary and G. Sarrabayrouseb

[1] Electronics Laboratory of Physics Department, Aristotle University of Thessaloniki, Aristotle University Campus, 54124 Thessaloniki, Greece

[2] LAAS – CNRS, University of Toulouse, 7 Avenue du colonel Roche 31077 Toulouse cedex 4, France

ABSTRACT

In this chapter the most important flicker noise models that have been proposed in the literature are analytically presented and discussed. The available flicker noise simulation models in software packages like SPICE and HSPICE and SPECTRE are also examined and selection criteria in relation to the type of MOSFET and the operating regime are proposed. The impact of scaling down technologies is discussed and extra analysis is presented in relation to the type of the bias in CMOS transistors. In particular the switched biasing effect in the MOSFET low frequency noise performance is presented. Additionally, low frequency noise measurement set-ups are presented and measurement techniques are suggested.

1.1. INTRODUCTION

CMOS transistors are required as "building blocks" of analog applications circuits. They are preferred for dense, low-voltage and low-power applications. While the advantages of

---

* A version of this chapter was also published in *MOSFETs: Properties, Preparations and Performance*, edited by Noah T. Andre and Lucas M. Simon, Nova Science Publishers. It was submitted for appropriate modifications in an effort to encourage wider dissemination of research.

† Corresponding author: E-mail address: tnoul@physics.auth.gr.

CMOS technology are well established, there is a major drawback for analog applications: MOSFETs are "noisy". On the other hand, the conventional bipolar semiconductor devices such as BJT or HBT with their superior noise characteristics are commonly used for such analog circuit applications requiring satisfactory noise performance. However, with the recent progress in noise reduction for MOS transistors (MOS-FETs), analog circuit configuration using MOSFETs is becoming possible in areas traditionally regarded as difficult.

In particular, the noise behaviour of CMOS devices is primarily dominated by two sources, thermal and flicker noise. Other sources that are sometimes present in the noise spectrum are shot noise, generation/recombination noise, and "popcorn" noise. Of these noise sources, thermal and shot noise are physically fundamental to the operation of the device and are always present. Thermal noise is the voltage fluctuations due to the random Brownian motion of electrons in the resistive medium. It is broadband white noise, and it gets larger with increasing resistance and temperature. Shot noise is caused by the non smooth current flowing across a junction and it is comprised of individual electrons arriving at random times. This non-uniform flow gives rise to broadband white noise. Generation/recombination (G/R) noise is due to trapping of carriers in the bulk or at the interfaces of the device. The trapping of carriers causes fluctuation in their number, and thus fluctuation in the resistance. The G/R noise creates a Lorentzian noise spectrum. Popcorn noise, sometimes called burst noise or random-telegraph-signal (RTS) noise is due to a discrete modulation of the channel by the capture and emission of a channel carrier.

Regarding flicker noise phenomenon, it appears through both quality-dependent and fundamental noise processes. It seems to be large in MOSFETs and therefore sets the lower limit of the signal level that can be processed by VLSI devices and circuits. It is also commonly called $1/f$ noise because the noise spectrum varies as $1/f^{\alpha}$, where f is the signal frequency and α, an exponent very close to unity (α=1±0.2) (see Figure 1). In contrast to the channel thermal noise and the shot noise which are quite understood, flicker noise is not yet fully known, although its presence is surprisingly universal in all types of semiconductor devices [1]. A lot of studies have been done so far in order to explain and describe the flicker noise phenomenon [2]-[19]. However, controversy still exists in the modelling of $1/f$ noise in CMOS transistors.

In this chapter, the most important $1/f$ noise models proposed in the literature are presented and discussed. The available flicker noise simulation models in software packages like SPICE and HSPICE are examined and selection criteria in relation to the type of MOSFET and the operating regime are also proposed. Extra analysis is presented in relation to the type of the bias in CMOS transistors. In addition, low-frequency noise measurement set-ups are also presented and measurement techniques are suggested.

1.2. Flicker Noise Modelling and Simulation

In integrated circuits design, noise simulation is a key point of the process. To evaluate circuit noise performance, the designer uses one of the noise models available in the simulator. The selection of the noise model is very important in flicker noise evaluation depending upon the type of MOSFET (n or p) and the operating mode. So any information or rule about the model to select is very essential and useful to the designer.

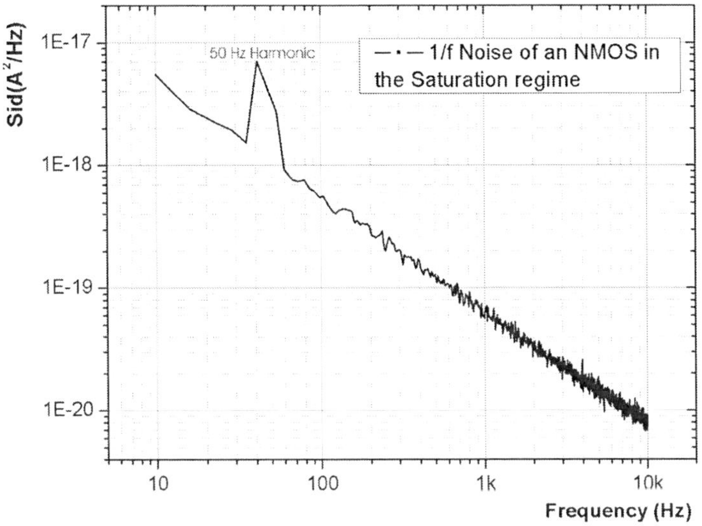

Figure 1. Power Spectral Density of Flicker Noise is inversely proportional to frequency

1.2.1. The Empirical and the Typical 1/*f* Noise Models

A. Simple empirical model

The most common flicker noise model used in the IC analog designer's community was proposed by Tsividis [20]. Tsividis says that "in a common bulk CMOS fabrication process," one may find an input-referred voltage due to flicker noise of:

$$v_f^2(f) = \frac{K_F}{C_{OX}^2 \cdot W \cdot L \cdot f} \tag{1}$$

where K_F is the empirical flicker noise coefficient equal to 5×10^{-9} fC^2/μm^2 for NMOS devices, and 2×10^{-10} fC^2/μm^2 for buried channel PMOS devices. W and L represent the transistor's channel width and length respectively and C_{ox} *the* gate capacitance per unit area and f the frequency.

Many of the empirical models, including the above model, given by textbooks are simple and easy for hand calculations. Unfortunately they are also questionable. In particular, this equation is an empirical relation, and K_F is not entirely independent of the operating point (and is not the same in weak and strong inversion). This model was implemented in early versions of SPICE, and is generally considered to be oversimplified.

B. Number fluctuation model and mobility fluctuation model

According to the Number Fluctuation model (or McWhorter Model) the flicker noise phenomenon is primarily a surface effect [2]. In particular, the McWhorter number fluctuation (Δn) theory states that flicker noise is generated by fluctuations in the number of carriers due to charge trapping in surface states. McWhorter obtained the necessary $1\triangleleft f$ spectrum by assuming that the time constant τ of the surface states varied with a $1/\tau$

distribution. Christensson et al. [3, 4] were the first to apply the McWhorter theory to MOSFETs, using the assumption that the necessary time constants are caused by the tunnelling of carriers from the channel into traps located inside the oxide.

A number of other applications of the McWhorter theory to MOSFETs have been done. Das and Moore [5] reviewed and compared a number of these theories and found that the basic assumptions behind any explanation, affects the interpretation of the experimental results. They said, "theoretical calculations of the MOSFET drain noise current have been performed by many investigators, which have led to different results, mainly due to the difference in the method of attack and nature of assumptions".

Reimbold [6] developed the McWhorter model further, taking into account all the capacitive components of the small-signal equivalent circuit, so as to account for all transistor operating regimes. This development was done to fit Reimbold's measurements in weak inversion. Ghibaudo [7] presented a shortcut through Reimbold's work and came up with the same result.

According to the above model the spectral density of the drain current I_{DS} is:

$$i^2_f = \frac{K_F q^4 I^2_{DS}}{n^2 kTWLf} \cdot \frac{N_{ST}}{(C_{OX} + C_{SS} + C_D + C_I)^2} \tag{2}$$

where N_{ST} is the surface trap density, C_{SS} is the interface state capacitance, C_D is the depletion capacitance, and C_I is the inversion layer capacitance. T is the temperature, k the Boltzman constant, q the electron charge.

On the other hand according to the mobility fluctuation model (or Hooge model), $1/f$ noise is not associated to surface effects. F. N. Hooge proposed that $1/f$ noise is essentially a bulk phenomenon. Working with metal films he extracted an empirical relation for $1/f$ noise in terms of resistance fluctuations [8-11]. In particular, the Hooge model suggests gate voltage dependence in the input-referred noise. The model is based on empirical experimental observations of the noise in homogeneous samples, and the input referred noise shows strong gate bias dependence.

1.2.2. Combined Models

A lot of research has focused on resolving the above two theories regarding the $1/f$ noise in CMOS technology. However, it has not provided any satisfactory resolution. Inconsistent experimental results were proposed for both n- and p- type channel devices.

Specifically, in an effort to fit as much published data as possible, some authors have combined the number fluctuation and mobility fluctuation theories into a single model. The main idea is that charges, when trapped, cause correlated surface mobility fluctuations due to inversion layer carriers scattering on the trapped charges. Jayaraman and Sodini [14] were the first to propose this model. They also showed that oxide band bending due to a non-uniform oxide trap distribution gives rise to a gate voltage dependence of the flicker noise coefficient, and of α. However, from their data, they were unable to determine if the gate voltage dependence they observed was due to correlated mobility fluctuations, or to non-uniform oxide trap distributions (with a corresponding higher trap density), or both.

Trying to improve the flicker noise model in circuit simulators such as SPICE, the unified model idea was developed further by Hung et al [15, 16]. By examining the channel current modulation due to the trapping of a single electron (popcorn noise in a very small device), they were able to quantify the contribution of the correlated surface mobility fluctuations. However, their model is geared towards numerical simulation.

Literature supports number fluctuations as the primary source of flicker noise in MOSFETs. Responding to this evidence, Hooge [17] has suggested that in semiconductors there is always a mobility-correlated 1◁*f*noise but that other types of 1◁*f* noise may be present and may be dominant.

Other authors have suggested that 1◁*f*noise in NMOS devices is caused by carriers number fluctuations while in PMOS devices the noise is due to mobility fluctuations [18, 19]. This dichotomy makes sense if the NMOS transistors are surface channel devices where the carriers interact more readily with Si-SiO_2 surface states, and the PMOS transistors are buried channel devices where the carriers interact only with the lattice (however in modern deep-submicron technologies (0.35μm and below), both NMOS and PMOS transistors are surface channel devices, so that this dichotomy may no longer exist). Furthermore, differences in the energy trap density between NMOS and PMOS transistors could also account for the differences in observed bias dependence [21].

1.2.3. Noise Simulation Using the BSIM MOSFET Model in Software Packages

Flicker Noise Simulation Models

While conclusions and design criteria concerning noise are extracted using simulations in different types of typical analog integrated circuits and while other issues like the MOS transient response were studied in respect to the simulation model and Level, few extended research has been performed on the validity of the available $1/f$ noise simulation models in relation to the MOS transistor model that is developed in the simulator. In particular BSIM4 model, which is one of the most commonly used Spice MOS transistor simulation models, develops two distinct models, SPICE-Flicker and BSIM-Flicker, to calculate flicker noise.

In particular, when $1/f$ noise performance of MOS transistors or MOSFET circuits has to be simulated there are different flicker noise models that might be used. The simulation of MOSFET $1/f$ noise performance is differentiated in relation to the used flicker noise model. HSPICE simulator incorporates in its algorithms specific flicker noise models. On the other hand BSIM4.X (and BSIM3V3.X) develops its own flicker noise models. The above model selector parameters and the respective equations are listed in Table I.

For all flicker noise models incorporated either in HSPICE or in BSIM algorithms, there are serious inaccuracies in terms of MOSFET noise estimation. It seems that one single model can not simulate accurately flicker noise performances of MOSFETs in all operating regions and for both transistor types, P and N.

As it can be seen in Table I, in the HSPICE simulator there are three different equations and so three different models that can be used for flicker noise estimation depending on the value of the selector parameter NLEV. Also, in case of BSIM3/BSIM4 two different equations can also be chosen depending on the selector parameters fnoimod/noimod.

Compatibility between the BSIM3/BSIM4 flicker noise models and the HSPICE flicker noise models is presumed when fnoimod is 0 in BSIM4 (noimod 1,4 in BSIM3) and when NLEV is 0 in HSPICE (Level 49).

Regarding noise parameters, they are listed in Table I and have the following meaning: *F*licker noise coefficient *KF*, is a fitting parameter and depends critically on processing, g_m is the transconductance, W_{eff} and L_{eff} are the effective channel width and length, respectively, *AF* (flicker frequency exponent) and *EF* (flicker exponent) characterize the power dependence of the measured $\overline{i_{1/f}^2}$ on I_{DS} and frequency respectively. *h* is a function of the SPICE parameters *NOIA, NOIB, NOIC* and g is a function of the SPICE parameter *EM. NOIA, NOIB* and *NOIC* can be treated as fitting parameters, like *KF* in the SPICE-Flicker model. *NOIA*, *NOIB*, and *NOIC* have different default values in NMOS and PMOS transistors [24]. In addition these parameters are unitless in BSIM3V3, but have some rather complicated units in BSIM4. *EM* represents the field at which the carrier velocity saturates. *EM* and *KF* retain the same meaning in both BSIM3V3 and BSIM4. *EM* is a parameter that affects only the noise models but not the I-V or the C-V characteristics [24].

The unified BSIM-Flicker noise model of BSIM4 also follows the framework of BSIM3V3, but with the slight improvements of using smoothing functions and considering the bulk charge effect [24] – [27].

Until recently two works that examine which flicker noise simulation model is more appropriate have been published [28], [29]. They concern NMOS devices operating from subthreshold to strong inversion and PMOS transistors only in the saturation region. Noise comparisons refer to old versions of the BSIM MOSFET model. 1/*f* noise performance is examined in relation to the gate-source bias, the transconductance and its linearity, and the MOSFET channel length. Different 1/*f* noise models are considered suitable for different MOS operating regions and the accuracy of each 1/*f* model is examined at only two particular frequencies (1 Hz and 1 kHz).

Table I. Flicker noise models

Selector simulation parameter	Noise model	
NLEV = 2 or 3 (HSPICE)	$\frac{\overline{i_{1/f}^2}}{\Delta f} = \frac{KF \cdot g_m^2}{C_{ox} W_{eff} L_{eff} f^{AF}}$	(3)
NLEV = 1 (HSPICE)	$\frac{\overline{i_{1/f}^2}}{\Delta f} = \frac{KF \cdot I_{DS}^{AF}}{C_{ox} W_{eff} L_{eff} f}$	(4)
NLEV = 0 (HSPICE) fnoimod = 0 (BSIM4.X) noimod = 1 or 4 (BSIM3V3.X)	SPICE-Flicker in BSIM4 (Spice2 in BSIM3), $\frac{\overline{i_{1/f}^2}}{\Delta f} = \frac{KF \cdot I_{DS}^{AF}}{C_{ox} L_{eff}^2 f^{EF}}$	(5)
fnoimod = 1 (BSIM4.X) noimod = 2 or 3 (BSIM3V3.X)	BSIM-Flicker in BSIM4 (BSIM in BSIM3), $\frac{\overline{i_{1/f}^2}}{\Delta f} \approx h(NOIA, NOIB, NOIC)\left[\frac{I_{DS}}{C_{ox} L_{eff}^2 f^{EF}} + \frac{g(EM) I_{DS}}{W_{eff} L_{eff}^2 f^{EF}}\right]$	(6)

According to the above works, for long channel NMOS devices (L=5 μm), the input-referred gate voltage noise is independent of gate bias, which shows the surface state model and the noise model (NLEV=2,3) in SPICE are more appropriate. SPICE can predict the long channel NMOS noise performance very well in saturation regime, but not in the sub-threshold region, and also cannot be used to simulate short-channel NMOS noise performance. For short channel NMOS devices (L=0.6 μm) the input-referred gate voltage noise is independent of the gate-source voltage V_{gs}, increasing proportionally as V_{gs} increases which is due to the non linearity of the transconductance g_m variation with gate bias. These devices can be modeled with the BSIM3.3 model and the same model is also appropriate. However these models are applicable to NMOS devices only [28]. Regarding PMOSFETs, for long channel devices, the flicker noise is due to the bulk effect [29]. For short channel PMOS devices, however it is due to the surface state effect. NLEV=0 is the most appropriate value for the simulation of long channel devices and NLEV=2,3 for short channel devices.

Contrary to the above results the authors have extracted different selection criteria considering the accuracy of the available 1/f noise simulation models [30],[31]. Particularly, in order to assess the flicker noise models described above, both n- and p-type MOS transistors were used. MOSFETs were fabricated in the 0.6 μm CMOS process by Austria Mikro Systeme (AMS) and their dimensions were 400μm/1.2μm for the NMOS and 1000μm/1.2μm for the PMOS transistor. Bias conditions varied as to measure MOS flicker noise performance from the threshold of linear operating region to saturation. The measurements were obtained for a frequency range of 10 Hz to 10 kHz. Respective simulations were performed using built-in noise models in the HSPICE simulator and the BSIM4 MOS transistor model. Comparison between measurements and simulations was made in relation to the same MOSFET drain current as to achieve accuracy. Temperature was constant at 295 K.

In particular, for an NMOS transistor operating in the linear or saturation region, BSIM-Flicker is the best simulation model, since it appears to be closer to the experimental results. In both operating regions, and especially in saturation, the selected model provides an accurate description of the NMOS flicker noise behavior. A low frequency measurement and the respective simulations using all the available $1/f$ noise models in the saturation and in the triode regime are given in Figure 2

Additionally, for a p-type MOS transistor, SPICE-Flicker noise model (and NLEV=0 for HSPICE) is more preferable in the saturation region and BSIM-Flicker in the triode region. In both operating regions, the selected models overestimate flicker noise. However, the best noise model for a PMOS transistor is not as accurate as it is for an NMOS transistor. In all operating regions, for an NMOS and a PMOS transistor, simulation noise models used by HSPICE for NLEV = 1 and NLEV = 2 or 3 underestimate largely $1/f$ noise and appear to be inaccurate. A low frequency measurement and the respective simulations using all the available $1/f$ noise models in the saturation and in the triode regime are given in Figure 3

The above study implies that NMOS transistor flicker noise performance can be simulated with reliability using the BSIM-flicker noise model of BSIM4.X instead of SPICE-flicker or HSPICE $1/f$ models. However, all the above noise models appear to be quite problematic for a PMOS transistor. This suggests the necessity to develop a new model to calculate PMOS flicker noise performance and to implement it in the simulator.

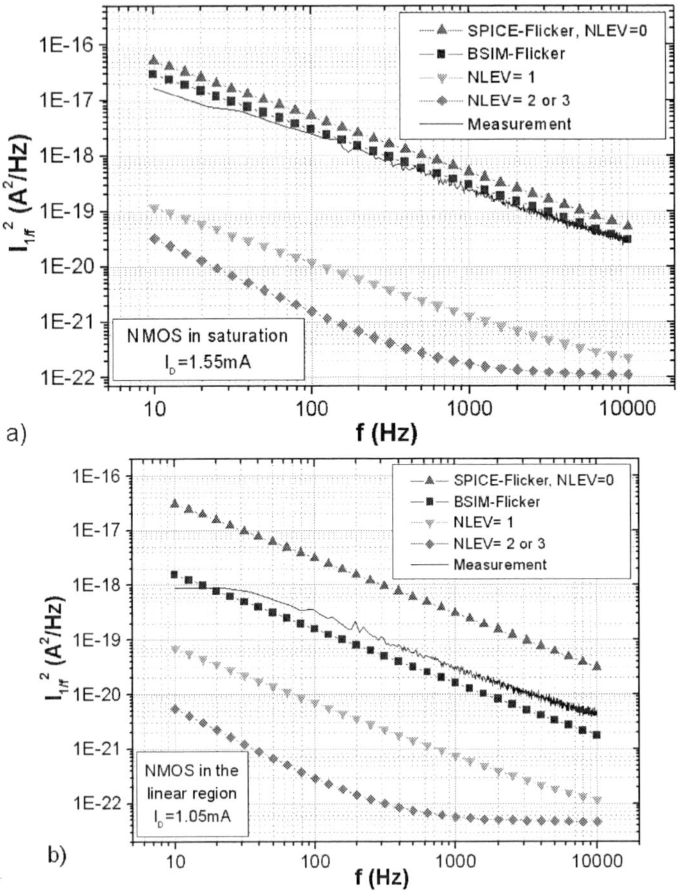

Figure 2. Noise measurement and simulations of an NMOS transistor operating (From [30] with permission) a) in the saturation region for a drain current of 1.55 mA, b) in the linear region for a drain current of 1.05mA

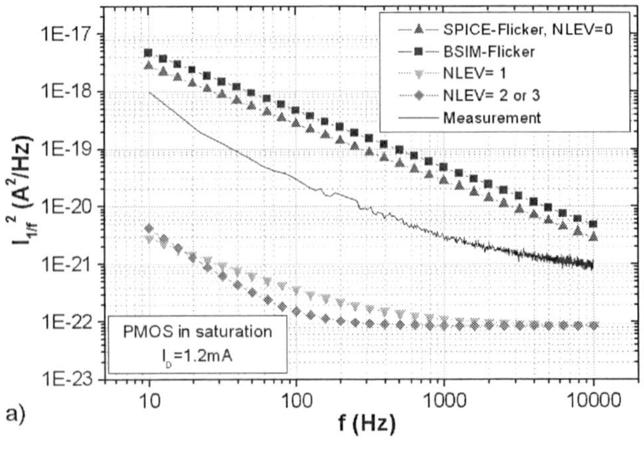

Figure 3. Continued on next page

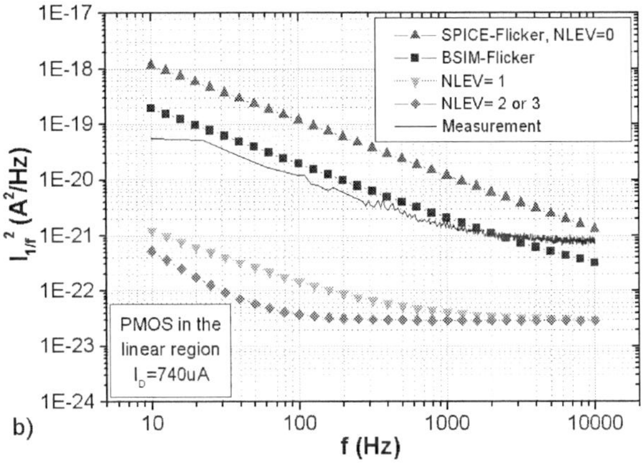

Figure 3. Noise measurement and simulations of a PMOS transistor operating a) in the saturation region for a drain current of 1.2 mA, b) in the linear region for a drain current of 740 μA (From [30] with permission)

Table II. Usability of all models for a particular device type under particular bias conditions

Device type	Operating Region	Optimum model	Accuracy
NMOS	Saturation	BSIM-Flicker	Good
	Linear region	BSIM-Flicker	Good
	Threshold	BSIM-Flicker	Good
PMOS	Saturation	SPICE-Flicker	Medium
	Linear region	BSIM-Flicker	Medium
	Threshold	-	-

An overall view of the usability of all models for a particular device type under particular bias conditions is given in Table II. Finally, the results about the reliability of the specific $1/f$ noise simulation models can be considered as valid for other typical analog CMOS scaled processes since, relatively long channel MOSFETS in analog circuit applications are used. However the accuracy of the noise magnitude each model provides is process dependent, since the majority of the model parameters depend critically on processing. Therefore the models reliability is also conditioned by an effective noise characterization of the used CMOS process.

Analysis was also performed in terms of the $1/f$ noise exponent [31], [32]. In order to find out the exact flicker noise dependence on frequency variations and compare it to the respective simulation parameter (*EF*), which is common to both models, $1/f$ noise measurements were exponentially fitted. The flicker noise measurements of an NMOS and a PMOS FET in the linear and saturation region are shown in Figure 4 and Figure 5.

Regarding the description of an NMOS transistor noise behavior in linear and saturation regions, the experimental values of the flicker noise exponent, varied from 0.91 to 0.98, quite close to that of the simulation parameter *EF*, which is unity. Therefore, both models can describe with accuracy the noise spectral density dependence on frequency. The flicker noise exponent of an N-MOSFET at the threshold of the linear region, was an exception, since it was found to be 0.85 due to harmonics in the kHz range.

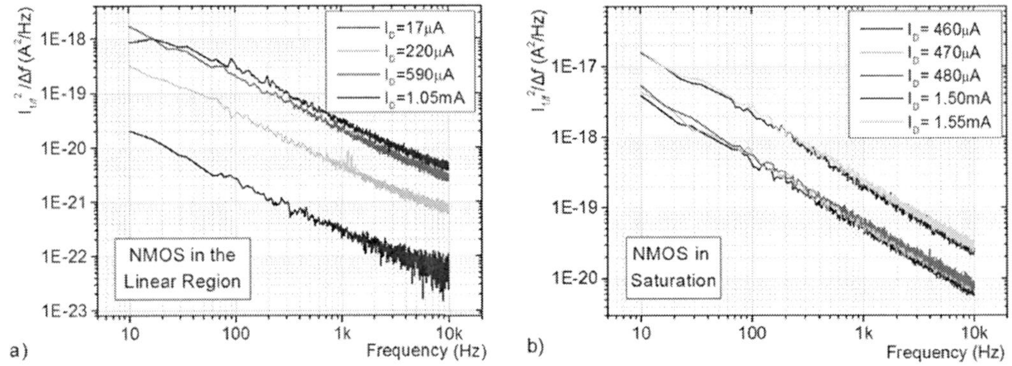

Figure 4. NMOS transistor noise performance a) in linear and b) in saturation region for various values of drain current

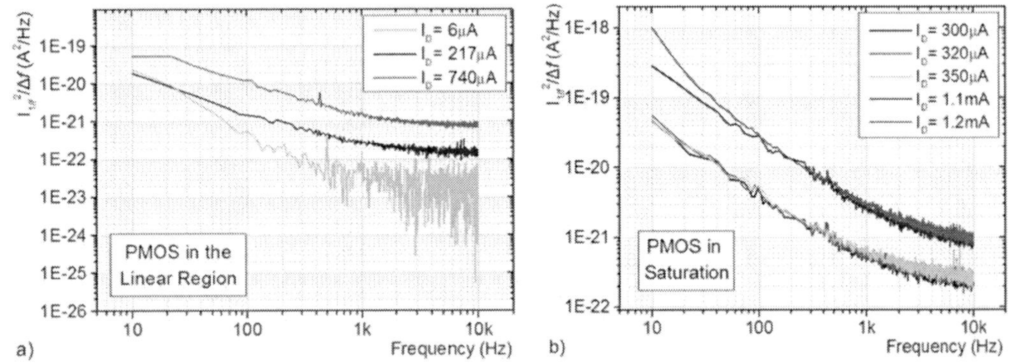

Figure 5. PMOS transistor noise performance a) in linear and b) in saturation region for various values of drain current

Contrary, P-MOSFET $1/f$ noise measurements showed a different dependence on frequency. Particularly, in the saturation region, the noise parameter *EF* ranged from 0.63 to 0.79, whereas in the linear region, from 0.53 to 0.58. These values, in all operating regions and especially in the triode region, appear to be quite lower than unity. This is due to the fact the generation-recombination (GR) noise sources also contribute to the total noise performance. The respective error on the flicker noise exponent in the PMOS transistor is much higher than that on the NMOS. This implies even more difficulty in the PMOS noise estimation using the above simulation models in the specific frequency region.

J. C. Vildeuil et al [33] proposed an appropriate noise parameters (*NOIA, NOIB, NOIC*) extraction methodology in order to optimise the BSIM3 $1/f$ noise model. Specifically, n and p-channel LDD MOSFETs issued from a 0.8 μm CMOS technology and noise measurements were analyzed versus gate bias to identify the origin of the $1/f$ noise and then BSIM3 noise parameters were deduced using an appropriate extraction methodology.

Particularly, they showed that BSIM3 model provided good agreement between measurements and simulations for $1/f$ intrinsic channel noise developing an extraction methodology for the respective noise parameters. For P-MOSFETs $1/f$ noise agreed with the Hooge model and in that case *NOIA, NOIB* were constant and provided relatively accurate simulation from subthreshold to strong inversion. For N-MOSFETs $1/f$ noise agreed with Mc

Whorter's model. *NOIA* was still constant whereas in strong inversion good agreement between the measurements and modelling was obtained only if the *NOIB* parameter varied as $(V_{gs}-V_T)^{-1}$. For both n- and p-channel transistors the *NOIC* parameter was negligible.

Critical Issues on the Unified 1/f Noise Models

The above unified models that are available in SPICE, assumes implicitly that the mobility, limited by Coulomb scattering at trapped interface charges, does not depend on the inversion carrier density. However recent experimental results indicate that the mobility, limited by Coulomb scattering, is proportional to the square root of the carrier inversion density [34]-[36]. These results are in agreement with a theoretical model based on the work of Nicollian and Brews [37], [38]. E.P. Vandamme and L. K. J. Vandamme [39] showed that the correlated mobility fluctuations can be neglected compared to the noise contribution from the carrier number fluctuations, if the correct dependence of the Coulomb scattering limited mobility on inversion carrier density is taken into account. As a result the unified 1/*f* noise model can not predict the experimentally observed 1/*f* noise as a function of gate bias in p-type MOSFETs. Specifically the unified 1/*f* noise model predict the same gate bias dependence of the noise parameter α as the Δn model (α is defined as the noise parameter to compare the noise modeling results with experimental ones from different types of materials [17] and is given by: $a = f\mathrm{N}\dfrac{S_{I_d}}{I^2{}_d}$ where N is the total number of free carriers), which is not in agreement with the experimental results on PMOSFETs that show α values independent of $|V_{gs} - V_T|$ at low gate bias. Besides the unified 1/*f* noise models show weaker dependence of the current noise and input referred noise on gate bias than experimental results, and largely underestimate the noise level in strong inversion. Thus the present unified noise models cannot predict the 1/*f* noise in p-type MOS transistors.

1.2.4. Impact of Scaling down Technologies on 1/*f* Noise

Low frequency noise (LFN) in submicron MOSFETs is an important parameter for analog and RF applications. In complex mixed-signal and low-power integrated devices maintaining a high signal-to-noise ratio becomes more difficult as the power supply voltage and signal amplitude decrease. Assessing the impact of process scaling on 1/*f* noise is therefore of great interest. As LFN is strongly technology dependent, the introduction of advanced processing steps may lead to drastic changes in performances that are not easily predicted. As a result, opposite trends of the 1/*f* noise behaviour with downscaling in CMOS technology were reported [40–44]. The basic reason is that the noise is mainly governed by interface or border traps density related to the gate oxide thickness t_{ox} [42]. Marin et al [45] focused their study only on the impact of silicon dioxide thickness, leaving out the influence of any additional change in the oxidation process steps introduced into advanced CMOS technology.

Particularly, in their experiments, they studied three sets of surface channel p-MOS devices originated from splits processed in 0.25, 0.18 and 0.13 μm dual-gate CMOS technology nodes from STMicroelectronics (Crolles-France). The essential features of the device fabrication were: electric gate oxide thickness t_{ox} = 6, 3.5 and 2 nm respectively for the 0.25, 0.18 and 0.13 μm technologies. Transistors were designed for a 2.5 V (0.25 μm technology), 1.8 V (0.18 μm) and 1.2 V (0.13 μm) power supply voltage. The channel width was equal to W =12.5 μm for 0.25 μm technology and 10 μm for 0.18 and 0.13 μm technologies. Several gate lengths were available.

Measured spectra are predominantly $1/f^{\gamma}$ like for large area devices. An overall average frequency index of $0.96 < \gamma < 1.06$ was obtained by least-squares regression of the log–log plot of S_{ID} versus f. Moreover, it should be mentioned that when the drain current noise spectral density includes a Lorentzian spectrum behavior in the subthreshold regime or for small area devices, the $1/f$ noise component was extracted assuming that $S_{ID}(f)$ results in a superposition of $1/f$ noise and G/R noise components. Typical drain current noise spectra taken on L =0.22 μm devices are depicted in Figure 6. The signal limitation of noise power spectral density above 10 kHz was determined by the parasitic RC constant of the low noise current amplifier set-up in the subthreshold regime [45].

According to their work, the $1/f$ channel noise studied in the subthreshold, linear and saturation regimes can be attributed to the carrier number with correlated mobility fluctuations. The slow oxide trap density N_t and the Coulomb scattering parameter have been deduced. At sufficient high gate overdrive voltages, the increase in the noise level observed for the longer transistors is explained by the impact of the series resistance noise. The theoretical curves of S_{ID}/I^2_D calculated by considering the contribution of the access resistance noise to the intrinsic channel noise have shown a good agreement in relation to experimental data (see Figure 7).

In addition, three PMOS device technology generations (0.25, 0.18 and 0.13 μm) were compared with respect to noise performances. It is observed that as the oxide thickness is reduced, the interface trap density N_t increases, the Coulomb scattering parameter α_s stays almost constant and the Hooge parameter α_H increases. These parameter variations contribute to an input-referred noise $S_{VG}(t_{ox})$dependence lower than t^2_{ox} in weak inversion as previously predicted by the 1999 International Technology Roadmap for Semiconductors and to a non $S_{ID}/I^2_D \propto t_{ox}$ dependence in the linear regime (Figure 8 [45]).

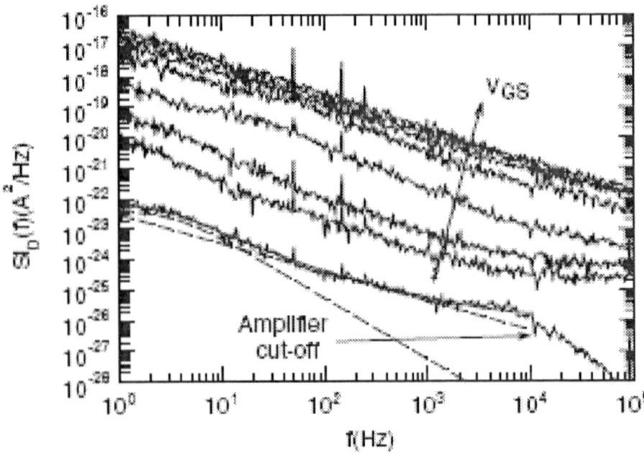

Figure 6. Typical variations of the drain current noise S_{ID} spectra in p-MOSFETs (10 × 0.22 μm^2), measured from weak to strong inversion at V_{DS}= -25 mV. Dotted lines represent the $1/f$ and G/R extracted noise components, solid line represents the total noise at $V_{GS} < V_T$. (From [45] with permission)

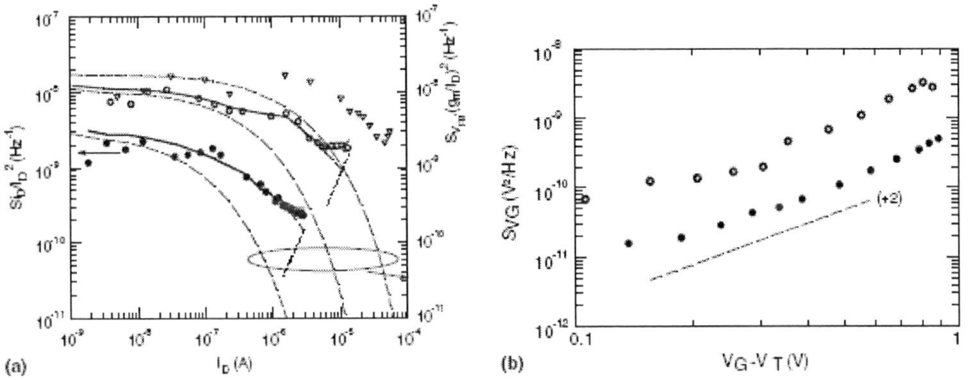

Figure 7. For PMOS devices (a) Variation of the normalized drain current spectral density S_{ID}/I^2_D and $S_V(g_m/I^2_D)$ ratio (dashed lines) versus drain current ID in all operating regimes at V_{GS}=-25mV (b) variations of S_{VG} as a function of (V_{GS}-V_T) in strong inversion.(From [45] with permission)

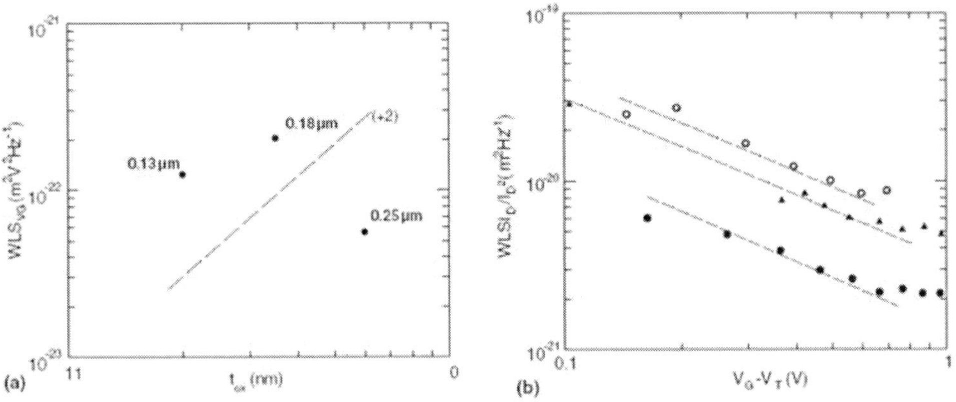

Figure 8. Input-referred noise spectral density S_{VG} vs t_{ox} for three pMOS device generations (0.13, 0.18, 0.25 µm) biased in subthreshold region at V_{DS}=-50mV and f=1 Hz. Dashed line represents $S_{VG} \propto t^2_{ox}$ dependence as predicted by the ITRS roadmap (b) Normalized current noise spectral density WLS_{ID}/I^2_D vs the gate overdrive voltage for t_{ox}=6 nm, 3.5 nm and 2nm respectively (PMOS devices in linear region at f=1 Hz and V_{DS}=-50 mV) (From [45] with permission).

Their study clearly demonstrated that to leave out a number of important noise parameters and considering only gate oxide thickness in order to predict trends in S_{VG} versus technology scaling, leads to inaccurate or too optimistic results.

1.3. Noise Reduction

1.3.1. 1/*f* Noise under Switched Bias Conditions

A. P. van der Wel et al performed low frequency measurements with switched biased MOSFETs (Figure 9) [46]. The aim was to compare switched bias and constant bias noise power for the same measurement bandwidth. For 50% duty cycle, a low-frequency noise power that is reduced by 6 dB compared to the constant-bias situation was expected and indeed, a gap of about 6 dB was observed. Further noise reduction was observed when the gate-source voltage in the "off" state was decreased. A more or less flat plateau below the switched frequency, which gradually changes to a noise spectrum with reduced power density was also observed at low frequencies. The switching frequency was varied from 100 Hz up to 1 MHz (Figure 9b). Even at 1 MHz, where the settling of the output voltages becomes incomplete, this noise reduction was also found. Below, a quasi-plateau has been observed, while a noise spectrum appeared again at more than two decades lower frequencies. This can be explained by assuming different noise generating processes, as suggested in [47]: 1) a process that *is* affected by switching resulting in the plateau (e.g., oxide trap noise with associated RTS's, that effectively disappears when switching between inversion and accumulation [47]); and 2) other processes that are not affected by the switching and render the residual low frequency noise (e.g., mobility fluctuations) [46].

In order to study the noise reduction for different CMOS processes, devices from six different manufacturers were tested by A. P. van der Wel et al [46]. An intrinsic noise reduction was found in all cases, with values of 5–8 dB. The effect occurs in all devices, and was also observed for submicron MOS transistors [47]. It was also shown that phase noise in CMOS oscillators [48]–[50], operating at switching frequencies in the lower MHz region, benefits from switching.

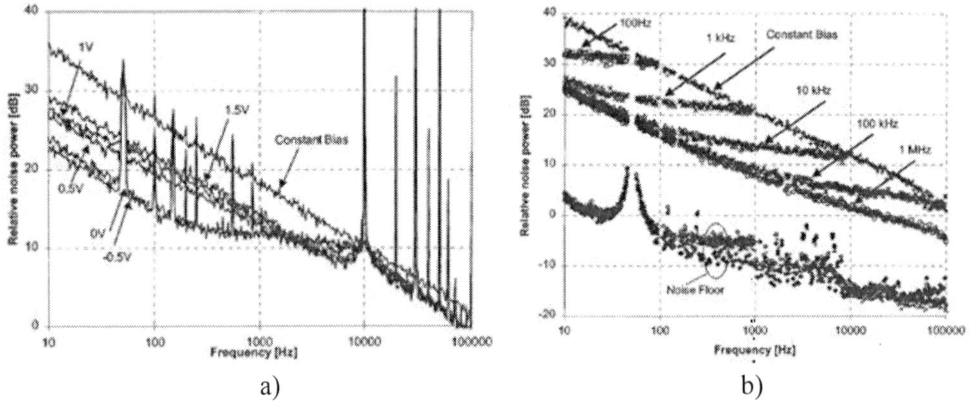

Figure 9 a) Noise reduction as a function of the "off" voltage: HEF 4007 NMOS V_{gs}=2.5V, V_t=1.9V f_{switch} = 10 kHz duty cycle = 50% (taken from [46] with permission ©2000 IEEE) b) Noise reduction while switching at different frequencies (indicated in the figure with different symbols), HEF4007 NMOS, V_{GSon} = 2.5 V, $V_{GS,off}$ = 0V, duty cycle = 50%, all spectra are shown up to 0.9 * fswitch and points within 5 Hz of multiples of 50 Hz are omitted (taken from [46] with permission ©2000 IEEE)

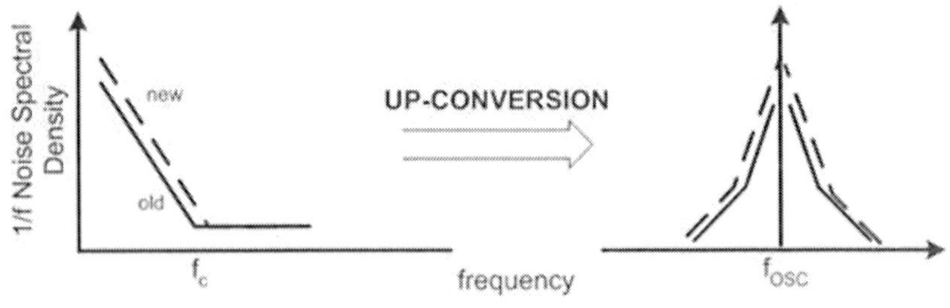

Figure 10. The 1/*f* corner frequency *f* of minimum size transistors in newer CMOS processes tends to increase. Upconversion, e.g., in oscillators renders also additional noise at high frequencies

Applications of the Switched Bias Technique

$1/f$ noise is of increasing interest as minimum size transistors in newer CMOS processes tend to have higher corner frequencies, typically well above 1 MHz. Moreover, as shown in Figure 10, the effect of noise is not limited to low frequencies: noise is also upconverted to high frequencies [51], e.g., in active mixers, VCO's [52], and frequency dividers [53]. This is one of the important difficulties in implementing analog CMOS circuits with similar noise performance than their bipolar counterparts. Circuit techniques like chopping and correlated double sampling reduce the effect noise has on circuits. In contrast, Eric A. M. Klumperink at al. [54] have proposed a technique that tackles the problem at its physical roots: it reduces the intrinsic noise itself. It exploits an intriguing physical effect: Cycling a MOS transistor between strong inversion and accumulation reduces its $1/f$ noise.

From a designer's perspective, a pragmatic view is usually taken, concentrating on the question: given a certain IC process, what can be done to reduce the $1/f$ noise? The main controllable parameters are W and L of the transistor and its biasing.

Changing the biasing only helps for transistors showing significant bias dependence (as for the mobility fluctuation model). However, biasing also affects many other important circuit performance aspects like transfer function, signal swing, speed, linearity and current efficiency.

Due to the many tradeoffs, changing the bias is rarely practical. Increasing the gate area WL at constant W/L and constant bias has much less impact on most of the performance aspects mentioned. The main disadvantage in this case is the increase in gate-capacitance resulting in speed penalty. Lowering the impedance level, the loss in speed can be compensated. However, a lower impedance level requires higher transconductance values of the transistor, resulting in more power consumption. It is concluded that the only practical way to reduce $1/f$ noise is by increasing device gate area. However, this comes at the cost of speed or power consumption [54].

1.3.2. Available Methods

Known circuit techniques for noise reduction will be reviewed briefly, in order to make a distinction between these techniques and the switched bias technique [54].

A. $V_{DS} = 0$:

$1/f$ current becomes zero for a MOS operating in the triode region at V_{DS}=0V. This property is useful in resistive circuits, for instance in current dividers [55] or transconductors [56]. However, if V_{DS} is increased, the $1/f$ noise comes up. Furthermore, almost all transistor circuits require voltage amplification, while a MOS transistor in the triode region has a maximum "intrinsic voltage gain" of 1. The majority of MOS circuits exploit MOS transistors operating in saturation, i.e., $V_{DS}\neq 0$.

B. DC Offset and Drift Reduction Techniques

DC offset and drift are basic problems in analog circuits, and several techniques have been developed to reduce the resulting low frequency errors. Examples of such techniques are chopper stabilization [57], auto-zeroing techniques (correlated double sampling [57]), dynamic element matching [58], dynamic current mirrors [59] and current copiers [60]. As $1/f$

noise at low frequencies is indistinguishable from drift, a reduction of low frequency $1/f$ noise is also achieved. The reduction factor is typically limited by device mismatch, timing errors and charge injection. All these techniques are limited to use at low frequencies. They do not help to reduce up-converted $1/f$ noise.

C. Reduce the Upconversion of 1/f Noise

Hajimiri proposed a theory that is useful to quantify the upconversion effect of $1/f$ noise in oscillators [51]. The oscillator is modeled as a linear time variant system, and an impulse sensitivity function is defined that characterizes the sensitivity to upconversion. Based on this theory, it can be shown that symmetry in the oscillator waveform helps to minimize the upconversion [61]. However, the achievable symmetry is limited especially in cases where complementary devices are used.

D. Phase-Locked Loop (PLL)

In a PLL the phase of a voltage controlled oscillator (VCO) is locked to the phase of a reference signal by means of a phase detector with feedback loop. If the reference is clean, the PLL suppresses the phase noise of the VCO by an amount determined by the loop gain [24]. Although this is a very useful commonly used technique, it has its limitations. For frequencies larger than the loop bandwidth, the phase noise of the (effectively free running) VCO still determines the phase noise of the PLL. Hence, $1/f$ noise can be a problem, especially for applications requiring a low loop bandwidth. Within the loop bandwidth, the noise contribution of the phase detector and the frequency divider, that is often present between the VCO and the phase detector, is a source of phase noise contributions that are not attenuated by the loop gain [61]. Hence, noise improvements are wanted in these blocks.

E. The switched biasing approach

Cycling between Strong Inversion and Accumulation

Figure 11 illustrates the basic idea of cycling a transistor between strong inversion and accumulation for an NMOS transistor. A voltage source with a square wave signal switches the gate-source voltage of the NMOS between two bias values. The high level, V_{gson}, is larger than the threshold voltage, so that the transistor is biased at a constant voltage in strong inversion (NMOSFET is "on"). The low level, V_{gsoff}, is equal or lower than the threshold voltage (NMOSFET "off") and is variable. Depending on V_{gsoff}, the low state of the square wave corresponds to biasing in moderate inversion, weak inversion or finally accumulation (holes in the p-bulk accumulate under the gate-oxide).

Having a switched bias the noise power is roughly 6 dB lower. This is what we can expect for 50% duty cycle, as the switching operation can be represented as a multiplication of the $1/f$ noise current with a square-wave signal with 50% duty cycle, $m(t)$, as follows:

$$m(t) = \frac{1}{2} + \frac{2}{\pi}\sin\omega_{sw}t + \frac{2}{3p}\sin 3\omega_{sw}t + \frac{2}{5p}\sin 5\omega_{sw}t + ..., \tag{7}$$

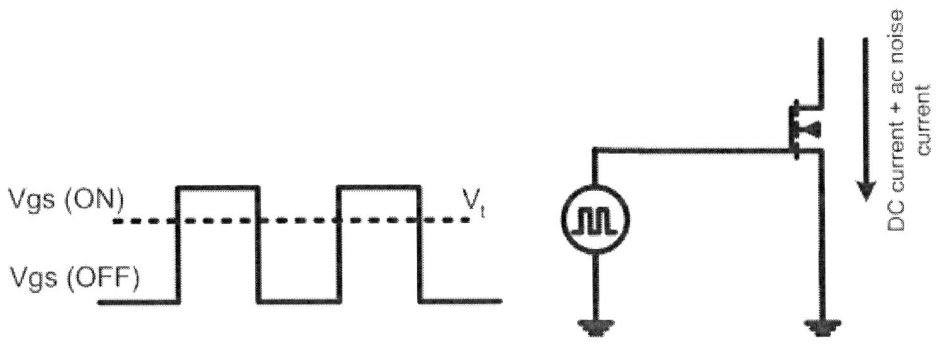

Figure 11. Varying V_{gs} cycles a MOSFET between strong inversion and accumulation.

In the frequency domain this corresponds to a convolution of the *PSD* of the *1/f* noise with a spectrum with delta functions at dc, ω, 3ω, 5ω, etc. The dc-term determines the resulting noise power in baseband, which is -6 dB (or $(1/2)^{1/2}$) compared to the original *1/f* noise power.

The general concept of the switched biasing technique is demonstrated below:

Many applications where the specific technique was applied in order to reduce the 1/*f* noise, are given in [54].

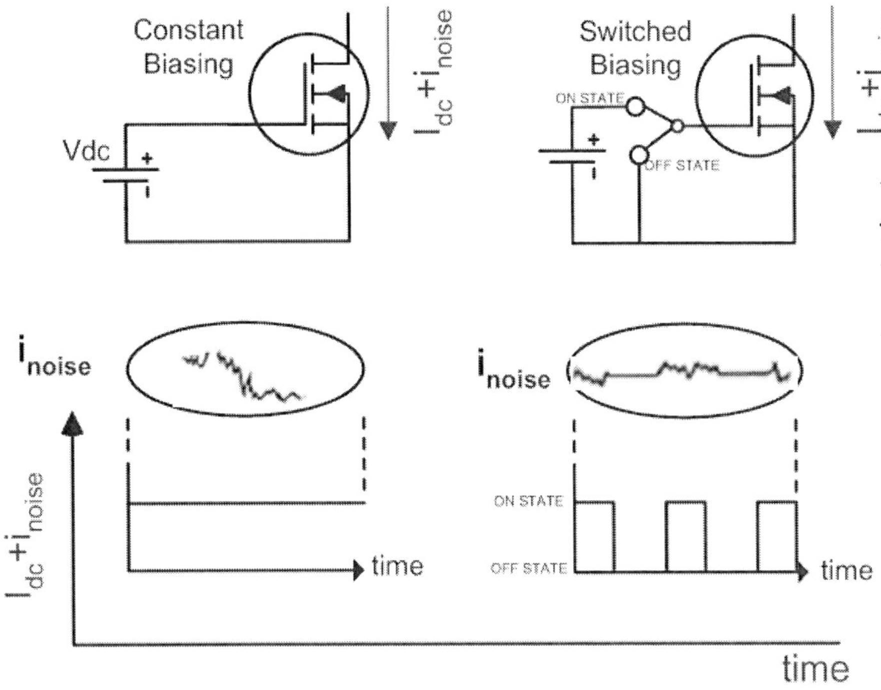

Figure 12. Concept of the switched biasing technique (V_T>0)

1.4. Measurement of Low Frequency Noise in Mosfets

1.4.1. Low Frequency Noise Measurement Technique under Constant Bias

It is well known that the extrinsic noise behaviour of a transistor is fully described by two noise generators, together with their correlation. Two different characterization techniques can be used to measure these noise sources: the first one, called "multi-impedance technique" consists to measure the noise sources with only one low noise voltage amplifier and several different resistors at the input impedance of the device under test (DUT. Then the two input noise generators, voltage v_G and current i, and their coherence γ' (Figure 13b) are extracted with a numerical procedure [62]. The second one, called "direct technique", is based on two low noise amplifiers at the input and the output of the DUT in order to directly measure all the noise sources. This configuration usually required two low noise current amplifiers (transimpedances) [63] or one voltage amplifier associated to a transimpendance amplifier [64]. Thus by using two transimpedance amplifiers, the noise sources are directly measured like on Figure 13c: the noise generators are at the input (i_G) and at the output (i_D) terminals of the DUT, together with their coherence γ. It is possible from one noise representation to derive to another one due to the equivalence between all the noise representations.

During some previous studies of bipolar transistors, we already had the opportunity to compare between two LFN set-ups based on these two different techniques [63]: the direct technique is rather utilized than the multi-impedance one due to a smaller characterization time without a numerical procedure.

A very simple view of the set-up is on Figure 14, the current noise sources (i_G and i_D) are directly AC short-circuited at the transistor input and output terminals with the low input impedance of two transimpedance amplifiers.

The study of the gate noise source i_G and of the coherence can give some important physical information [64]. However, as the gate leakage from the transistor is small and as the input impedance of the transistor is typically much more than the driving source impedance generally used, the LFN measurement can be reduced to the study of the noise source i_D as a first approximation in order to obtain the LFN behavior from a circuit simulation [65, 66]. A more complete view of the set-up is given on Figure 15 for the drain LFN measurements.

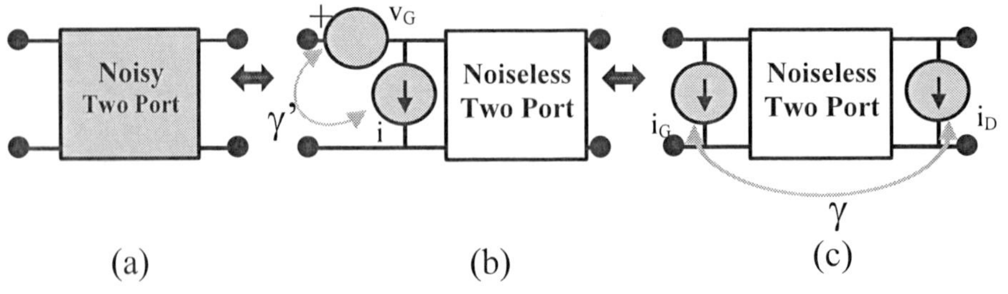

Figure 13. Equivalent LFN noise representations

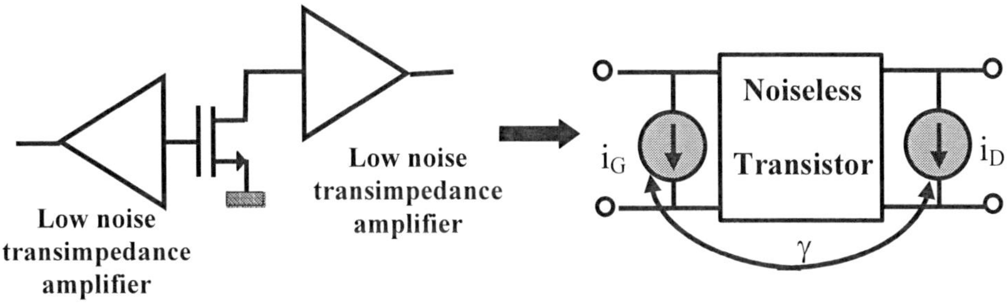

Figure 14. LFN set-up based on transimpedance amplifiers.

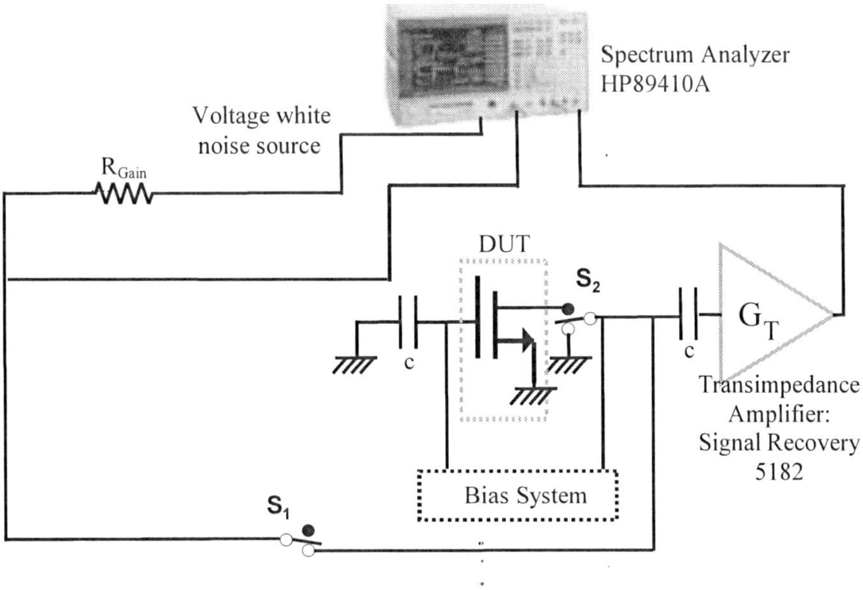

Figure 15. LFN set-up based on transimpedance amplifier for the S_{ID} measurement

The Several S_1 and S_2 are used in order to configure the set-up: to measure either the noise of the set-up or the amplifier gain or the drain current spectral density S_{ID} respectively (Table III).

The voltage white noise source and the resistance R_{Gain} are applied in order to obtain the gain G_T of the transimpedance. The capacitors *c* have been introduced to separate the DC part from the LFN part.

Table III. The different measurement configurations, the position of each switch is given by an open or filled dot, as shown in Figure 15

Switch	S_1	S_2
Measurement configuration		
Set-up noise	(like a floating position)	o
GTD Gain	o	•
SID	(like a floating position)	•

In order to obtain S_{ID}, it is necessary to divide the spectrum density measured at the output of the amplifier by G_T and to substrate the noise of the set-up.

The design of the set-up could be more direct by connecting the amplifier and the bias system in serial instead of the transimpedance with a capacitor in parallel of the bias system and of the DUT (Figure 15). Two points prevented this last configuration. The first and main point is that the amplifier that we utilized for the set-up has a virtual ground at the input: without a physical access of the ground, the serial connection is not possible. The second one is due to that the transimpedance is easily saturated at the drain terminal by the amplitude of the DC current amplitude.

By using the Eq.8, it is possible to obtain from the S_{ID} measurement the input noise voltage spectrum density S_{VG}.

$$S_{VG} = \frac{S_{ID}}{g_m^2} \tag{8}$$

It is important to notice that to characterize only one LFN source, S_{ID} or S_{VG}, a voltage amplifier can be used instead of the current amplifier at the output of the DUT: the measurement consists of dividing the voltage noise measured at the output of the amplifier by the gain of the system "DUT and Amplifier" and to take into account the noise of the set-up.

1.4.2. Low-Frequency Noise Measurement Technique under Switched Bias

In order to measure MOSFET noise under constant and switched bias conditions, the measurement setup of Figure 16 was proposed [46]. This setup exploits the high common mode rejection ratio of a differential probe to measure small differential noise currents superimposed on much larger common mode switched bias currents of two equal (DUT large-signal behaviour should be equal for minimal common to differential conversion) MOSFET devices under test. The bipolar cascode transistors supply the drains of the MOSFET's with an almost constant voltage and convey the MOSFET drain currents to resistors R_{D1} and R_{D2}, which act as I-V converters. To establish switched bias conditions, the gate-source bias voltages of the two DUT's are driven by a common square-wave signal. As a result, the input signal of the differential probe corresponds to two distinct bias conditions: in the first, both MOSFET's are "off" (below threshold), and in the second, they are "on" in strong inversion and saturation. The probe ideally only senses the differential voltage that is dominated by the noise current of the two DUT's in the "on" state. However, a large common signal also occurs at the probe input due to the switched biasing. Mismatches in the large-signal DUT characteristics and the large, but finite common mode rejection of the probe (>80 dB from 10 Hz to 50 kHz), result in a residual switching signal at the output of the probe. Resistive and capacitive trimmers are provided to equalise the common signals and thus minimize the differential residual to sufficiently low values so as not to exceed the input dynamic range of the spectrum analyzer. Furthermore, the periodic nature of the switching residual results in spectral peaks that can easily be distinguished from noise contributions.

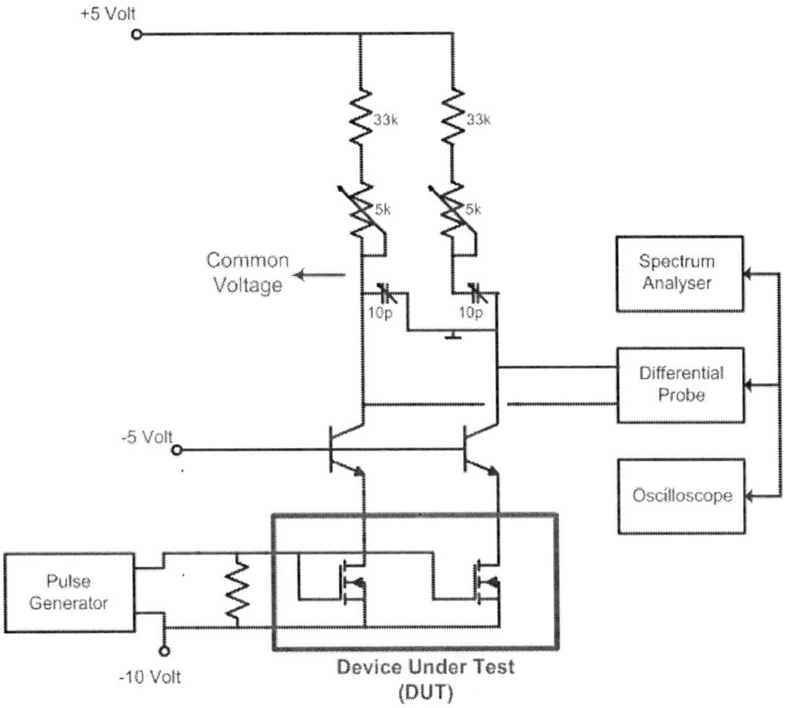

Figure 16. Low-Frequency noise measurement setup for MOSFETs under switched bias conditions. The large common signal is largely rejected by the differential probe

References

[1] Halford, D. A general model for f ^{a} spectral density random noise with special reference to flicker noise 1/*f*. *IEEE Proceeding,* 1968, 56(3), 251-258.

[2] McWhorter, AL. 1/*f* noise and germanium surface properties. In *Semiconductor surface physics*, R.H. Kingston, Ed.,University of Pennsylvania Press: *Philadelphia,* 1957, 207-228.

[3] Christensson, S; Lundström, I; Svensson, C. Low frequency noise in MOS transistors-I. Theory. *Solid-State Electronics,* 1968, 11(9), 797–812.

[4] Christensson, S; Lundström, I. Low frequency noise in MOS transistors–II. Experiments. *Solid-State Electronics,* 1968, 11(9), 813–820.

[5] Das Makunda, B; Moore, JM. Measurements and interpretation of low-frequency noise in FETs. *IEEE Transactions on Electron Devices,* 1974, 21(4), 247–257.

[6] Reimbold, G. Modified 1/*f* trapping noise theory and experiments in MOS transistors biased from weak to strong inversion – influence of interface states. *IEEE Transactions on Electron Devices,* 1984, 31(9), 1190-1198.

[7] Ghibaudo, G. A simple derivation of Reimbold's drain current spectrum formula for flicker noise in MOSFETs. *Solid-State Electronics,* 1987, 30(10), 1037-1038.

[8] Hooge, FN. 1/*f* noise is no surface effect. *Physics Letters,* 1969, 29(3), 139–140.

[9] Hooge, FN. Discussion of recent experiments on 1/*f* noise. *Physica,* 1972, 60(1), 130–144.

[10] Hooge, FN; Vandamme, LKJ. Lattice scattering causes 1/*f* noise. *Physics Letters,* 1978, 66(4), 315–316.

[11] Hooge, FN. 1/*f* noise. *Physica B,* 1976, 83(1), 14–23.

[12] Jindal, RP; Van Der Ziel, A. Phonon fluctuation model for flicker noise in elemental semiconductors. *Journal of Applied Physics,* 1981, 52(4), 2884–2888.

[13] Vandamme, LKJ; de Werd HMM. 1/*f* noise model for MOSTs biased in the nonohmic region. *Solid-State Electronics,* 1980, 23(4), 325–329.

[14] Jayaraman, R; Sodini, CG. A 1/*f* noise technique to extract the oxide trap density near the conduction band edge of silicon. *IEEE Transactions on Electron Devices,* 1989, 36(9), 1773–1782.

[15] Hung, KK; Ko, PK; Hu, C; Cheng, YC. A unified model for the flicker noise in metal-oxide-semiconductor field-effect transistors. *IEEE Transactions on Electron Devices,* 1990, 37(3), 654–665.

[16] Hung, KK; Ko, PK; Hu, C; Cheng, YC. A physics-based MOSFET noise model for circuit simulators. *IEEE Transactions on Electron Devices,* 1990, 37(5), 1323–1333.

[17] Hooge, FN. 1/*f* noise sources. *IEEE Transactions on Electron Devices,* 1994, 41(11), 1926–1935.

[18] Vandamme, LKJ; Xiaosong, Li; Rigaud, D. 1/*f* noise in MOS devices; mobility or number fluctuations? *IEEE Trans Electron Devices,* 1994, 41(11) : 1936–1945.

[19] Chang, J; Abidi, AA; Viswanathan, CR. Flicker noise in CMOS transistors from subthreshold to strong inversion at various temperatures. *IEEE Trans Electron Devices,* 1994, 41(11), 1965–1971.

[20] Yannis, P. Tsividis. *Operation and Modeling of the MOS Transistor*. McGraw-Hill, New York, 1987, 343.

[21] John, H. Scofield, Nick Borland, and Daniel M. Fleetwood. Reconciliation of different gate- voltage dependencies of 1/f noise in n-MOS and p-MOS transistors. *IEEE Transactions on Electron Devices*, 41(11), 1946-1952, November 1994.

[22] Xie, D; Cheng, M; Forbes, L. SPICE models for flicker noise in n-MOSFETs from subthreshold to strong inversion. *IEEE Trans Computer-Aided Design,* 2000, 19(11), 1293–1303.

[23] Zhou, J; Cheng, M; Forbes, L. SPICE models for flicker noise in p-MOSFETs in the saturation region. *IEEE Trans Computer-Aided Design,* 2001, 20(6), 763-767.

[24] Liu, W. *MOSFET models for SPICE simulation including BSIM3V3 and BSIM4*, New York: Wiley–Interscience; 2001.

[25] BSIM4.3.0 *MOSFET- User's manual.* Berkeley: The Regents of the University of California; 2003.

[26] *BSIM3v3.2.2 Manual.* Berkeley: UC Berkeley; 1999.

[27] Star-HSPICE manual. Fremont: Avant! Corporation and Avant! Subsidiary; 2001.

[28] Xie, D; Cheng Mengzhang, Forbes, L. SPICE models for flicker noise in n-MOSFETs from subthreshold to strong inversion. *IEEE Trans Computer-Aided Design,* 2000, 19(11), 1293-303.

[29] Zhou, J; Cheng Mark, Forbes, L. SPICE models for flicker noise in p-MOSFETs in the saturation region. *IEEE Trans Computer-Aided Design,* 2001, 20(6), 763-7.

[30] Noulis, T; Siskos, S; and Sarrabayrouse, G. 'Comparison between BSIM4.X and HSPICE flicker noise models in NMOS and PMOS transistors in all operating regions'. *Microelectronics Reliability,* 2007, 47,. 1222-1227.

[31] Noulis, T; Siskos, S; Sarrabayrouse, G. 'Analysis and selection criteria of BSIM4 flicker noise simulation models', *Intenational Journal of Circuit Theory and Applications* (accepted for publication).

[32] Noulis, T; Siskos, S; Sarrabayrouse, G; Bary, L; Hatzopoulos, AA. 'Detailed study of BSIM3v3 flicker noise models in NMOS and PMOS transistors from threshold to saturation', *Proc. 20th DCIS,* Portugal, Lisboa, 2005.

[33] Vildeuil, JC; Valenza, M; Rigaud, D. 'Extraction of the BSIM3 1/f nosie parameters in CMOS transistors', *Microelectronics Journal*, 1999, 30, 199-205.

[34] Koga, J; Takagi, S; Toriumi, A. ' A comprehensive study of MOSFETelectron mobility in both weak and strong inversion regimes', *IEDM Tech. Dig*., 1994, 475-478.

[35] Takagi, S; Toriumi, A; Iwase, M; Tango, H. 'On the universality of inversion layer mobility in Si MOSFETs: Part I-effects on substrate impurity concentration' *IEEE Trans Electron Devices,* 1994, 41, 2357–2362.

[36] Mujtaba, A; Takagi, S; Dutton, R. 'Accurate modelling of coulombic scattering and its impact on scaled MOSFETs' in Symp. *VLSI Tech. Dig*., 1995, 99-100.

[37] Soppa, WM; Waagemann, HG. 'Investigation and modelling of the surface mobility of MOSFETs from -25 to +150°C' *IEEE Trans Electron Devices,* 1988, 35, 970–977.

[38] Nicollian, EH; Brews, JR. *MOS Metal oxide Semiconductor Physics and Technology,* New York: Wiley, 1982.

[39] Vandamme, EP; Vandamme, LKJ. 'Critical discussion on unified 1/f noise models of MOSFETs', *IEEE Trans Electron Devices,* 2000, 47, 2146–2152.

[40] Simoen, E; Claeys, C. 'On the flicker noise in submicron silicon MOSFETs' *Solid-State Electronics,* 1999, 43, 865-882.

[41] Ghibaudo, G; Roux-dit-Buisson, O; Brini, J. *Phys Status Solidi,* (a) 1992, 132, 501.

[42] Kolarova, R; Skotnicki, T; Chroboczek, JA. 'Low frequency noise in thin gate oxide MOSFETs', *Microelectronics Reliability,* 2001, 41, 579.

[43] Brederlow, R; Weber, W; Schmitt-Landsiedel, D; Thewes, R. *'Fluctuations of the low frequency noise of MOS transistors andtheir modeling in analog and RF-circuits'* International Electron Devices Meeting (IEDM), 1999, 159-162.

[44] Knitel, MJ; Woerlee, PH; Scholten, AJ; Zegers-Van Duijnhoven ATA. In: *Technical Digest International Electron Devices Meeting.*

[45] Marin, M; Akue Allogo, Y; De Murcia, M; Llinares, P; Vildeuil, JC. 'Low frequency noise characterization in 0.13 μm p-MOSFETs. Impact of scaled-down 0.25, 0.18 and 0.13 μm technologies on 1/f noise', *Microelectronics Reliability,* 44, 2004, 1077-1085.

[46] Van der Wel, AP; Klumperink, EAM; Gierkink, SLJ; Wassenaar, RF; Wallinga, H. 'MOSFET 1/f Noise Measurement Under Switched Bias Conditions', *IEEE Electron Devices Letter*, 2000, 21(1), 43–46.

[47] Dierickx, B; Simoen, E. "The decrease of 'random telegraph signal' noise in metal-oxide-semiconductor field-effect transistors when cycled from inversion to accumulation," *J. Appl. Phys.*, vol. 71, 2028–2029, Feb. 1992.

[48] Gierkink, SLJ; Van der. Wel, A; Hoogzaad, G; Klumperink, EAM; V. Tuijl, AJM. "Reduction of the 1/f noise induced phase noise in a CMOS ring oscillator by increasing the amplitude of oscillation," in *Proc. 1998 Int. Symp. Circuits and Systems (ISCAS'98)*, CD-ROM Paper MPA9-8, Monterey, CA, May 31–June 3 1998.

[49] Gierkink, SLJ; Klumperink, EAM; Van derWel, AP; Hoogzaad, G; Van Tuijl, AJM; and Nauta, B. "Intrinsic 1/f device noise reduction and its effect on phase noise in CMOS ring oscillators," *IEEE J. Solid- State Circuits*, vol. 34, 1022–1025, July 1999.

[50] Gierkink, SLJ; Klumperink, EAM; Van Tuijl, E; Nauta B; "Reducing MOSFET 1/f noise and power consumption by "Switched Biasing"," in *Proc. 25th Europ. Solid-State Circuits Conf.*, Duisburg, Germany, Sept. 21-23, 1999, 154-157.

[51] Hajimiri, A; Lee, TH. "A general theory of phase noise in electrical oscillators," *IEEE J. Solid-State Circuits*, vol. 33, 179-194, Feb. 1998.

[52] Craninckx, J; Steyaert, MSJ. "A fully integrated CMOS DCS-1800 frequency synthesizer," *IEEE J. Solid-State Circuits*, vol. 33, 2054-2065, Dec. 1998.

[53] Krishnapura, N; Kinget, P. "A 5.3 GHz programmable divider for HiPerLAN in 0.25 _m CMOS," in *Proc. 25th Eur. Solid-State Circuits Conf.*, Duisburg, Germany, Sept. 21-23, 1999, 142-145.

[54] Klumperink Eric, AM; Gierkink Sander, LJ; Van der Wel Arnoud, P. Nauta Bram, 'Reducing MOSFET 1/f Noise and Power Consumption by Switched Biasing' *IEEE J. OF Solid-State Circuits*, vol. 35, No. 7, July 2000.

[55] Bult, K; Geelen, GJGM. "An inherently linear and compact MOST-only current division technique," *IEEE J. Solid-State Circuits*, vol. 27, 1730–1735, Dec. 1992.

[56] Mensink, CHJ; Nauta, B; Wallinga, H. "A CMOS "soft-switched" transconductor and its application in gain control and filters," *IEEE J. Solid-State Circuits*, vol. 32, 989-998, July 1997.

[57] Gregorian, R; Temes, GC. *Analog MOS Integrated Circuits for Signal Processing*. New York, NY: Wiley, 1986.

[58] Plassche, RJVD. "Dynamic element matching for high accuracy monolithic D/A converters," *IEEE J. Solid-State Circuits*, vol. SSC-11, 795-800, Dec. 1976.

[59] Wegmann, G; Vittoz, EA. "Very accurate dynamic current mirrors," *Electron. Lett.*, vol. 25, 644-646, May 1989.

[60] Daubert, SJ. *et al.*, "Current copier cell," *Electron. Lett.*, vol. 24, 1560-1562, Dec. 1988.

[61] Hajimiri, A; Limotyrakis, S; Lee, TH. "Jitter and phase noise in ring oscillators," *IEEE J. Solid-State Circuits*, vol. 34, 790-804, June 1999.

[62] Escotte, L; Plana, R; Graffeuil, J. "Evaluation of noise parameter methods," *IEEE Trans. Microwave Theory Tech.*, vol. 41, 369-381, Mar 1993.

[63] Bary, L; Borgarino, Plana, MR; Kovacic, S; Lafontaine, H; Feng Zhou, Z. and Graffeuil J; "Transimpedance amplifier based full low noise characterization set-up for Si/SiGe HBT's," *IEEE Trans. Electron Devices,* vol. 48, no. 4, 767-773, April 2001.

[64] Giusi, G; Crupi, F; Ciofi, C; Pace, C. "Instrumentation Design for Gate and Drain Low Frequency Noise Measurements," *Instrumentation and Measurement Technology Conference (IMTC) 2006, Pro. of the IEEE,* 1747-1750, April 2007.

[65] Van der Wel, AP; Klumperink, EA; Kolhatkar, JS; Hoekstra, E; Snoeij, MF; Salm, C; Wallinga, H; Nauta, B. "Low-Frequency Noise Phenomena in Switched MOSFETs," *IEEE Journal of Solid-State Circuits*, vol. 42, 540-550, March 2007.

[66] Scholten, AJ; Klaassen, DBM. *"New 1/f noise model in MOS Model, Level 903,"* Philips Research Nat. Lab., Unclassified Report 816/98, 1998 [Online], Available: http://www.nxp.com/Philips_Models/mos_models/model9/.

INDEX

G

H

I

J

K

L

M

N

O

P

Q

R

S

T